WITHDRAWN

Challenges of Astronomy

W. Schlosser   T. Schmidt-Kaler   E.F. Milone

Challenges of Astronomy

Hands-on Experiments
for the Sky and Laboratory

With 190 Illustrations

Springer-Verlag
New York Berlin Heidelberg London
Paris Tokyo Hong Kong Barcelona

W. Schlosser
Universität Bochum
Bochum, FRG

T. Schmidt-Kaler
Universität Bochum
Bochum, FRG

E.F. Milone
Department of Physics and Astronomy
The University of Calgary
2500 University Drive, NW
Calgary Alberta T2N 1N4
Canada

Library of Congress Cataloging-in-Publication Data
Schlosser, W. (Wolfhard)
    [Astronomische Musterversuche. English]
    Challenges of astronomy : hands-on experiments for the sky and
laboratory / W. Schlosser, Th. Schmidt-Kaler, and E.F. Milone.
        p.   cm.
    Rev. translation of: Astronomische Musterversuche.
    Includes bibliographical references and index.
    ISBN 0-387-97408-3
    1. Astronomy—Experiments.   I. Schmidt-Kaler, Th. (Theodor),
1930–   II. Milone, E.F., 1939–   III. Title.
    QB62.7.S3713   1991
522′.078—dc20   90-43848
                                                CIP

Printed on acid-free paper.

Typeset by Publishers Service of Montana, Bozeman, Montana.
Printed and bound by Edwards Brothers Inc., Ann Arbor, Michigan.
Printed in the United States of America.

9 8 7 6 5 4 3 2 1

ISBN 0-387-97408-3 Springer-Verlag New York Berlin Heidelberg
ISBN 3-540-97408-3 Springer-Verlag Berlin Heidelberg New York

Contents

Appendices

Introduction

This book began as a series of laboratory exercises in astronomy and related physical sciences by the authors on both sides of the Atlantic. The bulk of it is translated from the German work, *Astronomische Musterversuche*, a compendium of useful and interesting laboratory exercises developed mainly for teachers in the German public schools. Its revision and translation arose from the recognition by the third author of the unique character of the exercises, which resemble a series of challenges to the dedicated amateur and student as well as to the professional (observer, experimentalist, or theoretician), rather than the rote activities of most laboratory manuals. Accordingly, encouraged by Theo Schmidt-Kaler, I sought a publisher with connections in Germany who might be sympathetic to the enterprise. In Springer-Verlag I found such a publisher, and in its (former) physics editor, Jeffrey Robbins, both a kindred soul and a strong supporter of the project. The present physics editor, Tom von Foerster has maintained that strong support.

The work commenced in 1987 and was completed except for suggested revisions, figures, and tables, in 1989. Its timely completion is due for the most part to the award of a Killam Resident Fellowship in the fall, 1988, for which my co-authors and I are grateful. The final result provides not only a rich resource for teachers and instructors in planning their own laboratory courses, but a work more directly accessible to the student and to the amateur astronomer and physicist as well.

The exercises are arranged in chapters. They are intended to be self-guiding but not self-standing: the terminology of later chapters depends strongly on familiarity with astronomy or on a thorough working of the earlier chapters. The progression is from ancient astronomy, emphasizing phenomena, to more modern astronomy, involving physical principles; it finishes with Olbers's paradox and a glimpse of the revelations of cosmology. The exercises tend to become more demanding as one progresses. Thus the first six chapters or so are suitable for bright high school students while later chapters usually require a fuller background. The chapters on celestial mechanics (14), photometry (23), spectroscopic analyses (24), and the Sun (28) are exceptionally detailed and are suitable for advanced astronomy undergraduates, as well as the dedicated amateur instrumentalist. Chapter 14 will be especially interesting to students with computer programming skills; but those with less interest in the calculational procedure will find a shortcut early in that chapter. The topics table provides a rough guide to the suitability of the material in a given chapter for a particular learning situation. In short, the early chapters stress mechanics and trigonometry; the later chapters involve light and atomic physics as basic tools.

All three authors have had the opportunity to review the manuscript at intermediate stages and all chapters have been field-tested, either at the University of Bochum or at the University of Calgary. It is a pleasure to acknowledge the help and advice given by three colleagues at the University of Calgary, Dr. Hans Laue, Dr. David Fry, and Prof. T. Alan Clark, and the encouragement of Prof. Harlan J. Smith of the University of Texas. University of Calgary students Gaston Groisman and Steven Griffiths provided several computer graphics figures. Finally, the readability of the manuscript has been much improved by the sharp eye of my wife, Helen. Notwithstanding all this assistance, errors in the final text are the responsibility of the third author alone.

E.F. Milone
Calgary, 1989 Sept. 29

Plate I.1. The Temple of Heaven, section of ceiling and upper wall, Beijing, China.

1

The Vault of Heaven

1.1 The Shape of Heaven

Many observers of the sky in early civilizations, such as ancient China and Greece before the Christian era, depicted the sky as a great spherical chamber on which the stars were emblazoned and within which the Sun, Moon, and planets moved. But not all ancient observers agreed on that perception. The shape of the sky is a subjective notion, conditioned by cultural, psychological, and physiological factors. It has sometimes been depicted as a flat concave bowl, as in Fig. 1.1.

The perceived shape of the sky is connected to estimated angles of elevations (astronomical *altitudes*) from the horizon.[1] The summer midday Sun (or winter full moon) is typically estimated to be close to the zenith, or point overhead, even at high-latitude locales such as northern Europe or Canada, where the latitude is in the range 50 to 65°. Over this range of latitude, the maximum altitude of the midsummer Sun is about 63 to 48°. The midpoint altitude of any *vertical circle* arc from the zenith to the horizon will be underestimated by many people by as much as 15° from the true value of 45°. A related perception is that of the impressively large disk of the rising or setting Sun or Moon. It is striking, unmistakable, and yet purely illusory. In fact, the low-altitude disks are slightly flattened in the altitude direction (i.e., toward the zenith) because atmospheric refraction bends light up toward the zenith, and the bending is stronger for the lower limb than for the upper.

---

[1]Altitude, azimuth, zenith, etc., are defined and illustrated in Appendix A.

1.2 Determining the Shape of the Sky

In a series of measurements made at a rooftop location just south of the University of the Ruhr at Bochum (in the industrial heartland of Europe), at 1430 Local Civil Time (LCT),[2] July 14, 1981, the results listed in Table 1.1 were obtained for the measured altitude of the estimated halfway point of vertical circle arcs made at various azimuths. The mean altitude was 30.2° and its uncertainty—the standard deviation of the mean[3]—was 1.3°. The uncertainty is several times smaller than the largest deviations from the mean; the largest deviations in this case possibly reflect the difficulties of making the eye estimate for the arc midpoint and of setting a hand-held sextant to such a point. Nevertheless, the experiment was successful, and the precision clearly was improved by the use of several measurements. In this particular experiment, as in others, the observer, sighting along a line at 45° altitude, was startled at how close to the zenith this line appeared to run.

The measured result suggests that the horizon appears to be twice as distant as the zenith. Figure 1.2 is a sketch of the cross section of the perceived sky in the plane defined by the observer and a particular azimuth direction. The arc *ZMH* is bisected at the point *M*, which is therefore the midpoint of that arc. The mean measured altitude $<h>$ of that

---

[2]This is the time in effect at that location. It is usually a type of mean solar time, e.g., EST, EDT, GMT (UT).
[3]Appendix B reviews error analysis and defines terms such as mean, standard deviation of a single observation, and standard deviation of the mean.

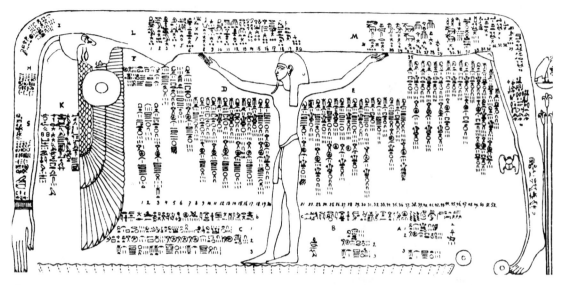

FIGURE 1.1. An Egyptian depiction of the sky as a some-what flattened shell. Seen in cross section is the person of the sky goddess, Nut, raised up by her father, Shu, god of the atmosphere, whose action separated the sky from the Earth and thus created the present universe. After a frieze at Luxor, 1300 B.C., courtesy of Fr. Becker (1980). As with all cosmic myths, a literal interpretation of this depiction is not necessarily what the depictors had in mind.

point is indicated also. The observer is at point O. The ratio $r \equiv OH/OZ = 2.33$ is a measure of the apparent distortion of the sky. Note that if $OZ = OH$, the altitude would measure $45°$.

1.3 Activities

The perception of the sky as a flattened bowl varies from person to person. Let us find out how this shape appears to you.

A cross-staff is a simple instrument consisting of essentially two sticks, a long one along which you sight and a shorter one held perpendicular to the first and slid along the first until the object to be measured is matched in angular size. Make a cross-staff (or sextant, a more elaborate but perhaps more familiar instrument) of your own design, or follow the instructions for making one in Appendix C, and calibrate it. If you are hopeless as an instrument maker, you may use the increments of an equatorial telescope's declination setting circle, or even a blackboard protractor.

With such a device, you can make your own determination of the shape of the sky by replicating the experiment described previously. It is important to try to make the observations at a level site to avoid excessive horizon dip or obscuration. Keep a record of the location, sky condition, date, and time at which the observations are made, the azimuth (bearing from the north, through the east directions), as well as the measured altitude of your best estimated halfway point, for each observation. It would be best to make several observations at the same azimuth before moving to another (if this is possible at the observing site), and arrange your data in tables to facilitate analysis.

After the data have been collected, and you are in a warm, well-lighted place with calculator at hand, compute the altitude means and their uncertainties for each azimuth. Then by geometric construction determine the quantity r, and describe the apparent shape of the sky. Table 1.2 will be needed.[4] We will

TABLE 1.1. Sample measurements of midarc altitudes.

Measure	Altitude (°)
1	27
2	28
3	30
4	34
5	32
Mean	30.2 ± 1.3

[4]Similar tables were developed in antiquity by, for example, Ptolemy.

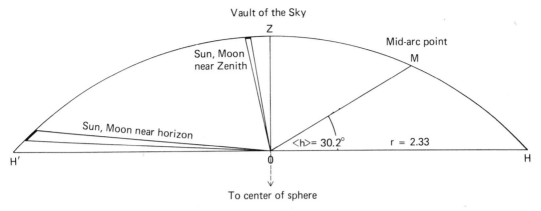

FIGURE 1.2. A constructed cross section of the perceived vault of the sky. The right side illustrates the shape of the sky determined from the measured altitude of the apparent vertical circle arc midpoint. The left side illustrates the effect of the shape of the sky on the apparent angular size of the Moon or the Sun.

make a simplifying assumption in this procedure that the arc of the sky that we see (and will draw) is merely a portion of a circle whose center is not at the observer. The quantity R which appears in column 2 of Table 1.2 is the *assumed* radius of this circle, while the chord length HH', given in units of R, is seen in Fig. 1.2. Select a large piece of graph paper and, with a compass, lightly draw a circle with a radius of 10 cm (the reason for this particular length will be clear in a moment). Now draw a horizontal line to represent your visible horizon, HH', as the chord of the circle you just drew. Place it at the location where its length, in centimeters, is equal to 10 times the relative chord length (column 2) of Table 1.2 for the particular value of $<h>$ you have determined; for example, for $<h> = 30°$, draw a line 14.37 cm long. Place a dot at the midpoint of this line, to represent your observing position, O. Draw a line perpendicular to HH' through O, to represent the zenith–nadir directions. The line intersects the circle at Z, your local zenith. Now from O, construct the angle $<h>$ by means of a protractor, and draw a line from O in that direction to the point where it intersects the circle. The intersection should mark M, the approximate midarc point. Verify that arc $ZM \approx$ arc MH, and that the ratio HH'/R is equal to the ratio indicated by Table 1.2. Finally, measure the line segments OH and OZ and compute $r = OH/OZ$.

A second interesting activity is to measure the angular size of the full moon near the horizon. Unless you are extremely lucky, you cannot do this experiment and the sky shape activity on the same day. However, this will give you a chance to follow the Moon's motions, to observe where it rises or sets, and this may lead to another interesting activity. The Sun can be observed more readily, but it is not safe to observe it without proper filters in place so we do not recommend this procedure. When you find the full moon, measure its largest angular extent both near the horizon and at other altitudes, measuring the altitude of the Moon at the same time. Record the date and sky condition, and keep a table of the time, azimuth, altitude, and angular diameter of each observation. As before, it is a good idea to make several observations at each azimuth to improve the precision of the results.

TABLE 1.2. The half-arc angle $<h>$ vs. chord length relation.

$<h>$ (°)	HH'/R
15	0.705
20	0.946
25	1.191
30	1.437
35	1.677
40	1.888
45	2.000

References

Becker, Fr. (1980) *Geschichte der Astronomie*. Bibliographisches Institut, Mannheim.

2

Astronomy of the Spheres

2.1 The Sky as a Sphere

Despite the apparent flatness of the sky canopy, explored in Chapter 1, the sky has often been depicted as a great sphere. The diurnal motions of the stars are along circular arcs, as the photographed paths of stars near the North Celestial Pole show (see Fig. 2.1). Each stellar diurnal circle, regardless of angular diameter, takes the same length—a sidereal day—to complete. The stars also maintain the same positions relative to each other. In the past when people assumed the Earth to be motionless, the easiest way to account for such a phenomenon was to imagine that the stars were embedded on a great rotating sphere.

Today we have little difficulty picturing the Earth rotating beneath the vast vault of the universe—with each object in it at a different distance from us—thereby creating the diurnal motions. Yet even as late as the early seventeenth century, this point of view was greeted skeptically by many people. The commonsense view of ancient and medieval times was that since we neither see nor feel any effects of the Earth's high speed required by rotation, it was not taking place. The modern view is that since we are moving continuously with the Earth, there is no *relative* motion of objects near us, therefore no inordinate winds constantly whipping past us, or objects careening off tables, or other phenomena attributable to relative motions. Although the geocentric celestial sphere model is physically incorrect, we can still use it to find objects in the sky. Thus we make use of star charts to locate stars among constellations and planispheres of various kinds to locate constellations in the sky; we use catalogue coordinates of an object to point a telescope to it, if the telescope has computer control or setting circles; and we can compute the time of day, our latitude, perhaps our longitude, or even the catalogue positions of observed objects, by making use of the trigonometry of the sphere. Figure 2.2 depicts part of the constellation of Orion.

2.2 Charting the Skies

A star chart, showing objects in their correct relative positions on a portion of the entire sky, is indispensable for observational work. Such a chart often superposes a coordinate grid on the star fields. The coordinates provided are the right ascension and declination. These are the celestial analogues of our terrestrial longitude and latitude, respectively. The *right ascension* (abbreviated α, or RA) is expressed in hours, minutes, and seconds of time; the declination (δ or DEC) is given in degrees. The right ascension is measured along a great circle called the celestial equator and increases eastward from a point called the March or (more commonly) vernal equinox.[1] *Declination* is measured along the shortest arc to one of the two celestial poles, either the North or South Celestial Pole (NCP or SCP). Whereas the RA of an object can be anywhere between 0 and 24 h (spanning 360°), the declination must be between +90° and −90°. Figure A.5 (Appendix A) shows how these coordinates could be imagined to appear on the celestial sphere, while Fig. A.4 depicts the terrestrial coordinate system for comparison. Notice, however,

---

[1]This point is the intersection of the north-moving Sun and the celestial equator on the first day of Northern Hemisphere spring. The September equinox marks the intersection of the south-moving Sun and the celestial equator.

FIGURE 2.1. Diurnal circle arcs on the celestial sphere caused by the Earth rotating beneath the canopy of the sky.

that the latter figures show the sky from the *outside*, whereas Fig. 2.2 is a view from the *inside*.

In Fig. 2.2, the RA increases to the left—as one faces the southern half of the sky. Especially useful are models of the celestial sphere that are made of clear plastic with all labeling on the inside of the sphere. The near side of the sphere can be looked through to see the stars as they appear on the sky on the far inside surface of the sphere. The planisphere and most star charts present the sky as it looks when the charts are held up against the sky, with the lower edge of the chart pointing to the same direction on the horizon (e.g., southwest).

To be seen at all, an object must be above the horizon. To determine if an object is above the horizon, it is useful to picture that portion of the celestial sphere that is seen by the observer. Figure A.7 shows a pair of coordinates in another system —the horizon or altazimuth system. The elements of this system were described briefly in Chapter 1. Figure 2.3 locates the RA and DEC coordinates in the sky of an observer at a particular instant of local time and at a particular latitude. Both the equatorial and the horizon systems are involved. The horizon system will be familiar to those who have attempted Chapter 1.

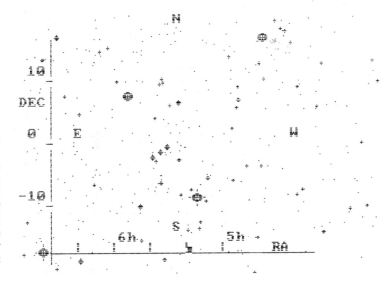

FIGURE 2.2. A portion of the constellation Orion displaying the stars of greater brightness as larger symbols. The star field is depicted as viewed by a Northern Hemisphere observer looking south as the field crosses the meridian. For earlier sidereal times, the star field will appear rotated to the left and will be east of the meridian; for later times, it will be rotated to the right and will be west of the meridian.

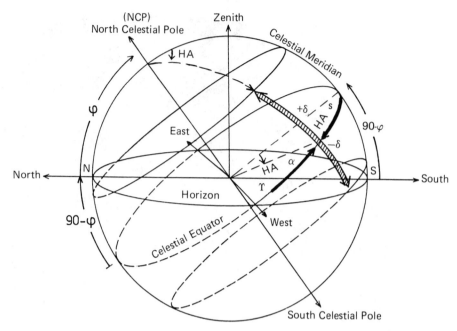

FIGURE 2.3. The celestial sphere as viewed from the outside. The orientation is arbitrary but is appropriate for someone in midnorthern latitudes. The location of the vernal equinox is marked by the symbol of the ram's horn (for the zodiacal constellation Aries, in which it was once located). The celestial meridian is a great circle through the NCP and the observer's zenith (point directly overhead). The celestial meridian cuts the horizon at the north and south points (N and S). The other cardinal points are removed from N and S by 90° of bearing, or azimuth, measured along the horizon. Azimuth may be measured from the north point eastward or from the south point westward; the sense of direction is the same in either case. The altitude is measured from the horizon up toward the zenith. Other quantities are defined in the text and in Appendix A.

2.3 Time and Place

Anyone who has tried to find a star in the sky given its coordinates immediately discovers that the star's visibility depends critically on the time of year and on the time of night. It also depends on the latitude, as one can imagine by tilting the NCP in Fig. 2.3 closer to the horizon (low latitudes) or raising it closer to the zenith (high latitudes). Close to the North Pole, for example, most of the southern stars (at negative declinations) will always be invisible. In addition to redefining a number of terms common to both horizon and equatorial systems, Fig. 2.3 defines a measure of *time*—the hour angle (HA). The hour angle of an object is the number of hours, minutes, and seconds since an object has been on the celestial meridian. When the object in question is the vernal equinox, the hour angle is called the local sidereal time (LST).

At any given instant the hour angle of any object in our sky would be different from that of any other observer, at another place, by the difference in longitudes of the observers' locations. With the sidereal time at some other place whose longitude we know, by measuring the local hour angle of a star we could compute the longitude of our location.

A more usual problem is that we have (or can find) the sidereal time and want to find the hour angle of an object. Notice from Fig. 2.3 that since

$$\text{local sidereal time} = \text{HA} \qquad (2.1)$$

we have for any star

$$\text{LST} = \text{RA*} + \text{HA*} \qquad (2.2)$$

Therefore,

$$\text{HA*} = \text{LST} - \text{RA*} \qquad (2.3)$$

where HA* is the hour angle of the star of interest. By the rules of spherical trigonometry, an arc measured along a great circle is equal both to the angle subtended at the center of the sphere and to the subtended angle at the pole of that great circle on the surface of the sphere. All three quantities involving the hour angle are depicted in Fig. 2.3. Sometimes for hour angles near 24 h the HA is given as negative, for example, -03^h29^m. This means that the object is east of the celestial meridian and will be on the meridian in 3 hours and 29 minutes of sidereal time. Unfortunately, LST is not the same as EST, CST, MST, PST, or any other kind of mean solar time because the Sun runs slow, on average, over the year, compared to the stars. The average rate of this difference is 03^m56^s per day. Note that the hour angle must be in hours, minutes, and seconds of sidereal time.

2.4 Gauging the Sky

Once you have become familiar with the notations and orientations of star charts, they can be used like any road map. The following activities will provide practice:

1. Given the name of a particular star or constellation, you should be able to locate it on a star chart and then determine its approximate RA and DEC. Using a star atlas—a list of bright stars in an astronomical almanac or other observing aid will do nicely—you can check your estimate.
2. Given the RA and DEC of a celestial object, you should now be able to locate it on a star chart.
3. To determine if the object is visible, one needs the LST. This can be obtained readily by locating the celestial meridian in the sky and (if you are in the Northern Hemisphere), facing south,[2] noting which constellations are on meridian. The RA of those constellations will tell you the approximate LST. There is nothing sacred about facing south, but if you face north, note that objects on the meridian below the pole will have an RA 12 h different from those above it. Check your time with a sidereal clock if one is available.
4. Once an object is found in the sky, its azimuth and altitude should be recorded, along with the time. For this purpose, it is advisable to maintain an observing log. Entries should include the date(s), object, RA, DEC, LCT, LST, hour angle (HA), azimuth, altitude, magnitude, and other estimates and comments. The HA must be calculated from LST and the RA. Some of these data are actually redundant. We will return to this point later.

---

[2]In the Southern Hemisphere, you would face north. To follow the subsequent discussion, Southern Hemisphere observers should replace the word "south" with "north" and vice versa.

Table 2.1 lists objects on which this exercise can be tried. Some of the objects are moderately faint and may require an observing site well away from city lights. Not all of the double stars are resolvable as such to the unaided eye and will require a small telescope. For detailed descriptions, a pair of binoculars or a telescope is highly desirable. The list includes what is probably the most beautiful object in the sky: the κ Crucis star cluster, also known as "the Jewel Box." Note, however, that not all objects are visible from all latitudes.

The list of observing targets can be supplemented by a few planets, clusters, nebulae, galaxies, and any of several thousand stars. The sky is of course the limit: the limiting magnitude of the sky due to haze, cloud, or sky brightness will determine how deeply one can reach into the sky. However, even under unfavorable conditions, if it is not actually overcast, one can still observe the brightest objects (one of the authors can attest that astronomical observing is possible even in New York City, where he grew up).

For each object that is found, an entry in the log should be made along with the diameter and magnification of optical aids, if any; in addition, estimates may be made of the angular sizes of extended objects, the colors of stars, orientations of double stars, and the proximity to other objects in the sky. Angular sizes may be measured with a cross-staff or similar device (see Chapter 1 and Appendix C). The direction of the line separating the two components of a double star can be gauged relative to the direction toward the celestial pole. In a telescope or in a tripod-mounted pair of binoculars, the stars' diurnal motion will reveal the west direction and a slight motion of the telescope to the north will cause the stars seen through it to move south.

It was mentioned earlier that some of the data entered in the log were redundant. The rules of spherical astronomy can be used to find the hour angle and declination of an object given its azimuth and altitude, and the reverse. For both transformations, the latitude is needed. The relations between the horizon and equatorial systems are given as follows in terms of the sine and cosine functions:

$$\sin h = \sin \phi \cdot \sin \delta + \cos \phi \cdot \cos \delta \cdot \cos HA$$
$$(2.4)$$

$$\sin A = -\cos \delta \cdot \sin HA \, / \cos h \qquad (2.5)$$

where h is the altitude, ϕ is the latitude, δ is the declination, A is the azimuth, and HA is the hour

TABLE 2.1. A short list of interesting objects for naked eye viewing.

Type name	Position, 1988					V magnitude[a]
	α		δ			
Bright stars						
α Orionis (Betelgeuse)	05	34	+07	24		+0.50
α Canis Majoris (Sirius)	06	45	−16	42		−1.46
α Bootis (Arcturus)	14	15	+19	16		−0.04
α Lyrae (Vega)	18	36	+38	46		+0.03
Extended objects						
M31 (galaxy)	00	42	+41	12		+3.5
Pleiades (open star cluster)	03	46	+24	05		+1.5
Large Magellanic Cloud (galaxy)	05	24	−69	46		+0.1
M42 (gaseous nebula)	05	35	−04	49		~4
Melotte 111 (open star cluster)	12	25	+26	10		+2.9
NGC 4765 (κ Cru)[b] (open star cluster)	12	53	−60	17		+5.2
M13[b] (globular star cluster)	16	41	+36	29		+5.9
Double stars					Separation (arc·min)	
β Tucanae[b]	00	31	−63	01	0.5	+4.4, +4.4
ζ Ursae Majoris[b]	13	24	+54	59	0.2	+2.3, +4 .0
and 80 Ursae Majoris	13	25	+55	02	12	+4.0
ν Draconis[b]	17	32	+55	11	1.0	+4.9, +4.9
ε Lyrae	18	44	+39	37	3.5	+4.5, +4.7
β Cygni[b]	19	30	+27	56	0.6	+3.2, +5.4
α¹, α² Capricorni	20	17	−12	35	6.3	+3.6, +4.2
δ Cephei[c]	22	29	+58	21	0.7	+3.5, +7.5

[a]The unit of star brightness most used in astronomy is the *magnitude*, which is actually an index of faintness since it increases as the star brightness decreases. *V* refers to *visual* magnitude, i.e., in the yellow part of the color spectrum. A star of magnitude +5 is just visible to the unaided eye in a typical suburban sky. A difference of 2.5 magnitudes is equivalent to a ratio of 10:1 in brightness.
[b]Requires a telescope.
[c]Variable.

angle, all expressed here as degrees.[3] The formula for sin A gives the correct answer, but A may be anywhere between 0 and 360° and *two* values of an angle can be found from the same value of the sine function when this range is permitted. Both A and 180° − A should be examined; one of the two values will be correct, the other will not.

2.5 Further Challenges

The following questions, which involve these quantities, may prove challenging:

---

[3]HA or any quantity, say, Q, given in hours, minutes, and seconds can be converted with the formula

$$Q \text{ (degrees)} = \text{hours} \times 15 + \text{minutes} / 4 + \text{seconds} / 240$$

1. Determine the latitude of your site using two methods: measure or estimate as well as you can the altitude of the celestial pole (latitude $\phi = h_p$, pole altitude); measure or estimate the altitude of an object, the declination of which is known, as it crosses the meridian (i.e., has an hour angle of 0). The second method does not yield the answer quite so directly. See if you can argue the logic behind these methods. Compare your results with the known latitude of your site.
2. Verify the claim that an object with HA = 30°, $\delta = 10°$, and observed from a latitude $\phi = 40°$ will be seen at an altitude $h = 50°$ and azimuth $A = 230°$. Why is A *not* −50° (or 310°)? Now verify your observed values of A and H for the log entries.
3. The right ascension of the Sun near the autumnal equinox is ~12ʰ. At about what time of night will an object at $\alpha = 18^h$ and $\delta = 0°$ set at the beginning of autumn?

3

Early Astronomy

3.1 Introduction

One of the most important tasks of astronomy, since its beginnings, has been the determination of time and the closely related problem of the determination of position. The more we probe into the past, the more these activities appear to have had very practical aims. The Greeks were probably the first to regard astronomy as a topic of pure research.

The most important measurements of time—in antiquity as in the present—involve the continuous rotation of the Earth and the motions of the Sun and the Moon through the sky. The Earth's rotation determines the length of the day; the Moon's motion relative to the Sun in our sky is the basis for the month (as the name indicates), and the Sun's circuit among the stars, the year. While the length of the seven-day week and the modern names of the days of the week derive from the seven "planets" of antiquity,[1] it is also true that the seven-day week is about a quarter of a month.

The faithful recurrence of astronomical phenomena never ceases to amaze; it may account for the repeated attempts to find procedures that mark the progress of time. Just as the rising points of the Sun and Moon in Stone Age Britain had calendrical significance, the first sighting of the crescent Moon in the contemporary Middle East has calendrical importance today. One of these procedures is the use of the directions of rise and set of the Sun

and Moon on the horizon. Stonehenge was in use for a millennium or more; at least three major phases of construction are recognized. The several different site lines, as well as foresights and backsights (i.e., points and objects of reference far and near, respectively), the most likely purpose of which must have been to identify on the horizon certain critical rising points of the Sun, and perhaps the Moon, indicate both continual rediscovery and reaffirmation of these activities. This is confirmed by the broad geographic range of the sites. In addition to a large number of megalithic sites in Europe, and elsewhere in the Old World, at which there is some evidence for deliberate astronomical alignments, at least some of the medicine wheels in North America have shown evidence of solar, possibly also stellar, alignments.

The existence of prehistoric astronomical alignments requires that the early observers discovered that the fixed stars continually rise and set at the same points of the horizon when they are viewed from the same place and that the rising and setting points and daily paths of the Sun and Moon depend on the time of year.

To be sure, the phenomenon of the precession of the equinoxes does in fact change the rise and set points of stars slightly over generations because it slowly changes the declinations of stars not near the ecliptic poles. The earliest known discovery of this phenomenon is that of the Greek astronomer Hipparchus (fl. 190–127 B.C.). While the Sun and Moon rise points are not affected by precession in the same way, the Moon's motions are more complicated than just seasonal variation. Because the Moon's period is only ~27⅓ days, its rise and set points vary with the time of month, and because its orbit is tilted with respect to the *ecliptic* (i.e., the Sun's annual path in the sky), they vary from

---

[1] In ascending order from Earth: the Moon, Mercury, Venus, the Sun, Mars, Jupiter, and, in the "seventh heaven," Saturn. In early European astrology, each planet is associated with a particular hour, in cyclic order. The planet "dominating" the first hour of a day gave its name to that day.

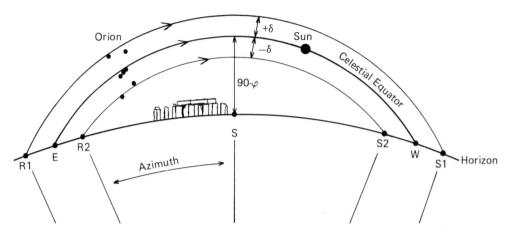

FigURE 3.1. Diurnal circle arcs. The rise and set positions and diurnal motions of the stars of Orion and of the
Sun across a two-month interval. Note that the diurnal
paths of the stars are the same from night to night. Note
also that the astronomical altitude of an object is greatest
at meridian crossing, above the south point of the horizon.
The altitude $h = 90 - \phi + \delta$, where ϕ is the latitude of
the observer and δ is the declination of the object.

month to month as well. Moreover, the orbit swings
around, or regresses, over a period of 18.61 y, so
that the rise and set points vary from year to year as
well. The Moon's daily path across the sky thus
varies with time of month, with time of year, and on
a longer time scale, an 18.61-y cycle. There is much
evidence that megalithic Europe studied the Moon's
motions intently and may well have discovered all
three dependences. The discoveries noted earlier
are the most basic, however, and they are illustrated
in Fig. 3.1, a wide-angle view of the southern
horizon. The celestial equator is shown passing
through the east and west points of the horizon.
Though it is an imaginary arc, it is nevertheless the
most important arc in the sky, for it traces the daily
path of the spring and fall equinox sun through the
sky. The diurnal or daily paths of the Sun at $\pm 10°$
declination are also shown. The extent of the variation of the sun's declination during the course of the
year is greater than the $\sim 20°$ extent of the constellation Orion. The maximum variation amounts
to $\pm\varepsilon$, where $\varepsilon = 23.5°$ (this angle was slightly
greater in megalithic times). As noted in Chapter 2,
the diurnal arcs are entirely due to the earth's rotation beneath the celestial objects and should not be
confused with the slower, eastward motion of the
Sun and Moon against the star background. It is
important to notice that the rise and set points are
symmetrically located on either side of the south
point of the horizon.

It is easy to mark the directions of the rise and set
of objects like the Sun and Moon by setting large
stones in place or by constructing the architectural
lines of buildings to agree with the alignments.
With permanent markers in place, celestial phenomena such as summer solstice could be
predicted with precision year after year. From this
stage, it is conceivable that the visibility of the
Moon and planets and the possible occurrence of
eclipses could be predicted. The relations among
these phenomena could be determined with the
help of counting procedures alone, but we have no
evidence that this was actually done. Nevertheless,
there has been ample speculation. The opinion of
some scientists is that Stonehenge was a "Stone
Age computer" (although "Stone Age *abacus*"
would seem to be a more appropriate expression of
its alleged functioning). In the next section we will
show how Stonehenge might have been used to calculate eclipses.

3.2 History of Stonehenge

Stonehenge, the most famous example of megalithic construction in the world, is located in the
vicinity of Salisbury Plain in southern England.
The grounds have been heavily restored, and so in
gauging its working we must trust that the reconstruction is in accord with its ancient designers'
plans. Basically the construct consists of four or
five concentric rings surrounded by a circular
mound and a ditch. The outermost ring is 88
meters in diameter and contains the Aubrey holes,

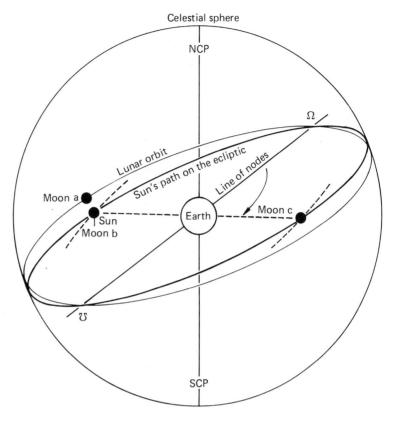

FIGURE 3.2. The lunar orbit. The orbit of the Moon shifts west against the ecliptic, causing the Moon's node crossing to occur earlier each year.

56 pits which long ago were filled in with chalk and are today marked in concrete. The innermost "ring" (it is actually horseshoe shaped) consists of massive stones, called trilithons because they are arranged in pairs, each of which is capped by a third stone, 10 m across. Archaeologists differentiate several periods of construction, the first (Stonehenge I) occurring around 2800 B.C. The Aubrey holes belong to this stage. In the next major construction phase, Stonehenge II, ~2150 B.C., the heel stone was set in place. The last phase, with the Sarsen Circle, the final bluestone configuration, and the innermost trilithons (Stonehenge III), was completed ~2000 B.C. Evidently Stonehenge was constructed from the outside in. The existing avenue to the northeast was extended ~1100 B.C., completing the site. Probably the site was considered sacred for millennia. Aerial photographs have revealed that the surrounding area was not used for agricultural purposes until about A.D. 500, well into the Christian era.

In 1906, the English astronomer and amateur archeologist Norman Lockyer showed that the summer solstice Sun rose approximately over an erect heelstone when viewed from across the center of the site. This phenomenon has long attracted tourists, and access to the site has had to be restricted in recent years. Many of the original standing stones of the construct had fallen, over the preceding centuries, and have had to be reerected. Nevertheless, the numerous solar alignments are not considered artifacts of misplacements of stones; many involve the Aubrey holes, mounds, and other features the placement of which there is no doubt.

The astronomical orientation of Stonehenge is not itself unusual. There are hundreds of other megalithic structures throughout Europe and elsewhere oriented toward the rising and setting horizon positions of the Sun and Moon. Few are as elaborate as Stonehenge, however.

What may differentiate Stonehenge from most of the others is its possible utilization as a kind of Stone Age abacus for eclipse calculation. Unfortunately, proof of this function of Stonehenge is not available due to the absence of written evidence from the prehistoric society that produced it. On the other hand, the basic motions of the Sun and

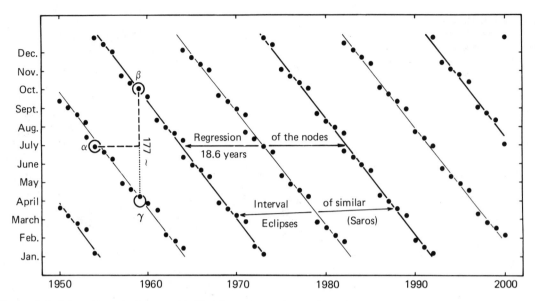

FIGURE 3.3. Solar eclipses in the second half of the twentieth century. The darker lines show eclipses in series that are separated by the interval known as the Saros. Lighter lines show that series of eclipses also occur at intervals of about 9 y, although the characteristics of the eclipses in those intervals are not as closely matching.

Moon are sufficiently clear that it is believed they were recognized even in megalithic times. From these motions, "eclipse warning" tables could be constructed. In the mid-1960s, Gerald Hawkins drew attention to circumstances indicating that in addition to marking the course of the Sun, Stonehenge could have been used to mark the Moon's motion, and with this the occurrence of solar and lunar eclipses. The conjecture has sparked much controversy.

3.3 The Astronomy of Eclipses

The Moon's motions and the possibility of eclipses are certainly knowable, but they are not easy to obtain. If the Moon traveled the Sun's path, the ecliptic, there would be a solar eclipse each new moon and a lunar eclipse each full moon. Eclipses do not occur each month because the Moon's orbit in space is tilted to the ecliptic. The two points where the orbits cross are the *nodes*, an ascending node where the Moon moves from south to north of the ecliptic, and the descending node where it moves from north to south. Therefore, both Sun and Moon must be within certain limits of one of the nodes for an eclipse to occur. Figure 3.2 illustrates the effect of the nodes on the occurrence of

eclipses. The line joining the nodes (the line of nodes) regresses, or moves westward, making a complete revolution in 18.61 y. As the Sun moves eastward during the year, it encounters each node sooner than it would without the regression, creating *eclipse seasons*, which are only about 173^d apart.[2] This causes a variation in the time of year when eclipses can occur, a variation compounded by the fact that neither the orbit of the Moon nor that of the earth is in fact a circle. Astronomical objects move faster when they are closer to, and slower when they are farther from, the object they are orbiting; therefore, the speed at which both Sun and Moon travel on the sky varies with the place in the orbit. Despite these complications, and others, there are periodicities among eclipses. Figure 3.3 shows all the solar eclipses for the period 1950–2000. It demonstrates that eclipses are not random events but follow one another in definite cycles. The cyclicity can be summarized by several rules of thumb:

1. Eclipses occur at intervals of nearly half a year (∼177 days), for example, β − γ in Fig. 3.3.

[2]Note that 173^d is *not* an even multiple of a lunation, which is ∼29.5^d long; thus eclipses must occur at *other* intervals such as near 29.5^d, 148^d, or 177^d.

TABLE 3.1. Solar eclipses, 1950–2000.

Number	Gregorian calendar date	Julian day number	Universal time			Number	Gregorian calendar date	Julian day number	Universal time		
7523	Mar. 18, 1950	2433 359	15^h	20^m	4^s	7580	Nov. 3, 1975	2442 720	13	5	1
7524	Sept. 12, 1950	2433 537	3	29	1	7581	Apr. 29, 1976	2442 898	10	19	9
7525	Mar. 7, 1951	2433 713	20	51	6	7582	Oct. 23, 1976	2443 075	5	10	0
7526	Sept. 1, 1951	2433 891	12	50	0	7583	Apr. 18, 1977	2443 252	10	36	8
7527	Feb. 25, 1952	2434 068	9	16	7	7584	Oct. 12, 1977	2443 429	20	30	8
7528	Aug. 20, 1952	2434 245	15	21	5	7585	Apr. 7, 1978	2443 606	15	16	3
7529	Feb. 14, 1953	2434 423	1	11	0	7586	Oct. 2, 1978	2443 784	6	40	8
7530	July 11, 1953	2434 570	2	28	8	7587	Feb. 26, 1979	2443 931	16	46	6
7531	Aug. 9, 1953	2434 599	16	10	3	7588	Aug. 22, 1979	2444 108	17	10	9
7532	Jan. 5, 1954	2434 748	2	21	8	7589	Feb. 16, 1980	2444 286	8	52	1
7533	June 30, 1954	2434 924	12	26	9	7590	Aug. 10, 1980	2444 462	19	10	6
7534	Dec. 25, 1954	2435 102	7	34	0	7591	Feb. 4, 1981	2444 640	22	14	3
7535	June 20, 1955	2435 279	4	12	0	7592	July 31, 1981	2444 817	3	53	1
7536	Dec. 14, 1955	2435 456	7	8	4	7593	Jan. 25, 1982	2444 995	4	56	9
7537	June 8, 1956	2435 633	21	30	0	7594	June 21, 1982	2445 142	11	52	9
7538	Dec. 2, 1956	2435 810	8	12	5	7595	July 20, 1982	2445 171	18	56	4
7539	Apr. 29, 1957	2435 958	23	54	8	7596	Dec. 15, 1982	2445 319	9	18	6
7540	Oct. 23, 1957	2436 135	4	43	8	7597	June 11, 1983	2445 497	4	38	2
7541	Apr. 19, 1958	2436 313	3	24	0	7598	Dec. 4, 1983	2445 673	12	26	4
7542	Oct. 12, 1958	2436 489	20	51	8	7599	May 30, 1984	2445 851	16	48	0
7543	Apr. 8, 1959	2436 667	3	29	7	7600	Nov. 22, 1984	2446 027	22	57	5
7544	Oct. 2, 1959	2436 844	12	31	5	7601	May 19, 1985	2446 205	21	42	2
7545	Mar. 27, 1960	2437 021	7	37	1	7602	Nov. 12, 1985	2446 382	14	20	1
7546	Sept. 20, 1960	2437 198	23	13	2	7603	Apr. 9, 1986	2446 530	6	8	5
7547	Feb. 15, 1961	2437 346	8	11	0	7604	Oct. 3, 1986	2446 707	18	55	2
7548	Aug. 11, 1961	2437 523	10	35	9	7605	Mar. 29, 1987	2446 884	12	45	5
7549	Feb. 5, 1962	2437 701	0	11	1	7606	Sept. 23, 1987	2447 062	3	8	8
7550	July 31, 1962	2437 877	12	24	5	7607	Mar. 18, 1988	2447 239	2	3	0
7551	Jan. 25, 1963	2438 055	13	42	8	7608	Sept. 11, 1988	2447 416	4	49	7
7552	July 20, 1963	2438 231	20	42	7	7609	Mar. 7, 1989	2447 593	18	19	4
7553	Jan. 14, 1964	2438 409	20	44	7	7610	Aug. 31, 1989	2447 770	5	45	2
7554	June 10, 1964	2438 557	4	23	1	7611	Jan. 26, 1990	2447 918	19	21	1
7555	July 9, 1964	2438 586	11	30	5	7612	July 22, 1990	2448 095	2	54	1
7556	Dec. 4, 1964	2438 734	1	18	6	7613	Jan. 15, 1991	2448 272	23	50	8
7557	May 30, 1965	2438 911	21	13	8	7614	July 11, 1991	2448 449	19	6	4
7558	Nov. 23, 1965	2439 088	4	10	8	7615	Jan. 4, 1992	2448 626	23	10	5
7559	May 20, 1966	2439 266	9	42	9	7616	June 30, 1992	2448 804	12	18	6
7560	Nov. 12, 1966	2439 442	14	26	6	7617	Dec. 24, 1992	2448 981	0	43	3
7561	May 9, 1967	2439 620	14	56	8	7618	May 21, 1993	2449 129	14	7	6
7562	Nov. 2, 1967	2439 797	5	47	9	7619	Nov. 13, 1993	2449 305	21	34	0
7563	Mar. 28, 1968	2439 944	22	48	4	7620	May 10, 1994	2449 483	17	7	4
7564	Sept. 22, 1968	2440 122	11	9	2	7621	Nov. 3, 1994	2449 660	13	35	6
7565	Mar. 18, 1969	2440 299	4	52	3	7622	Apr. 29, 1995	2449 837	17	36	2
7566	Sept. 11, 1969	2440 476	19	56	1	7623	Oct. 24, 1995	2450 015	4	36	8
7567	Mar. 7, 1970	2440 653	17	43	2	7624	Apr. 17, 1996	2450 191	22	48	6
7568	Aug. 31, 1970	2440 830	22	2	6	7625	Oct. 12, 1996	2450 369	14	15	2
7569	Feb. 25, 1971	2441 008	9	49	1	7626	Mar. 9, 1997	2450 517	1	15	5
7570	July 22, 1971	2441 155	9	15	1	7627	Sept. 1, 1997	2450 693	23	52	2
7571	Aug. 20, 1971	2441 184	22	54	0	7628	Feb. 26, 1998	2450 871	17	27	0
7572	Jan. 16, 1972	2441 333	10	53	3	7629	Aug. 22, 1998	2451 048	2	3	4
7573	July 10, 1972	2441 509	19	39	5	7630	Feb. 16, 1999	2451 226	6	39	7
7574	Jan. 4, 1973	2441 687	15	42	9	7631	Aug. 11, 1999	2451 402	11	8	4
7575	July 30, 1973	2441 864	11	39	1	7632	Feb. 5, 2000	2451 580	13	3	9
7576	Dec. 24, 1973	2442 041	15	8	1	7633	July 1, 2000	2451 727	19	20	8
7577	June 20, 1974	2442 219	4	55	6	7634	July 31, 2000	2451 757	2	24	7
7578	Dec. 13, 1974	2442 395	16	25	5	7635	Dec. 25, 2000	2451 904	17	22	3
7579	May 11, 1975	2442 544	7	5	7						

Source: In Th. von Oppolzer (1887), Trans. by O. Gringerich, Canon of Eclipses, Dover, New York, 1962.

TABLE 3.2. Lunar eclipses, 1950–2000.

Number	Gregorian calendar date	Julian day number	Universal time		Number	Gregorian calendar date	Julian day number	Universal time	
4886	Apr. 2, 1950	2433 374	20^h	44^m	4923	May 25, 1975	2442 558	5	46
4887	Sept. 26, 1950	2433 551	4	15	4924	Nov. 18, 1975	2442 735	22	24
4888	Feb. 11, 1952	2434 054	0	40	4925	May 13, 1976	2442 912	19	50
4889	Aug. 5, 1952	2434 230	19	49	4926	Apr. 4, 1977	2443 238	4	21
4890	Jan. 29, 1953	2434 407	23	50	4927	Mar. 24, 1978	2443 592	16	25
4891	July 26, 1953	2434 585	12	19	4928	Nov. 16, 1978	2443 768	19	3
4892	Jan. 19, 1954	2434 762	2	34	4929	Mar. 13, 1979	2443 946	21	10
4893	July 16, 1954	2434 940	0	22	4930	Nov. 6, 1979	2444 123	10	54
4894	Nov. 29, 1955	2435 441	17	6	4931	July 17, 1981	2444 803	4	48
4895	May 24, 1956	2435 618	15	31	4932	Jan. 9, 1982	2444 979	19	56
4896	Nov. 18, 1956	2435 796	6	47	4933	July 6, 1982	2445 157	7	30
4897	May 13, 1957	2435 972	22	32	4934	Dec. 30, 1982	2445 334	11	26
4898	Nov. 7, 1957	2436 150	14	28	4935	June 25, 1983	2445 511	8	25
4899	May 3, 1958	2436 327	12	11	4936	May 4, 1985	2446 190	19	57
4900	Mar. 24, 1959	2436 652	20	17	4937	Oct. 28, 1985	2446 367	17	43
4901	Mar. 13, 1960	2437 007	8	30	4938	Apr. 24, 1986	2446 545	12	44
4902	Sept. 5, 1960	2437 183	11	23	4939	Oct. 17, 1986	2446 721	19	19
4903	Mar. 2, 1961	2437 361	13	32	4940	Oct. 7, 1987	2447 076	3	59
4904	Aug. 26, 1961	2437 538	3	8	4941	Aug. 27, 1988	2447 401	11	6
4905	July 6, 1963	2438 217	22	0	4942	Feb. 20, 1989	2447 578	15	37
4906	Dec. 30, 1963	2438 394	11	7	4943	Aug. 17, 1989	2447 756	3	4
4907	June 25, 1964	2438 572	1	7	4944	Feb. 9, 1990	2447 932	19	12
4908	Dec. 19, 1964	2438 749	2	35	4945	Aug. 6, 1990	2448 110	14	7
4909	June 14, 1965	2438 926	1	51	4946	Dec. 21, 1991	2448 612	10	34
4910	Apr. 24, 1967	2439 605	12	7	4947	June 15, 1992	2448 789	4	57
4911	Oct. 18, 1967	2439 782	10	16	4948	Dec. 9, 1992	2448 966	23	43
4912	Apr. 13, 1968	2439 960	4	49	4949	June 4, 1993	2449 143	13	0
4913	Oct. 6, 1968	2440 136	11	41	4950	Nov. 29, 1993	2449 321	6	26
4914	Feb. 21, 1970	2440 639	8	31	4951	May 25, 1994	2449 498	3	28
4915	Aug. 17, 1970	2440 816	3	25	4952	Apr. 15, 1995	2449 823	12	17
4916	Feb. 10, 1971	2440 993	7	42	4953	Apr. 4, 1996	2450 178	0	9
4917	Aug. 6, 1971	2441 170	19	44	4954	Sept. 27, 1996	2450 354	2	53
4918	Jan. 30, 1972	2441 347	10	53	4955	Mar. 24, 1997	2450 532	4	41
4919	July 26, 1972	2441 525	7	18	4956	Sept. 16, 1997	2450 708	18	47
4920	Dec. 10, 1973	2442 027	1	48	4957	July 28, 1999	2451 388	11	36
4921	June 4, 1974	2442 203	22	14	4958	Jan. 21, 2000	2451 565	4	44
4922	Nov. 29, 1974	2442 381	15	16	4959	July 16, 2000	2451 742	13	55

Source: Th. Oppolzer (1887), trans. by O. Gingerich, *Canon of Eclipses*, Dover, New York, 1962.

2. After 18 years and 11 days—an interval known as the *Saros*—eclipses with similar characteristics occur.

3. The dates of eclipses are displaced from year to year by about −11 days for each of several years, then by about −40, −11, and so on. The accumulated result over 5 y, to which the 177-day interval (which shifts the date close to the next eclipse season) is added, results in a +19 day/y average for the changing date of eclipse.

Still more detail can be gleaned from Tables 3.1 and 3.2 (from Oppolzer, 1887), which list all the solar and lunar eclipses, respectively, from 1950 to 2000. For example, after an interval of 19 y, eclipses occur on very nearly the same date. This interval is known as the *Metonic* cycle and was known to early Greek astronomy as the interval at which the solar and lunar cycles were reconciled.

Next we turn to the question of observability. That an eclipse occurs somewhere on earth is one thing; that it is visible at a particular place is another. Lunar eclipses are visible to all observers for whom the Moon is above the horizon at the time of the eclipse. This will be true for a given observer approximately every other time. It is also the case that many lunar eclipses are not particularly conspicuous. For solar eclipses, the observer must be within a 6000-km-wide strip for a partial eclipse to be seen, and within a strip about 300 km wide for a total eclipse. For Saros interval eclipses, the longitudes of successive eclipses are roughly

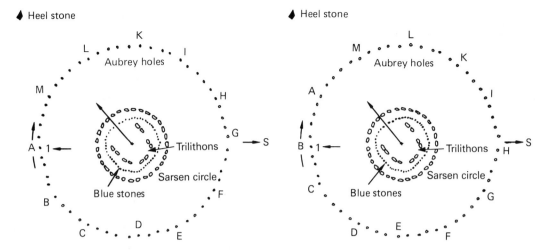

FIGURE 3.4. Planar views of Stonehenge which illustrate the procedure for eclipse prediction. The plan on the left shows the placements of the marker stones *A*–M among the Aubrey holes at a particular date (e.g., at the time of the solar eclipse of late June 1954). The diagram on the right shows the placements after five years of movement of the stones (e.g., at the time of the eclipse of early October 1959).

120° apart (there are also shifts in latitude). Only after three such intervals would the eclipse in this particular series be seen again at nearly the same site. At best, a Stone Age person would have been able to issue only an eclipse warning, that is, a warning that an eclipse might occur. Eclipse prediction, or even warning, requires record keeping. Were records kept at Stonehenge?

3.4 Activities with the Stonehenge "Computer"

According to Hawkins, a possible key to the mystery of Stonehenge is the westward motion of the line of nodes. Three nodal regression cycles total 55.83 y, the nearest integer to which is the number of Aubrey holes. The strategic placement of 12 distinctive heavy stones (Hawkins suggested six, three white and three black) and their movement at, say, summer solstice could provide the ledger needed to keep track of potential eclipse dangers. Figure 3.4 shows how these markers could have been placed among the Aubrey holes.

The possible use of the Aubrey holes as a ledger—actually a counting device or abacus—can be illustrated with present-day eclipses. Consider the solar eclipse of June 30, 1954. This eclipse was visible from the midwestern United States through eastern Canada and across the Atlantic to northern Europe and central Asia. Beginning on that date, place stone *A* in Aubrey hole 1 and the other markers as indicated. If each year the markers are advanced one hole in a specific direction (e.g., clockwise)—and always in the same direction—five years later, in 1959, stone *B* would then appear in hole 1. In that year another solar eclipse occurred. The specific date would require application of the rules of thumb. Adding 95 days to the previous date produces Oct 3, 1959, as the eclipse warning date (the rules yield other dates as well: April 9, 1959; June 19, 1955; June 8, 1956; May 27, 1957; May 16, 1958). By carrying out several such rules of thumb, several warning dates could be constructed for both solar and lunar eclipses.

A useful exercise is to examine the various series of single-node solar eclipses shown connected by solid lines in Fig. 3.3 and find the precise eclipse dates for the eclipses in Table 3.1. Show that the *average* interval between the eclipses (excluding the occasional 29.5-day intervals) noted along the diagonal lines of Fig. 3.3 is ~347 days but that the most likely interval is 354 days. Can you suggest what significance these intervals have?

Now for the most interesting questions. Stonehenge is regarded as a great megalithic indicator of the seasons and possibly also of the Moon's 18.61-y nodal regression cycle. How does the season affect the rise and set points of the Sun and Moon, and why should the Moon's rising and setting points change from year to year?

For more details on how Stonehenge could have been used as a computer, see Hawkins (1965) or Hoyle (1977).

References and Bibliography

Hawkins, G. S. (1965) *Stonehenge Decoded*. Dell, New York.

Hoyle, F. (1977) *On Stonehenge*. W. H. Freeman, New York.

von Oppolzer, Th. (1887) *Canon der Finsternisse*. Kaiserliche Akademie der Wissenschaften, Mathematisch-Naturwissenschaftliche Klassse, Denkschriften 52, Wien (Vienna). Translated by O. Gingerich, under the title *Canon of Eclipses*. Dover, New York, 1962.

4

The Earth's Radius

4.1 Introduction

The shape of the Earth, to a good approximation, is a sphere. To the extent that the density of the Earth depends only on the distance from the Earth's center, the arrow of gravity will point inevitably toward the Earth's center. With the help of a spirit level, a camera can be directed toward the zenith. A change in the latitude of the observer will show up as a shift in the displacement of the diurnal paths of the stars in the field of view of the exposure. For a displacement of 2 mm on an exposure taken with a camera lens of 135-mm focal length, a north–south displacement of 100 km is needed.

More than any other astronomical quantity, the *size* of the planet has determined the trade patterns of nations. Whereas geographic longitude and latitude can be determined by astronomical means without involving the Earth's radius, the *distance* in, say, kilometers between two locations requires it. Moreover, the Earth's size plays a role in everyday life. For example, it determines the amount of fuel required by intercontinental jet transports just as it determined the sailing times for the clipper ships of a bygone era. The spherical shape of the Earth was already known in antiquity, although like much classical learning this was largely lost during the Middle Ages.

The first attempts at determining the size of the Earth were carried out more than 2000 years ago. Best known are the procedures used by Eratosthenes (283–200 B.C.) and Posidonius (140–50 B.C.), which gave for the first time, as far as we know, the correct order of magnitude of the Earth's circumference. The basic method is illustrated in Fig. 4.1. A person moving a distance ℓ along a

meridian[1] will see a corresponding change in altitude of the North Celestial Pole (NCP) in the Northern Hemisphere (SCP in the Southern Hemisphere). This is also indicated by the change in altitude, h, or its complement, the *zenith angle, $z = 90° - h$*, of a star transiting the celestial meridian. Writing the changes in latitude and zenith angle as $\Delta\phi$ and Δz, respectively, we have

$$\Delta\phi = \Delta z \pm z_1 - z_2$$

Then, from the geometry of the sphere,

$$\Delta\phi / 360° = \ell / 2\pi R$$

we find the Earth's radius:

$$R = 360° \cdot \ell / (2\pi \cdot \Delta\phi)$$

The methods of the two ancient astronomers differed in that Posidonius, working between Alexandria in lower Egypt and the island of Rhodes in the Mediterranean, used the meridian crossing of the star Canopus, whereas Eratosthenes used the noonday Sun overhead at Syene (Aswan) at the same instant that its shadow indicated a nonzero zenith angle at Alexandria. Table 4.1 shows the results of these investigations and the modern values of R that they imply.

4.2 Measuring the Earth

In this exercise the classical experiment is replicated with the help of a modern camera, with a long focus lens in place of the eye and cross-staff.

---

[1]A north–south great circle of the terrestrial coordinate system. See Appendix A and refer to Chapter 2 for spherical astronomy designations.

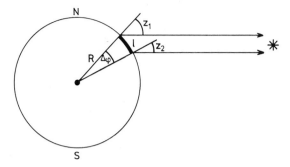

FIGURE 4.1. The angular measurement method of Eratosthenes and Posidonius.

The Sun is not especially suitable for our purposes, however. It may be visible at one site but cloud-enshrouded at the second site. If a measurement is then attempted at the second site on a later date, the declination of the Sun will have changed, vitiating the experiment. It makes more sense, therefore, to follow Posidonius and to work with stars, the declinations of which (for our purposes) are invariable. This makes it possible to make measurements at two different sites on different nights.

First, two suitable sites must be selected. Preferably they should be more than 100 km apart, along a north–south line. It would be best if they are remote from bright city lights to avoid excessive fogging of the film during exposure.[2] Second, the type of camera selected should have telephoto lenses. The combination of site separation and camera focal length will determine the shift of the star field on the film. The camera should be able to rest squarely on an adjustable platform that can be made horizontal with the help of a spirit level (Fig. 4.2). Once the camera is properly leveled, the zenith should lie squarely in the field of view and be centered on the exposures. Color film is not

necessary; black and white will do nicely. Either unmounted slide film or negative film can be used. The film speed is not particularly important, but a higher speed (high ISO or ASA numbers) means a "deeper" exposure, that is, fainter stars will register on the film. The camera will be less likely to be jarred at the opening and closing of the shutter if a cable release is used.

Two more aspects should be considered before work is started on this project. First, the plate scale ought to be considered, to ensure that there is sufficient scale to detect a shift; second, the observer should be well enough organized so that all critical information can be recorded at the time of the exposures.

The image scale, in degrees of sky per millimeter on the negative, is determined from the focal length of the camera lens and the use of the tangent function. The focal length can be drawn to scale for both the field of view and the plate scale, as in Fig. 4.3. The length of the exposed part of the film frame (revealed by the fogging due to the sky background) can be measured and drawn perpendicular to the optical axis (the line through the center of the lens). The angle subtended at the lens by the length of the frame is the angular distance on the sky which is captured on the film. This angle can be measured with a protractor or similar device. In the example in Fig. 4.3, the focal length is 135 mm and the film frame is 11.8 mm wide, so that the angle measured at the lens is 5°. This means that the image scale is 5°/11.8 mm, or 0.424°/mm. Mathematically, we have

$$\tan \theta = \text{film length/focal length}$$

Slightly more precise is the relation

$$\tan (\tfrac{1}{2} \cdot \theta) = \tfrac{1}{2} \cdot \text{film length/focal length}$$

Alternatively, one can determine the image scale by direct measurement of the separation of star trails of known declination difference.

The exposures made at the first site should be time exposures of a few minutes or more (multiple

---

[2]This condition is not essential, however; the Bochum site in West Germany used as an example in this exercise has a lot of light pollution.

TABLE 4.1. Evaluating two ancient methods of measuring the Earth.

Source	Method	Δz	ℓ[a]	R[a]
Eratosthenes	Summer solstice sun at celestial meridian	7.2°	890	7100
Posidonius	Canopus on the celestial meridian	7.5°	600	4600
Modern (equatorial)				6378

[a]ℓ and R are given in kilometers. The exact equivalent of the unit of antiquity, the stadium (plural: stadia), is uncertain.

FIGURE 4.2. Apparatus for measurement of the Earth's radius.

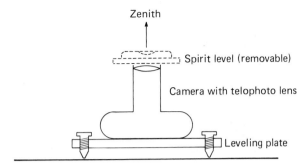

exposures and perhaps even multiple cameras would provide insurance against mistakes of one kind or another—especially if the camera is not a single-lens reflex camera). The correct times, exposure lengths, sky conditions, and other details (leveling procedure, stability of the platform, etc.) should be neatly recorded. The human memory is far too ephemeral to be trusted with material like this. Exposures at the second site should be made with the same camera or cameras and film, and the exposure times should be the same. If possible, exposures should be done within a few days of each other.

As mentioned in Chapter 2, sidereal time runs faster than the mean solar time by which societies regulate their activities. The difference rate is about 3^m56^s/day, so that for each day of delay, the exposures must be made 3^m56^s *earlier*. Since longitude too affects the local sidereal time at which exposures are made, the longitude difference between the sites should also be applied to the time of exposure. If the second site is east of the first one, the star field will be further west at the same standard time. Therefore, the exposure should be made earlier by an amount of time equal to the longitude difference. The combination of these two corrections with due regard for their signs should

give identical local sidereal times, so the images can be compared.

Once the film has been processed, the centers of each negative should be carefully marked with a fine-tipped pen, as in Fig. 4.4. One negative may be placed over the other, the star trails superimposed (if the start of exposure at the second site was correct), and the N–S shift of the zenith measured. With the established image scale, the angle in degrees, $\Delta z = \Delta \phi$, can be found. With ℓ known, the radius of the earth can be calculated.

If bad weather or a restricted travel budget prevents the experiment from being performed, one can always make use of the exposures shown in Fig. 4.4, bearing in mind that reproduction causes distortion and leads to larger than expected error. With a careful tracing of the zenith position and star trails of the negative print on the right of Fig. 4.4, the shift in z is obtained. In this case, $\Delta z = 3.2°$. The spatial N–S separation of the two sites is $\ell = 380$ km, leading to $R = 6800$ km.

4.3 Evaluation of the Errors

Errors may arise through misalignment of the camera axis and slight variations in the camera objective focal length from the nominal value. If one

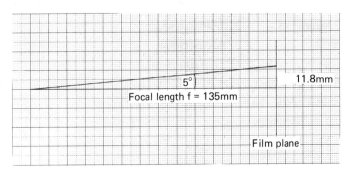

FIGURE 4.3. Basic geometry of the camera. The plate scale can be computed from the angular field of view, θ, and the size of the image on the film, s. To find θ, one can use the focal length, f.

North

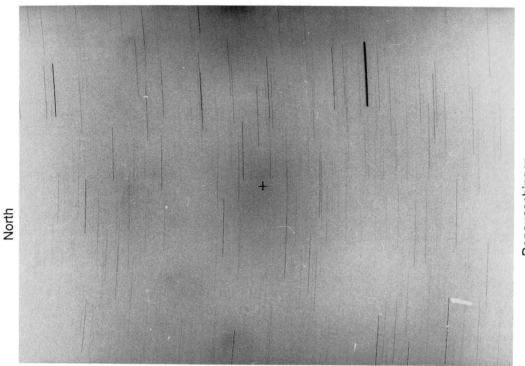

Donaueschingen

North

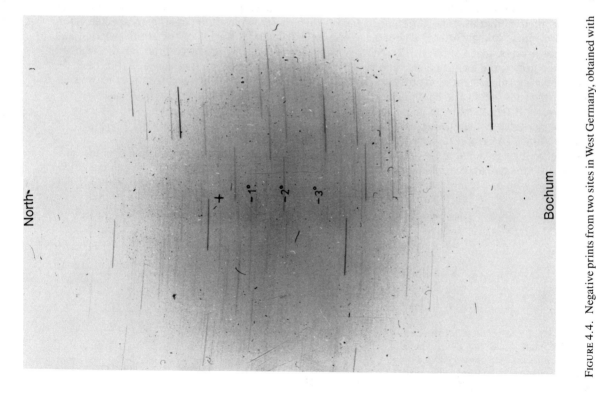

Bochum

FIGURE 4.4. Negative prints from two sites in West Germany, obtained with two different cameras, but both having focal lengths of 135 mm. Both exposures were made at $f/2.8$ for 10 min on black and white film with a speed of 17 DIN (40 ASA or ISO). The left photograph was taken at Bochum on March 5, 1978, 1950–2000 LCT; the right was taken at Donaueschingen on February 4, 1978, 2139–2149 LCT.

camera is used, the significance of these errors mimimized because the same error will presumably apply to both sets of photographs. Depending on which method was used to calculate the image scale, a slightly incorrect value for the focal length may affect the accuracy of the image scale. The authors estimate that the error arising from the example given here amounts to no more than about 10 arc min. This, then, is the uncertainty in Δz.

Since 10 arc min is equal to $0.167°$, the relative uncertainty in R is then given by

$$\Delta R/R = e_{\Delta z}/\Delta z = 0.167°/3.2° = 0.052$$

so that the uncertainty in R is 5% of R.

The greater the separation of the two sites, the greater the value of Δz. Given the same error in Δz, the error in R is proportionately reduced. Bear this in mind when you plan your next vacation!

5

The Distance of the Moon

5.1 Introduction

The Moon is the only heavenly body whose distance from the Earth can be directly determined by a simple geometric technique applied across a baseline of a few hundred kilometers or less.

The Moon's distance from the Earth is nearly two orders of magnitude greater than the Earth's radius (see Chapter 4). Traversing the distance between the Earth and the Moon, an astronaut becomes conscious of the decreasing angular size of the Earth. As the Earth diminishes, the Moon becomes a credible world. The consideration of even this nearest of all the (natural) heavenly bodies may help to remind us of the universe beyond the limited planetary domain that dominates the thinking and attention of almost everyone on Earth.

If the distance from the Earth to the Moon is determined and the Moon's angular size is measured, the actual diameter of the Moon can be calculated. It is, in fact, a good fraction of the Earth's diameter.

The values of the lunar distance and its diameter were already known in antiquity. In Ptolemy's *Almagest* (*ca.* A.D. 150), the distance was recorded as 59 Earth radii. The Earth radius was then known to a precision of only 10–20% and the lunar distance and diameter were known to about the same proportional precision. Aristarchus (ca. 320–250 B.C.) had described how to find the distance to the Sun in terms of the Earth–Moon distance. His method was to compare the angular distance between the Sun and Moon at the precise instant of a quarter Moon—when exactly half of the visible disk is illuminated. The departure from a right angle is so small (less than 9 arc min) that it is no wonder that his recorded result was greatly in error

(about 19 times too small). It is not clear how he performed the experiment, or if he did so at all. However the values were obtained, they implied for Aristarchus that the Sun was much farther from the Earth than was the Moon. Aristarchus was a geometrician and recognized that the angular sizes of the Moon and Sun, and the diameter of the Earth's shadow revealed during lunar eclipses, implied a hierarchy of sizes: Moon, Earth, and Sun in increasing order. Aristarchus was one of a very few astronomers of antiquity to favor the heliocentric hypothesis, probably on the basis of this hierarchy. Fig. 5.1 illustrates the Sun–Moon–Earth geometry during eclipses.

5.2 Lunar Parallax

The Moon's distance can be determined trigonometrically. Procedures of this kind are favored in astronomy, because their basic geometric nature, compared to some other techniques, makes them relatively free of special assumptions. The great disadvantage of the trigonometric method, however, is a sharply limited range over which precise distances can be determined. The principle involved in the method is illustrated in Fig. 5.2. The Moon is photographed at the same instant (or as close together in time as is practicable) from two locales (*A* and *B*) of known distance apart. On each photograph, the Moon appears against a background of stars, but the Moon's image will be shifted relative to the stellar background from one photograph to the other. The distances of the stars are so great that for present purposes they may be considered infinitely far away. The angular shift, p, is related to the lunar distance, r, the distance

FIGURE 5.1. (A)The size of the Earth's shadow exceeds the Moon's diameter. (B) The Sun and the Moon have nearly equal angular sizes as seen from the earth.

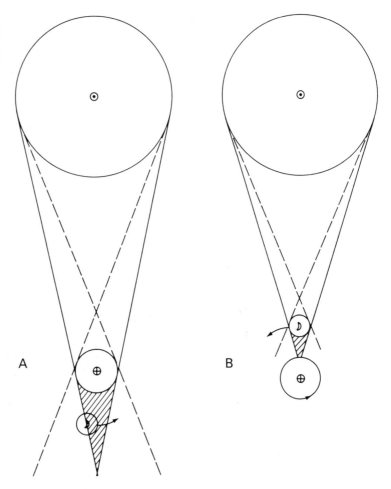

between the two sites, s, and the angle at site A between the Moon and the direction to site B, α, by the expression

$$r = s \cdot \sin \alpha \, / \sin p \qquad (5.1)$$

Figure 5.3 illustrates these quantities. Equation 5.1 follows directly from the law of sines of plane trigonometry:

$$\sin \alpha \, / \, r = \sin p \, / \, s$$

To carry out this experiment, an associate with an identical telescope–camera setup is required. One of you must observe at the second site and obtain the lunar photographs at the same time or times. The greater the site separation, the larger will be p: for every 110 km, p will change by around 1 arc min or about 3% of the Moon's

diameter. With a focal length of 1 m, this shift will be only $\sim \frac{1}{3}$ mm. This quantity *is* measurable, but just barely. The simultaneity of exposures is important. If this is not done, there will be added error due to the movement of the Moon among the stars in the intervening. Since the Moon takes about $27\frac{1}{3}$ days to complete its motion among the stars (a period known as the *sidereal month*), it moves on average

$$\frac{360° \cdot 60'/\mathrm{deg}}{27.3^{\mathrm{d}} \cdot 24^{\mathrm{h}}/\mathrm{day}} = 33 \text{ arc min/h}$$

among the stars. Therefore, to reduce this error well below 1 arc min, to, say, $\sim \frac{1}{2}$ arc min, the exposures should be made within 1 min of each other.

One difficulty is the determination of α, which is the angle between the direction to site B and the

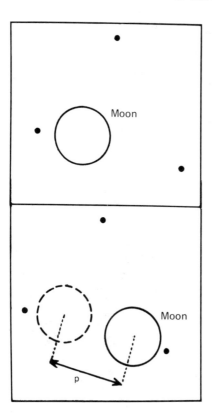

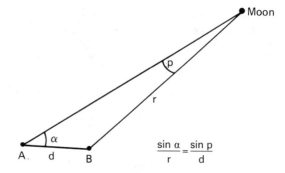

$$\frac{\sin\alpha}{r} = \frac{\sin p}{d}$$

FIGURE 5.3. The relations among the quantities p (angular shift), r (lunar distance), d (distance between observing sites A and B), and α (the angle between site B and the Moon as viewed from site A).

An error of 2° or less in the determination of α is tolerable since the precision of the distance result will be determined almost exclusively by the relatively larger error in the very small quantity p. To measure p, you will have to find the scale of your photographs. The technique described in Chapter 4 can be used if you know the effective focal length of your optical system. Another way to do it is to photograph, with the same apparatus, an untrailed image of a star near the celestial equator for a precise period of time. The lengths of the diurnal arcs of such stars will amount to 15 arc min for every minute of exposure. Therefore, if you can photograph the star field carefully, without jarring the camera, every 4 s, several times, with 4 s in between, you will create a kind of reseau, which can be used as a scale.[2] If disaster ensues, you can always use the data from §5.3!

FIGURE 5.2. The Moon and its parallactic shift among the stars. A shift in observing site causes the apparent shift in the position of the Moon. The angular shift, p, is also the angular separation of the two sites as viewed from the Moon (see Fig. 5.3).

moon. If both sites, the Earth's center, and the Moon happened to lie in the same plane, this would be the *altitude* of the Moon.[1] In general, the bearing of site B will be skewed from the lunar azimuth, however. One way to find α is to measure it with a protractor mounted on a tripod with a universal joint. The base can be made level to the horizon, while the protractor is moved into the plane of the Moon and the direction to site B. If site B can be seen, this is not difficult; it becomes more interesting if the view of site B from site A is obstructed or the distance is too great. In that case, you must obtain the true bearing from a geographic atlas or survey map or calculate it from the principles of spherical trigonometry.

5.3 Finding the Lunar Distance

On April 29, 1976, a partial solar eclipse was seen in Europe. The eclipse was photographed simultaneously at Bochum and at Donaueschingen. The images are shown in Fig. 5.4, and details of the exposures are given in Table 5.1.

The measuring scale seen in Fig. 5.4 was determined by the focal length of the telescope. By tracing one of the two eclipse figures, you will be able to superimpose the two, lining up the solar limb and sunspots, to reveal the shifted lunar limb.

---

[1]See Appendix A for definitions of terms like altitude, azimuth, and bearing.

[2]The use of a cable release can prevent camera shake. Also, a black cardboard held in front can function as a very effective shutter.

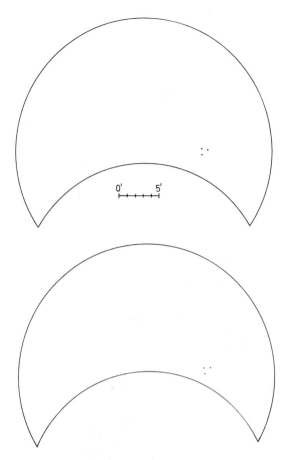

FIGURE 5.4. Images of the partially eclipsed Sun, made at the same instant, at the West German city of Bochum in the Ruhr valley (upper) and the village of Donaueschingen to the south (lower). Note the sunspot group in each photograph. In overlaying the tracing of one of the images on the other, use the limb of the Sun and the orientation of the sunspots as a frame of reference.

The shifted limb arc should be traced as dots or dashes to distinguish it. From the shift between the two arcs, the shift in the Moon's position can now be deduced. Find the centers of each lunar image by bisecting two or more chords drawn across each arc. Then use the scale to measure the shift between the determined centers.

Once you have traced, aligned, drawn, bisected, and measured, you should arrive at the quantity

TABLE 5.1. Exposure details.

Datum	Value
Time of exposure	1976 April 29^d 11^h 35^m LCT
Angle, α	52.7°
Separation, s	398 km

$p = 2.6$ arc min. From the tabular values and p, the lunar distance is found to be

$$r = 398 \text{ km} \cdot \sin (52.7°) / \sin (2.6') = 419,000 \text{ km}$$

The actual lunar distance on this date was 406,000 km. The chief source of error is in the measured shift in image center. This quantity, if obtained carefully, will be less than 0.3 arc min, or ~ 10%. The error in $\sin p$ (which at such small angles varies linearly with p), and thus in r, is therefore also ~ 10%.

As a further challenge, try your skill at spherical trigonometry. From the definitions and principles articulated in Appendix A and an examination of Fig. 5.5, show that

$$\cos \alpha = \cos h \cdot \cos (A_B - A_{\text{moon}})$$

where h is the altitude of the Moon, and A_B and A_{Moon} are the azimuths of the direction to site B and the Moon, respectively.

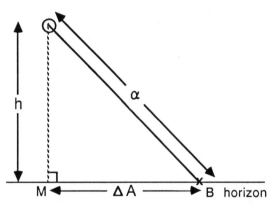

FIGURE 5.5. Finding α by spherical trigonometry. Can you solve the spherical triangle? See Appendix A and Chapters 1 and 2 for details on the definitions and usages of the altazimuth coordinate system.

6

Measuring Lunar Mountains

6.1 Introduction

Even a small telescope reveals many lunar surface features, a fact for which countless numbers of amateur astronomers have been grateful. The appearance of the lunar maria (dry lava beds, misnamed "seas"), walled plains, craters, and mountains testifies to the similarities between terrestrial and lunar landscapes. These features argue for the unity of nature and its laws throughout the universe, although cosmic matter can be seen under more exotic conditions than those prevailing on Earth or the Moon. Mountains on the Moon have varied origins: they may be blocks of lunar crust which were uplifted during the creation of huge impact basins; they may be ejecta created during crater impact; some may be volcanic domes. Therefore, they are found at the margins of the maria and in centers and raised walls of craters. The actual heights of lunar mountains are similar to those of Earth but as a percentage of global diameter, the lunar mountains are proportionately higher.

Galileo (1564–1642), the first person to record such features as mountains on the Moon, described a procedure for determining their heights. Later, as you will be challenged to do in this exercise, shadow lengths were used. The topography of much of the Moon's surface is known most accurately from spacecraft like the lunar orbiter. Although lunar spacecraft remain rare these days, we can readily attempt a ground-based measurement. First we require a first quarter moon, because at this phase, lunar shadows are still prominent and mid-disk features can be used.

6.2 Gauging Mountain Heights

It is best to select a feature near the center of the lunar disk to minimize effects of foreshortening, which complicate the calculations. A crater wall—the high crater rim formed during the crater-creating impact of a meteoroid—is a good choice, because one can use the crater as a standard of scale to measure the length of the shadow. There are numerous suitable craters near middisk and the quarter Moon condition brings out shadows of suitable length. Pick a crater that has a uniform depth, and in which the shadow extends to the middle. The eye is good at catching asymmetries in the shadow, which may indicate floor-level variation as well as variation in ridge height.

Figure 6.1 is a schematic of the crater region, showing the geometry. From the figure's right triangle,

$$\tan \phi = h \, / \, \ell \qquad (6.1)$$

Figure 6.2 places the crater region on a cross section of the Moon. The right triangle CDE involving the lunar radius, R, is similar to triangle ABC. This circumstance makes angle $CDE = \phi$. From triangle CDE,

$$\sin \phi = d \, / \, R \qquad (6.2)$$

Typically, $\phi \approx 10°$. The sine and tangent function values of an angle of this size are roughly equal. Using this approximation, we get

$$h \approx \ell \cdot d \, / \, R \qquad (6.3)$$

The quantities ℓ, d, and R on the lunar disk can be measured at the telescope by timing the interval

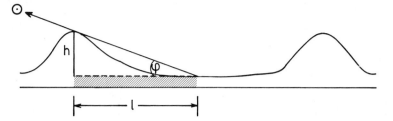

FIGURE 6.1. A cutaway crater showing the height of the wall, h, the length of shadow, ℓ, and the altitude of the Sun at the shadow edge, ϕ.

for these distances to cross a cross-hair or reticle marking in the telescope's eyepiece. If the Moon were exactly on the equator, the time for any arc length Δq to traverse the eyepiece would be

$$\Delta t \text{ (sec)} = \Delta q \text{ (arc sec)} / 15 \qquad (6.4)$$

In general, however, the Moon will have a declination other than zero, δ, the effect of which will be to alter the time it takes for one of these lengths to traverse the telescope eyepiece:

$$\Delta t \text{ (sec)} = \Delta q \text{ (arc sec)} / (15 \cdot \cos \delta) \qquad (6.5)$$

Solving the equation for Δq provides an angular quantity. Another way to obtain the quantities is to measure them on a lunar photograph, such as Fig. 6.3.

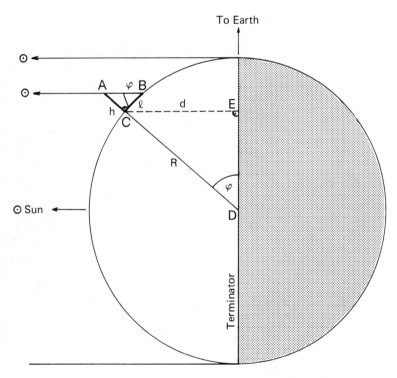

FIGURE 6.2. A cross section of the Moon in the plane defined by the crater, the Earth, and the Moon's center. R is the lunar radius and d is the distance from the plane defined by the terminator.

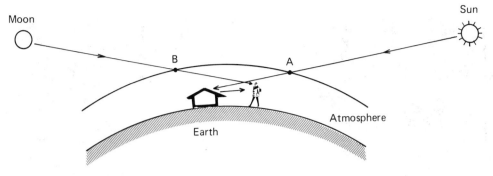

FIGURE 6.4. Conditions of illumination similar to those seen by John Herschel at Capetown, South Africa. Note that the important atmospheric path lengths of solar and lunar light rays are about equal. Sunlight traverses the distance from point A to the illuminated object on Earth, moonlight from point B to the observer's eye. This geometry assumes that the illuminated object is not so far away from the observer as to be appreciably affected by atmospheric extinction (see Chapter 19) and that the sky is clear from horizon to horizon.

The conversion to a linear quantity requires a scale factor. The easiest way to convert ℓ and d is to take the ratio of the angular radius to the linear radius and use this scale factor:

$$\Delta q \text{ (arc sec)} / \Delta q \text{ (m)} = R \text{ (arc sec)} / 1{,}738{,}000 \text{ (m)} \tag{6.6}$$

6.3 The Height of Horrocks's Rim

For illustration, the wall of the crater Horrocks, within the walls of the ringed plain Hipparchus, was selected. It casts a shadow to the crater's center. It is located near the center of the lunar disk, and therefore also near the terminator, which marks the line of sunrise. From an enlargement of Fig. 6.3, the quantities R, d, and ℓ, in measured units, were $R = 126$ mm, $d = 25$ mm, and $\ell = 1.1$ mm. From these,

$$h \approx 1.1 \cdot 25 / 126 = 0.22 \text{ mm}$$

Converting these units into real physical units, the scale factor becomes 1738 km / 126 mm = 13.79 km/mm. Therefore, the height of the crater wall becomes 0.22 mm $\cdot$ 13.79 km/mm, or 3.03 km. This is a typical height for lunar crater walls, which range from 2 to 6 km high. The diameter-to-depth ratio is an interesting property of craters. Since the diameter is $\sim 2 \cdot \ell$, the ratio is 2.2 mm / 0.22 mm = 10. This value indicates a rather flat crater, and not a semicircular hole as one might suppose from its telescopic appearance.

The major source of error in the experiment is in the measurement of ℓ, since the quantities d and R can be obtained with much greater relative precision. The error is

$$\Delta h / h \approx \Delta \ell / \ell \tag{6.7}$$

In the preceding example, $\Delta h \approx 0.1 \cdot h$. Therefore, $\Delta \ell \approx 0.1 \cdot \ell$ or 10%. In the case where ℓ is obtained from timing, a measure of Δt can be obtained from repeated measurements. The average time is

$$<t> = \Sigma t_i / n \tag{6.8}$$

where the symbol Σ indicates the sum of all individual measurements, t_i, and n is the total number of measurements. The standard deviation of the average is then

$$e_{<t>} = \sqrt{ \{ \Sigma \, [(t_i - <t>)^2] / [n \cdot (n - 1)] \} } \tag{6.9}$$

Now taking $\Delta t \pm e_{<t>}$, we have $\Delta \ell / \ell = \Delta t / <t>$ and Δh can be found as before.

FIGURE 6.3. An exposure of the first-quarter Moon taken on June 1, 1979, at 17^h30^m m LCT with an 11-cm refractor of 150-cm focal length. The exposure was ¼ s on ASA 25 (DIN 15) film. The selected crater is indicated by the white arrow.

FIGURE 6.5. Since the surface brightness of the Moon is independent of its distance from us, its surface must be darker than the observatory dome, corresponding more closely to the dark rocks in the foreground.

6.4 Further Challenges

1. One of the assumptions we made during this exercise is that the reference level from which we measured the height of the mountain (or rim, as in the present case) is the average radius of the Moon. In reality, the crater floor altitude is our local standard of measurement. There are many "ghost craters" on the moon, the interiors of which have been filled in by debris and lava flows. The determined height of the wall surrounding such a crater would be found to be too low (except in reference to the floor of the crater). Try to devise a ground-based observation that could determine whether the crater floor is at the same level as the region surrounding the crater.

2. Suppose you want to study a lunar feature that is not located at mid-disk. What qualitative effect would you expect to find on the height determination of a feature (a) close to the terminator but nearer to the north or south limb of the Moon; and (b) further away from the terminator but centered in a N–S sense?

3. When John Herschel (1792–1871), son of the famous discoverer of the planet Uranus, arrived at the Cape of Good Hope to study the stars of the southern sky, he noticed the full Moon rising over Table Mountain, a famous Capetown landmark, which was still in sunlight. Figure 6.4 recreates the geometry. Herschel noticed that the full Moon appeared darker than the rock of Table Mountain, although both were illuminated from approximately the same direction, and concluded that the surface brightness of the Moon was lower than that of Table Mountain. Now surface brightness is independent of the distance to the object, so this observation tells us something about the reflectivity of the Moon's surface. You can perform this experiment too. Figure 6.5 shows a gibbous Moon rising over a telescope dome at the European Southern Observatory at La Silla in Chile. To bypass weather problems, on any sunny day photograph a landscape and, on the same roll of film, photograph the Moon at about the same altitude the Sun had for the first photograph. Be sure that the photo of the Moon is taken at the same exposure time and f stop as the first picture. A more ambitious project is to study the Moon's surface brightness over the synodic period.

7

The Phases of Venus

7.1 Introduction

The phases of Venus have been known since Galileo (1564–1642) first reported them. They provided the clinching argument for the heliocentric theory, to any who would look through the telescope and believe the sight. In the Ptolemaic, or geocentric, view of the solar system, Venus revolved about Earth in an orbit *below* that of the Sun (hence the the term *inferior planet*). Since Venus's motion never carries it very far from the Sun in the sky, it should not be able to reveal more than a crescent phase. Figure 7.1 reveals why this is so and why all phases are visible if Venus actually revolves about the Sun.

After this, only Tycho Brahe's theory—that the planets revolved about the Sun but the Sun orbited Earth—remained (for a while) to challenge the heliocentric theory scientifically.

A detailed inspection of Venus's phases reveals a substantial atmosphere because of the great extension of the *cusps*, or horns, of the Venerian crescent when the planet is close to inferior conjunction. The atmosphere so thoroughly cloaks the visible surface of Venus that in *Perilandra*, part of a science fiction trilogy, C.S. Lewis was free to describe Venus as an ocean planet. Only later, when spacecraft revealed the high temperatures of the surface (as high as 730 K or ~460° C on the daytime side),[1] was it clear that Venus could have no liquid water on its surface. The vague markings that observers saw from earthbound telescopes have clearly been revealed to be circulation patterns in the high atmosphere. Only with Doppler-ranging radar in the latter part of the twentieth

century could the true rotation rate and the surface details finally be uncovered.

Here, the phases of Venus will help us establish another property of Venus—its distance from Earth. The distance of Venus can be determined in *astronomical units* (AU), that is, relative to the average distance of Earth from the Sun, $1.496 \cdot 10^{11}$ m. Through a study of the relative determination of Venus's distance from us, you can recreate a bit of the history of astronomy. In the eighteenth century, for instance, the relative distances of the planets were known far better than their absolute distances. Only later was the astronomical unit calibrated in terrestrial units. To this end, several eighteenth century expeditions set out to various spots around Earth to observe the transit of Venus across the disk of the Sun. Observed across a terrestrial baseline, a shift in time of the onset or ending of the event can be turned into an angular shift, and from the resulting trigonometric relation the distance from Earth to Venus can be found. Since the Earth–Venus distance was already known in AU from orbital study, the AU itself could be calibrated.

In this experiment, you will be able to use observations of Venus's position and phase to determine the planet's distance from Earth and from the Sun.

7.2 Distance to an Interior Planet

The planets Venus and Mercury are both inferior, or to use a more modern term, *interior planets*, and both undergo phase variations. Mercury's closeness to the Sun (0.387 AU, on average) makes it a difficult object to study from Earth. This makes Venus the inner planet of choice for such a study, but the intrepid—especially those who live in low-

---

[1] On the Fahrenheit scale the temperature is ~860°.

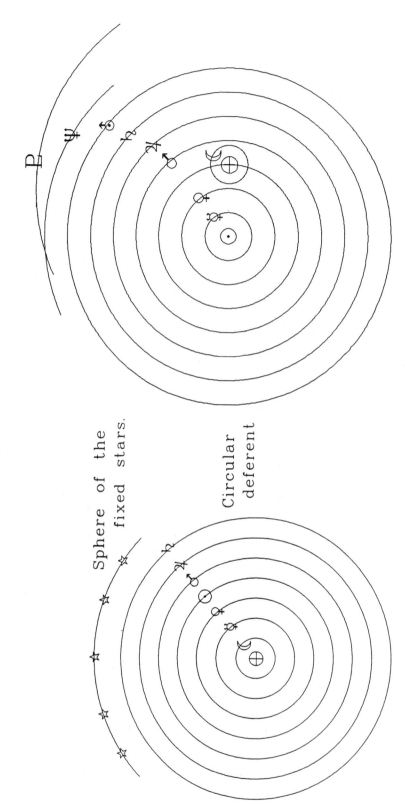

FIGURE 7.1. Two views of the solar system, the pre-Copernican geocentric system (*left*) and the heliocentric system (*right*), revealing how the discovery of gibbous phases of Venus refutes the geocentric theory. Notice that the heliocentric theory also accounts for the seemingly arbitrary limits to the maximum angle (elongation) from the sun that an inferior planet may attain. The orbits are not drawn to scale.

latitude sites—are encouraged to try their luck with our more elusive planetary neighbor.

The geometry of interior planetary configurations is seen in Fig. 7.2, which, like Fig. 7.1, is a view from north of the orbital planes. The angular distance of the planet from the Sun, α, the phase angle of the planet, β, the planet's distance from the Sun, r, and the Earth–Sun distance, $r_E \sim 1$ AU, are related by the sine law of plane trigonometry:

$$\sin \alpha / \sin \beta = r / r_E$$

Therefore, expressed in AU,

$$r \approx \sin \alpha / \sin \beta \qquad (7.1)$$

Figure 7.2 also shows the angular separation of the Earth and Venus as viewed from the Sun, γ, and the distance between the two planets, D. Because all the angles of a plane triangle sum to 180°, $\gamma = 180° - (\alpha + \beta)$. From another application of the sine law, and again setting $r_E \approx 1$ AU, the identity $\sin(180° - \theta) = \sin\theta$, lets us write

$$D \approx \sin \gamma / \sin \beta = \sin(\alpha + \beta) / \sin \beta \qquad (7.2)$$

also in AU. Equations 7.1 and 7.2 require evaluations of the angles α and β.[2] In the case of Venus, α is very nearly the difference in celestial longitude between the planet and the Sun because Venus is nearly (though not exactly) on the ecliptic. The tilt, or *inclination*, of Venus's orbital plane to the ecliptic is $\sim 3.4°$.

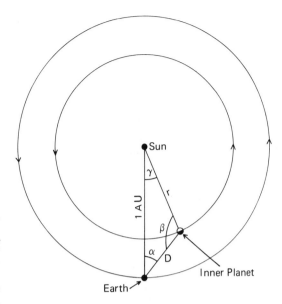

FIGURE 7.2. The geometry of the planetary configuration. Notice that although γ may be any angle between 0 and 360°, β cannot exceed 180° and α is more sharply limited, reaching a maximum value (maximum elongation) at $\beta = 90°$.

---

[2] The angle α should not be confused with the equatorial system right ascension coordinate (also written RA or R.A.) and the angle β should not be confused with the ecliptic system coordinate celestial latitude. These and other spherical astronomy coordinates are described in Appendix A and in Chapters 1 and 2.

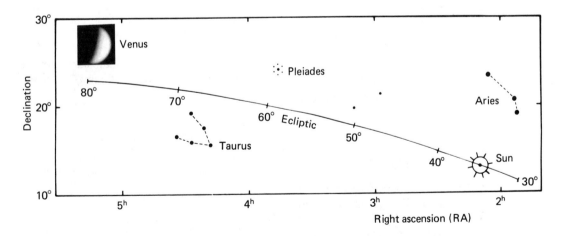

FIGURE 7.3. The positions of the Sun and Venus, April 24, 1972, superimposed on a star chart. The sizes of the Sun symbol and the photographic image of Venus have been greatly exaggerated.

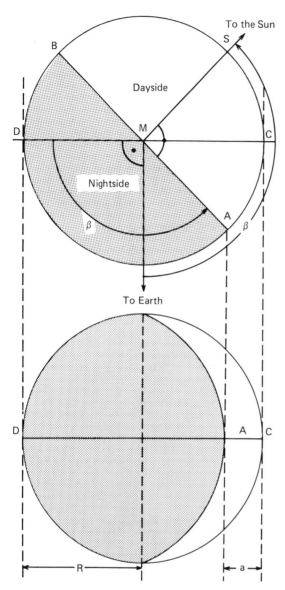

FIGURE 7.4. Interpreting β as a phase indicator for an interior planet. *Upper*: Cross section of the planet in the plane of Earth and the Sun. *Lower*: The disk of the planet as seen from Earth.

7.2.1 Finding α

There are two basic ways to find α: direct and indirect. The direct method involves measuring the angular separation between Venus and the Sun with a cross-staff,[3] sextant, or similar device. This is satisfactory for small angles. If the equipment is not available, or if the angle is large, an indirect method should be used.

There are basically two indirect methods to find α which do not require spherical trigonometry: an observational and a non-observational technique. Either way, you will need a detailed star chart which indicates the degrees of celestial longitude along the ecliptic as well as the RA and declination (DEC) coordinates. Such a chart is illustrated in Fig. 7.3.

It is possible to obtain the positions from observation – if you are fortunate enough to be able to spot Venus in the daytime (this is virtually impossible for Mercury). If observations can be carried out, and if you have a telescope equipped with accurate and finely divided setting circles, the RA or hour angle[4] and DEC coordinates of Venus and the Sun can be read off the setting circles once the objects are acquired. Do not under any circumstances look at the Sun in the telescope eyepiece; a *sunscreen* – a card held beyond the eyepiece – will tell you when the Sun has been acquired. Alternate the settings on the Sun and Venus and take several readings to improve the precision of your result. Aim for at least 1° of precision in each coordinate, noting that 1 min of RA measure is equal to ¼°. These data can be entered on your star chart and the longitude difference read off. This quantity is approximately equal to α.

Almanacs sometimes tabulate the celestial longitudes of these objects as well.[5] If you use these tabulations, calculate the difference in celestial longitude but plot the positions also; the plot will reveal if the difference you obtained by calculation makes sense or if, for example, you have to subtract the difference you obtained from 360°.

Finally, one may use spherical trigonometry to compute α (see §7.3).

7.2.2 Finding β

The angle β specifies the angular distance between Earth and the Sun at Venus. Figure 7.4 demonstrates that this angle can also be used to find the

---

[3]See Appendix C for its description and Chapter 1 for an example of its use.
[4]See Appendix A and Chapter 2 for definitions of these terms. If your setting circles give hour angle only, or if the RA circle must be reset each night, calibrate your circles by setting up on a star whose RA and DEC are known.
[5]The *Astronomical Almanac* is printed annually in hardbound form jointly by the U.S. Government Printing Office, Washington, D.C. 20402, and by Her Majesty's Stationery Office, PO Box 276, London, SW8 5DT, U.K. The U.S. Naval Observatory also produces a computer almanac on diskettes.

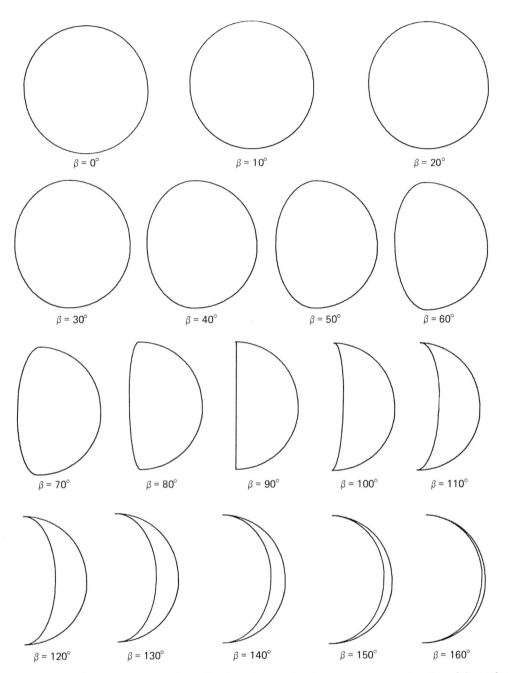

FIGURE 7.5. A succession of outlines of the illuminated areas of an interior planet as a function of the angle β.

phase of Venus: the smaller the angle β, the fuller the illumination on the Earth-facing part of the planet. If the radius of Venus is R, and a is the illuminated portion of R (measured at the widest point), it may be shown that

$$\cos \beta = (a - R) / R$$

The outlines of the illuminated part of Venus in Fig. 7.5 were computed on the basis of this equation. Table 7.1 relates β to several important phases of

TABLE 7.1. Relations among β, a, and Venerian phases.

$\beta(°)$	Phase	Orbital configuration	a
0	Full	Superior[a] conjunction	$2R$
90	Quarter	Greatest elongation	R
180	New	Inferior conjunction	0

[a]Beyond the Sun.

Venus. The value of β thus can be determined by estimation from the appearance of Venus in a telescope. Figure 7.5 illustrates the relation.

7.2.3 Finding r and D

The crescent of Venus which is superposed on Fig. 7.3 suggests that $\beta \approx 100°$. The relative positions of Venus and the Sun in Fig. 7.3 suggest also that $\alpha \approx 45°$. From these values,

$$r = \sin\alpha / \sin\beta = 0.72 \text{ AU}$$

Finally,

$$D = \sin(\alpha + \beta) / \sin\beta = 0.58 \text{ AU}$$

7.2.4 Find the Uncertainty

The angle α can be found more accurately than β. While the typical uncertainty in α may be $\sim 1°$, that in β may be 5–10°. Designate the uncertainties in these quantities as e_α and e_β, respectively, and set $e_\beta = 7.5°$. In radian measure this is

$$e_\beta = 7.5° \cdot \pi / 180° = 0.13 \text{ rad}$$

Error theory predicts that

$$e_r \approx r \cdot |\cot\beta| \cdot e_\beta \qquad (7.3)$$

where the effects of the error in α have been ignored (acceptable except when $\alpha \approx 0$). In the present example, $e_r \approx 0.02$ AU, so that $r = 0.72 \pm 0.02$. The minimum error occurs where $\cot\beta = 0°$ or when $\beta \approx 90°$. This is maximum elongation, where α reaches its maximum value and Venus appears farthest from the Sun. This means that when the inner planet is easiest to observe, it is also best for determining r.

In a similar way, the uncertainty in D can be determined. Expanding Eq. 7.2, and substituting Eq. 7.1 into it, we get

$$D = r \cdot \cos\beta + \cos\alpha \qquad (7.4)$$

Again ignoring the e_α term, substituting from Eq. 7.3, and making use of trigonometric identities, we may write

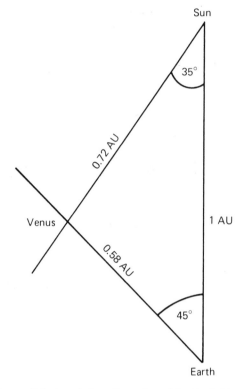

FIGURE 7.6. A scaled configuration of the Earth–Venus–Sun triangle which demonstrates the method of distance determination through construction.

$$e_D \approx r \cdot e_\beta / |\sin\beta| \qquad (7.5)$$

This quantity, too, has a minimum at $\sin\beta \approx 1$ or when $\beta = 90°, 270°$ – at maximum elongation. In the present example, $e_D \approx 0.10$ AU, so that we may write $D = 0.58 \pm 0.10$.

7.3 Further Challenges

1. One may avoid using trigonometric formulas by constructing an appropriate triangle and measuring the lengths of the unknown sides. Figure 7.6 demonstrates this. Scale the Earth–Sun distance as 10 cm, and use a protractor to draw the angles $\alpha = 45°$ and $\gamma = 35°$ from the ends of the baseline; then produce the lines the other two sides of the triangle. The intersection of the two produced lines is the position of Venus. Try it – and check the results.

2. With the aid of a spherical triangle, show that

$$\cos\alpha = \sin\delta_V \cdot \sin\delta_S + \cos\delta_V \cdot \cos\delta_S \\ \cdot \cos(\Delta RA) \qquad (7.6)$$

where δ_v and δ_s are the declinations of Venus and the Sun, respectively, and ΔRA is the difference in right ascension between them.

3. With the aid of a spherical triangle, show that

$$\cos \alpha = \sin h_v \cdot \sin h_s + \cos h_v \cdot \cos h_s \cdot \cos (\Delta A) \qquad (7.7)$$

where h_v and h_s are the altitudes of Venus and the Sun, respectively, and ΔA is the difference in azimuth between them.

4. Calculate the error that occurs as a consequence of assuming that α is the difference between the celestial longitudes of the planet and the Sun. (*Hint:* Look at Eqs. 7.6 and 7.7 and try to find the celestial longitudes and latitudes of your objects—either through tables or plot.)

5. As a final challenge for those who are up to the algebra, trigonometry, and error theory, derive equations Eqs. 7.3 and 7.5.

8

Saturn's Parallax and Motion

8.1 Introduction

The apparent motion of the planets in the skies of the Earth arise from the relative orbital motions of the planets. Figure 8.1 illustrates the changing orbital positions of an exterior planet—one beyond the Earth's orbit—and the Earth, and the consequent relative motion of the planet projected onto the sky. The looping or zigzag movement of a planet at opposition is seen to result from a combination of the planet's orbital motion and from the parallactic shifts caused by the motion of the Earthbound viewer.[1]

Figure 8.2 depicts how that movement would look after the mean motion of the planet has been subtracted. From a backdrop of distant stars, the parallax is obtainable. The parallax is formally defined as the half angle, p, and is related to the distance, r, by the expression $\sin p = 1$ AU $/ r$, so that one may write

$$r \text{ (AU)} = 1 / \sin p \qquad (8.1)$$

8.2 Finding the Parallax

In this exercise, you can determine the distance of an exterior planet by studying its motion. You will need two sheets of millimeter-division graph paper and at least one year's edition of the *Astronomical*

Almanac. We recommend selection of a planet beyond Jupiter so that you will have fewer points to plot and fewer almanac pages to look through. Pick one that is above the horizon and readily visible in your telescope (not Pluto, for example, unless you have a big telescope and deep sky charts!) so that you can compare its location to the tabulated value. Jupiter has an orbital period of ~ 12 y and so has a rough mean monthly motion of $\sim 360°/12$ y, or $\sim 30°$/y against the stellar background. Saturn, Uranus, and Neptune have smaller mean motions since it takes them longer to revolve around the Sun. Once you have selected your planet, note the range of right ascension (RA or α) and declination (DEC or δ) in the current year's almanac tabulation. Now select at least 1 point per month for plotting (more points will give you improved accuracy) and at least 1 point per 10 days if you select Jupiter. Label your graph paper to include the full range of motions of the planet, as in Fig. 8.3, which shows the movement of Saturn in the year 1976. Suitable scales would be 4 mm/degree of declination and 6 mm/minute of time of right ascension. As in Fig. 8.3, one of your plots should have the tabulated points, including the orbital motion; the other will have the (average) orbital motion subtracted out. First plot the observed (tabulated) motions of the planet on one of your sheets. For subsequent telescope work, you should also plot the positions of the bright stars in the field.

Representing the RA and DEC of the planet on a particular date of the year by α_t and δ_t, respectively, and indicating the difference between the values one year apart by $\Delta \alpha$ and $\Delta \delta$, we have

$$\Delta \alpha \pm \alpha_{t_2} - \alpha_{t_1} \quad \text{and} \quad \Delta \delta \pm \delta_{t_2} - \delta_{t_1}$$

where t_1 and t_2 represent the beginning and ending of the year, respectively (e.g., 1976 Jan 1 and 1977 Jan 1). After a full year's time the Earth is back to

---

[1]Normally a planet moves slowly eastward relative to the stars. This is called *direct* motion. However, an interior planet moving through inferior conjunction (from eastern to western elongation) and an exterior planet nearing opposition will appear first to stop and then to undergo a westward or *retrograde* motion in the sky before again stopping and resuming its direct motion. The locations where the planet appears to stop are called the *stationary points*.

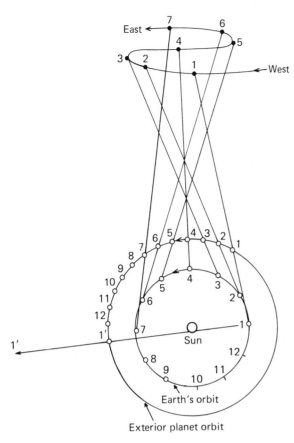

FIGURE 8.1. Retrograde (westward) motion arises from the relative motions of the Earth and planet.

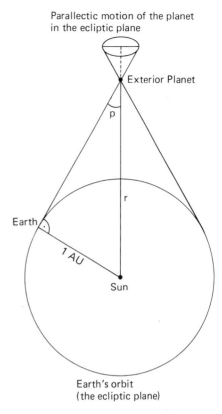

FIGURE 8.2. The parallactic shift of an exterior planet is revealed if the angular motion of the planet in its orbit is removed.

the same position in its orbit. From each of the positions α_i and δ_i for the n selected dates, the *mean motion* ($\Delta\alpha/n$ and $\Delta\delta/n$) can be subtracted to reveal the effect of the movement of the Earth. The subtraction will give the corrected positions, α_i' and δ_i':

$$\alpha_i' \pm \alpha_i - i \cdot \Delta\alpha/n \quad \text{and} \quad \delta_i' \quad \pm \quad \delta_i - i \cdot \Delta\delta/n$$

where i is the consecutive number of the point from the starting date. This requires a uniform length for the intervals. One way to start is to make a table with columns labeled i, date (t), α_i, δ_i, $i\alpha/n$, $i\delta/n$, α_i', and δ_i'. Arrange your entries in n rows separated by two or three blank lines so that you can interpolate data later at specific dates (e.g., around the stationary points).

With this technique, the mean motion of the planet will be eliminated; the orbital motion will be (more or less) frozen and the retrograde loop or zigzag will be transformed into an ellipse. This ellipse corresponds to the parallactic ellipses seen in the motions of the nearest fixed stars (but the latter are much smaller because the distances are so much larger). The semimajor axis (i.e., half of the major axis) is equal to the parallax, p. If the planet, or indeed a star, lay in the ecliptic plane (Earth's orbital plane), the ellipse would degenerate into a straight line. The ellipse's minor axis arises from the inclination of the orbital plane to that of the Earth and will be measurable when the average celestial latitude, β, over the year is not zero. Formally the minor axis is equal to $p \cdot \sin\beta$. Therefore, at the ecliptic pole, the parallactic ellipse for a star becomes a circle.

8.3 The Distance of Saturn

As an example, consider the motions of Saturn between $t_1 = 1976$ Jan 1 and $t_2 = 1977$ Jan 1. From tabulated positions in an almanac,

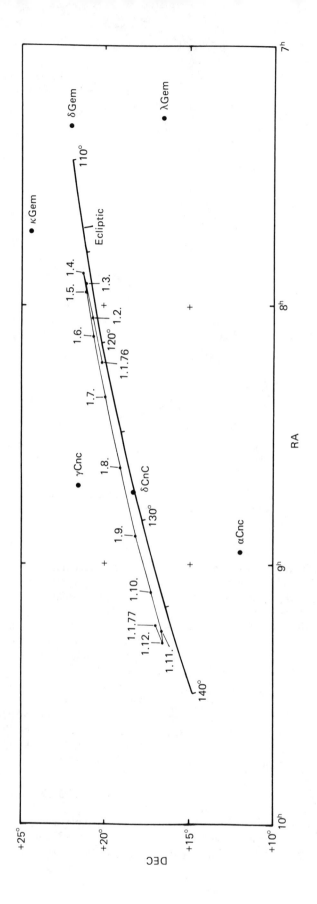

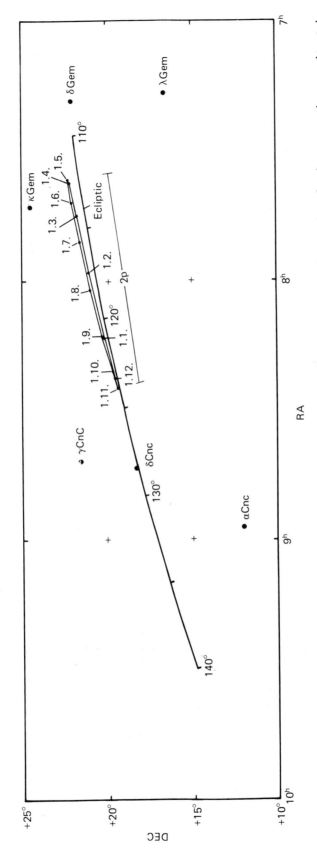

FIGURE 8.3. The motion of the planet Saturn in the sky during 1976. *Upper*: The tabulated (observed) motion. *Lower*: The same data after the mean motion was subtracted.

41

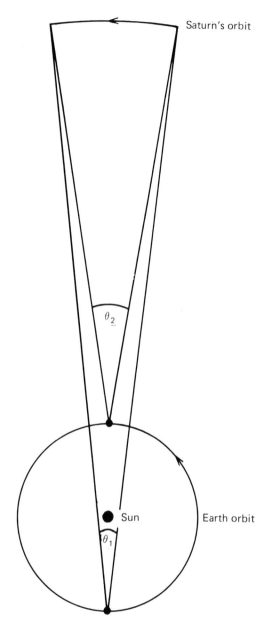

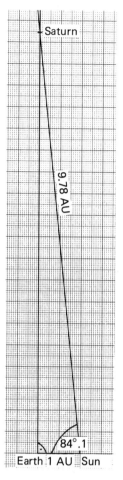

FIGURE 8.5. Graphical method of determining Saturn's distance from the Sun.

FIGURE 8.4. The angular motion of the planet is greater at opposition than at conjunction so that in the same time interval, $\theta_2 > \theta_1$. However, the difference in angle, $\theta_2 - \theta_1$, gets smaller as the distance to the planet increases.

t_1:

$\alpha_1 = 08^{\mathrm{h}}13^{\mathrm{m}}22^{\mathrm{s}}, \quad \delta_1 = +20°10'26''$

t_2:

$\alpha_2 = 09\ \ 14\ \ 15, \quad \delta_2 = +16\ \ 55\ \ 54$

$\Delta\alpha = 01\ \ 00\ \ 53, \quad \Delta\delta = -03\ \ 14\ \ 32$

Adopting $n = 12$, the mean motion then becomes

$\Delta\alpha/12 = 5^{\mathrm{m}}04^{\mathrm{s}}$ / month, $\Delta\delta/12$
$\qquad\qquad = -16'13''$/month

Calculation of the quantities $2 \cdot \Delta\alpha/12$, $3 \cdot \Delta\alpha/12$, ..., the corresponding δ quantities, and the α' and δ' values for each t_i provides the data necessary for the plotting of the lower panel of Fig. 8.3. The resulting ellipse has a major axis length of $2p = 11.74°$. Therefore, $p = 5.87°$. From Eq. 8.1, the distance of Saturn from the Sun then follows:

$$r\ (\mathrm{AU}) = 1/\sin(5.87°) = 9.78\ \mathrm{AU}$$

This value agrees reasonably well with the true value, 9.54 AU; the relative error is thus

$$(9.78 - 9.54)/9.54 = 0.025, \text{ or } 2.5\%$$

8.4 Sources of Error

What are the sources of error? A major source is poor time resolution of the planet's positions around the stationary points—where the planet changes direction. This source of error could be as large as 10%. To improve on the results, a finer spacing of data could be plotted at those points, with the intermediate values of the corrections interpolated from those already calculated. The finer grid treatment will result in a slightly increased value of p because the chances are overwhelmingly great that the course grid will miss the precise dates when the parallax is at maximum at both ends of the ellipse. Another source of error is the fact that planets do not have circular orbits about the Sun. When they are closer to the Sun they move faster than average; when farther away they move slower. Therefore, their rate of motion will not be uniform over the interval of a year selected for this exercise; the result will not be very large, however, because only a relatively small part of the orbit of the exterior planet is involved (unless Mars is selected!). The same variation in orbital velocity holds for the Earth too, however, and over the year it varies by $\pm 1.7\%$. Moreover, the apparent angular motion is a function of distance between the Earth and the planet. Figure 8.4 shows that the angular motion of the planet at opposition is much greater than that at conjunction, other effects being equal.

8.5 Further Challenges

1. Instead of using a calculator to evaluate the sine function, construct a triangle to scale, as in Fig. 8.5, and *measure* the distance of the planet from the Sun. This is not something to do all the time, but if you have never graphically tested a trigonometric function, this might be a good time to do so. A scale of 2 cm/AU will provide fair measuring precision. For the construction, remember that the sum of angles in a triangle is 180° and that in this right triangle, the angle at the Sun between the Earth and the planet must be equal to $180° - p$. Estimate the precision of your measurement and translate it into physical units (meters or kilometers) of planetary distance.

2. Compare the following quantities: $\sin p$, $\tan p$, and p [in radians, i.e., p (degrees) $\cdot \pi / 180°$]. For very small parallaxes, the three quantities approach the same value. This fact is used when stellar parallaxes involved.

3. If the weather permits, compare the sky with your tabulated planetary positions. If you are not using a computerized almanac, you may be comparing *geocentric* positions (i.e., referred to the Earth's center) instead of *topocentric* position (referred to the observer's location). Can you detect this difference in your positions? Illustrate how this effect arises, and show how it vanishes for an object observed at the zenith. Calculate the maximum effect for your object at opposition due to this *terrestrial parallax*.

9

The Astronomical Unit

9.1 Introduction

The mean distance between the Earth and the Sun is designated as the *astronomical unit* (AU).[1] This is perhaps the most basic unit of astronomy. Astronomers have been concerned about its determination for over two thousand years. By the early seventeenth century, the sizes of the orbits of the known planets *relative to the Earth's orbit* were obtainable with reasonably good precision. To translate these distances into miles, meters, or kilometers, however, the distance from the Earth to the Sun (or, more easily, the distance from the Earth to another planet) had to be determined in those units. Since the distance in AU was already known, the AU would then be calibrated. The transits of Venus were observed in previous centuries for just such a purpose[2]; in the twentieth century, to obtain a higher precision, the minor planet Eros and others were observed.[3] In the latter half of the twentieth century, the Doppler ranging technique, which employs the principles discussed in this chapter, was first applied to the planets.[4] From this method, the astronomical unit is known to about 2 km:

$$1 \text{ AU} = 1.495\ 978\ 70 \cdot 10^{11} \text{ m}$$

9.2 The Doppler Effect Phenomenon

In this chapter, the Doppler effect is used to calibrate the astronomical unit. In later chapters, this technique is applied to the stars and galaxies. Here we merely identify and make use of the phenomenon.

An observer will experience a rising pitch (or shortening wavelength) from the radiation emitted by an approaching object. The observer will experience a lowering pitch (or lengthening wavelength) from a receding object. The phenomenon of rising and falling pitch from a train whistle as a train first approaches and then recedes is a familiar illustration, especially when we consider that the whistle has a constant pitch when both the train and the observer are stationary.

The Doppler effect permits the relative velocity of approach or recession to be calculated if the shift in frequency or in wavelength can be measured. This shift is called the *Doppler shift*. The phenomenon is used in planetary astronomy to obtain the velocity of rotation of the planet. When coupled with radar pulse-ranging techniques, it can provide the distance and even permit the mapping of terrain relief on the surface of the planet. As we show later, the distance can also be obtained without the use of radar, which has a limited range but is capable of yielding much higher precision.

---

[1] A formal definition of the astronomical unit requires that there be a slight difference between this quantity and the mean Earth–Sun distance. The latter is actually 1.000 000 03 AU.

[2] For a remarkable account of some of those attempts and the results obtained from them, see Donald Fernie's delightful book, *The Whisper and the Vision: The Voyages of the Astronomers.*

[3] For a thorough summary of the celestial mechanical work involved, see the articles by Eugene Rabe (1950, 1954).

[4] An astronomy laboratory exercise by D. B. Hoff and G. Schmidt (1979) involves finding the rotation period of Mercury from radar pulses reflected from Mercury's surface.

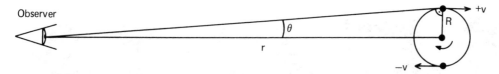

FIGURE 9.1. The basic trigonometry for the determination of the astronomical unit from the rotation of Saturn. In this view, the rotation axis is perpendicular to the line of sight of an Earth observer (and is facing the reader). The line features of the solar spectrum reflected from Saturn's equatorial limbs undergo a shift of $\pm 2v$, whereas spectral lines originating on Saturn would show a shift of only $\pm v$. Both θ and R have been greatly exaggerated for visibility.

The rotational velocity is obtained from the equatorial *limbs* (edges of the disk) where the Doppler shifts will be the largest. The rotational velocity coupled with the observed period of rotation of the planet yields the planetary radius. The linear and angular radii of the planets are related by the distance, which can then be obtained. The giant planets are ideal for this case, because their rotational velocities are high and they have perceptible disks.

If the rotation axis of a planet were perpendicular to the line of sight, the planet's equator would be in the same plane as the rotation axis and the equatorial limbs would be rotating at the full rotational speed. If, on the other hand, the rotational axis were facing the observer, all of the motion would be perpendicular to the line of sight, and no rotational Doppler shift would be seen. In Fig. 9.1 the conditions are closer to the former case than the latter, and the quantities +v and -v represent the equatorial speeds of the planet at the receding and approaching limbs, respectively. The speeds are taken with respect to the center of the disk. The speed, sometimes called the magnitude (as opposed to the direction), of the velocity is related to the planet's radius and rotation period:

$$|v| = 2\pi \cdot R / P \qquad (9.1)$$

Now the Doppler shift of a self-luminous source is given by

$$\Delta\lambda \pm \lambda_D - \lambda = |v| \cdot \lambda / c$$

where λ_D is the observed wavelength, λ is the wavelength of the same radiation from a stationary laboratory source, and c is the speed of light ($2.998 \cdot 10^8$ m/s $= 2.998 \cdot 10^5$ km/s). The Doppler shift of reflected sunlight from a planetary source is given by

$$\Delta\lambda = 2|v| \cdot \lambda / c = 4\pi \cdot R \cdot \lambda /(c \cdot P) \qquad (9.2)$$

after substitution for $|v|$ from Eq. 9.1.

9.3 Finding the Distance

Equation 9.2 may now be solved for the radius of the planet:

$$R = [\, | \, \Delta\lambda \, | \, / \, \lambda \,] \cdot c \cdot P / 4\pi \qquad (9.3)$$

The absolute value of the Doppler shift indicates that either the receding or approaching limb may be used. The planet's radius subtends an angle, say θ, at the earth, a distance r away. From Fig. 9.1, $R = r \cdot \sin\theta$. The angle θ is, however, very small, so that $\sin\theta \approx \theta$. Therefore,

$$r \approx [|\Delta\lambda| / \lambda] \cdot c \cdot P / (4\pi \cdot \theta) \qquad (9.4)$$

The units of r and $c \cdot P$ must agree. If c is in meters per second and P is in seconds, r is expressed in meters.

The distance r is in principle known in astronomical units already because the distances of planets from the Sun are known each moment in terms of their mean distances from the Sun and these are known from Kepler's third law.[5] The trigonometry of the orbits of Earth and the planet can be worked out so that the Earth–planet distance can be found in terms of planet–Sun and Earth–Sun distances and measurable angles. In the easiest case, the Sun, Earth, and planet are in line. Here, the Earth–planet distance $r = A - 1$ aU, where a is the planet–Sun distance. The calibration of the astronomical unit then follows:

$$AU\ (m)\ /\ 1\ (AU) = r\ (m)\ /\ r\ (AU)$$

---

[5]See Chapter 11 for a discussion of this law. Briefly it states that the square of the planet's sidereal period of revolution around the Sun is proportional to the cube of its mean distance from the Sun.

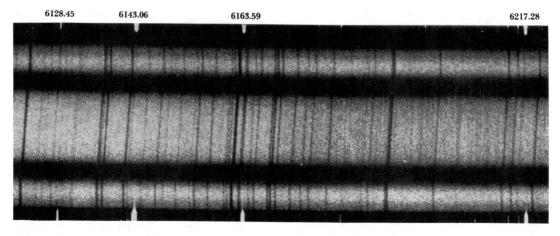

FIGURE 9.2. A spectrum of Saturn in the red spectral region. The white, labeled features at the edge of the figure are laboratory spectral lines with their wavelengths indicated. The large central band is the spectrum of the disk of Saturn (almost all of it from reflected sunlight); the narrower bands on either side are spectra of the rings of Saturn. The tilt of the spectral features shows that whereas the disk of the planet is rotating as a unit,[7] the rings are composed of many small bodies in different orbits. Orbital theory predicts that the bodies in inner orbits will revolve at higher speeds than those in outer orbits and thus will show higher Doppler shifts. In the planetary disk, the highest rotational speeds are displayed by material at the equator and at the very edge of the disk.

9.4 The Rotation of Saturn and the Astronomical Unit

Saturn offers a high equatorial velocity. Its rotation axis, although not always perpendicular to the Earth–Saturn line, is seen from different perspectives as Saturn revolves around the Sun, and the planet can be observed when the perpendicularity condition holds. A spectrum of the planet obtained at such a time will show the full rotational Doppler shift. The spectrum reproduced in Fig. 9.2 has a dispersion of 0.39 Å/mm or 0.039 nm/mm.[6] In the original, the shift between disk center and equatorial limb is 1.0 mm at a wavelength of 6170 Å. Therefore, $\Delta\lambda/\lambda = 0.39$ Å $/ 6170$ Å $= 6.32 \cdot 10^{-5}$. Observation of atmospheric features provides a rotation period, $P = 10^{h}14^{m}$, or 38,640 s. The sidereal period of revolution of Saturn around the Sun is 29.46^{y} and, by Kepler's third law, $a = 9.54$ AU. At opposition, it follows that Saturn is 8.54 AU from Earth. Now at opposition the angular radius of Saturn is found to be 9.76 arc sec, or 4.73 $\cdot$ 10^{-5} rad. With these data, we obtain

$$R = 5.56 \cdot 10^{7} \text{ m}$$
$$r = 1.18 \cdot 10^{12} \text{ m}$$

thus

$$1 \text{ AU} = 1.38 \cdot 10^{11} \text{ m}$$

Of all the potential sources of error in the calculations, ($\Delta\lambda/\lambda$, P, θ, and a), the largest source of error is in fact $\Delta\lambda/\lambda$. The uncertainty in the measured quantity $\Delta\lambda$ is $\pm$ 0.1 mm, or ~10%. If we ignore the much smaller uncertainties in the other quantities, the resulting uncertainty in the astronomical unit is also ~10%. Therefore our determination is

$$1 \text{ AU} = (1.38 \pm 0.14) \cdot 10^{11} \text{ m}$$

9.5 Further Challenges

1. Verify the calculations of §9.4 line by line so that you can trace both the logic and the solution.

---

[6]The proper term for this indication of scale on a photographic image of the spectrum is the *linear reciprocal dispersion*. The Ångstrom unit, Å, is 0.1 nm or $10 \cdot 10^{9}$ m. Therefore the wavelength of yellow-green light is expressible as 5500 Å or 550 nm or $5.5 \cdot 10^{-7}$ m.

[7]This statement is basically true for a belt or zone of gases at a particular latitude, but the rotational speed of an atmospheric feature in the giant planets varies with latitude, and is sometimes variable.

2. Examine Fig. 9.2, and with the aid of a diagram and the knowledge that the rings are in the equatorial plane of the planet, show how the planet may have looked when the spectrum was taken.
3. The equatorial plane of Saturn is tilted ~26° with respect to its orbital plane. Ignoring the inclination of Saturn's orbit to the ecliptic plane (Earth's orbital plane), calculate the maximum effect such a tilt can have on the measured Doppler shift.
4. Try measuring the Doppler shifts in Fig. 9.2 yourself with the help of a reticle. You have many lines to use, and the full diameter of Saturn across which to measure their shifts, for maximum precision. Make sure that you determine your own wavelength scale and interpolate the wavelengths of your selected features carefully. Compare your measured velocity to that in §9.4.

References and Bibliography

Fernie, D. W. (1976) *The Whisper and the Vision: The Voyages of the Astronomers*. Clarke, Irwin & Co., Toronto.

Hoff, D. B. and Schmidt, G. (1979) *Sky & Telescope*, **58**, p. 220.

Rabe, E. (1950) *Astronomical Journal* **55**, 112–126.

Rabe, E. (1954) *Astronomical Journal* **59**, 409–411.

10

The Moon's Orbit

10.1 Introduction

Orbital motion and the natural laws that cause it can be illustrated by the behavior of our nearest astronomical neighbor, the Moon. The Moon revolves around the center of mass of the Earth–Moon system in an elliptical orbit with a period of one sidereal month. The proximity of the Moon to the Earth and the shortness of its orbital period make it possible to deduce the basic character of its orbit. Even more can be obtained by careful study, since the Moon's orbit is further complicated mainly, though not exclusively, by the circumstance that it is in a three-body system. The Sun is the principal perturber of the lunar orbit; two of the other sources of perturbations are the nonspherical shape of the Earth and the gravitational attraction of the other planets.

In this chapter, the motion of the Moon is used to illustrate the validity of Kepler's first and second laws of planetary motion. Along the way, we discuss the degree to which the Moon's complicated motions cause departures from these relatively simple descriptions of solar system motion. You will need polar graph paper and a millimeter rule, or (better) a magnified reticle for measuring the Moon's diameter on photographs. If you wish to go further, you can take your own lunar photographs and follow the steps outlined here to reduce them.

10.2 Kepler's Laws

The motion of two massive objects, each under the influence of the other's gravity, is predicted by Johannes Kepler's three laws of planetary motion,[1] originally formulated for the Sun's retinue of planets:

1. The orbits of the planets are ellipses with the Sun at one focus.
2. The *areal* speed of a planet is constant, that is, the area swept out in a given interval of time by a Sun–planet line is constant anywhere in the orbit.
3. The square of the planet's sidereal period is proportional[2] to the cube of the mean distance from the Sun: $P^2 \propto a^3$.

These laws are derivable from Newton's law of gravity and his laws of motion.[3] The laws of motion are applicable to any two-body system and to the extent that we can identify the perturbing terms, to more complicated cases as well. Indeed, the solar system is actually a multibody system and careful prediction of the motions of each planet must take into account the effects of the other planets. While a full-blown treatment of the lunar orbit is beyond our scope, it is possible to discover Kepler's laws in the lunar orbit.

---

[1]Kepler's first and second laws were published in *The New Astronomy* in 1609; the third law appeared in *The Harmony of the Worlds* in 1619.
[2]By a judicious choice of units—years for time and astronomical units (AU) for distance—Kepler was able to write $P^2 = a^3$.

---

[3]The more general form of Kepler's third law becomes

$$(M_0 + M_p) \cdot P^2 = 4\pi^2 \cdot a^3 / G$$

where M_0 is the mass of the Sun, M_p is the mass of a planet, and G is the gravitational constant.

10.3 The Moon's Orbit and Kepler's First Law

The elliptical nature of the Moon's orbit is not difficult to demonstrate because we have a way to obtain the relative distance to this nearest natural neighbor in space: the angular or *apparent* diameter (α) of the Moon is inversely proportional to its distance (r). First, we may write

$$r \cdot \alpha = D$$

where D is the Moon's diameter, measured in the same units as r. Therefore,

$$\alpha \propto 1 / r \qquad (10.1)$$

If r_p is the distance at closest approach (*perigee*), and r_a is the distance at the farthest point (*apogee*) of the Moon in its Earthbound orbit, and if α_p and α_a are the angular diameters at those distances, then

$$r_p / r_a = \alpha_a / \alpha_p \qquad (10.2)$$

Figure 10.1 shows that $r_p + r_a = 2a$, where a is the mean distance of the Moon to the Earth. This quantity is also the *semimajor axis*; it defines the *size* of the orbit.

The *shape* of the orbit is defined by the *eccentricity*:

$$e \pm (a - r_p) / a$$

This equation can be rewritten:

$$e = (2a - 2r_p) / 2a = (r_a - r_p) / (r_a + r_p)$$

Thus

$$e = \frac{1 - r_p / r_a}{1 + r_p / r_a} = \frac{1 - \alpha_a / \alpha_p}{1 + \alpha_a / \alpha_p} \qquad (10.3)$$

Now examine the photographic images of the full Moon taken at different dates in Fig. 10.2. They were taken under roughly the same conditions with the same equipment. Notice that the images are not the same size even though they all have the same image scale. To make sense of the succession of photographs, we need a means to organize them according to their proximity to perigee. This is not the same thing as the *age* of the Moon, which is based on the *synodic* phase of the Moon (new, crescent, first quarter, gibbous, etc.). We will need a new type of phase, which we will call the *anomalistic phase*, ϕ. This term is used because the *true*

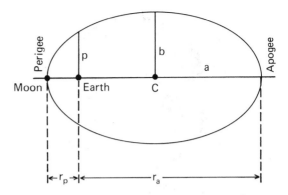

FIGURE 10.1. Some of the quantities that describe the lunar orbit. The eccentricity of the orbit is greatly exaggerated.

anomaly is the name given to the angle swept out by a body in its orbit starting from the perigee (or perihelion, or, more generally, pericenter for non-Earth orbits). The anomalistic phase can be computed as follows:

$$\phi \pm D\{(t - t_p) / P_{\text{anom}}\} \cdot 360° \qquad (10.4)$$

where t is the date of each exposure, t_p is a particular date when the phase was 0 (i.e., when the Moon was at perigee), and P_{anom} is the anomalistic period of the Moon—the time required for it to complete a revolution from perigee to perigee.[4] The symbology $D\{\cdots\}$ means the decimal part of. In this case we are interested not in the number of cycles the Moon has gone through since t_p but only the portion of the orbit traversed since the last perigee. To perform the phase calculation easily, use the table of Julian day numbers (JDN) provided (along with an illustration of its use) in Appendix D. The JDN is a running count of the number of days since a very remote date (JDN 0 $\pm$ January 1, 4713 B.C.). The JDN is reckoned from noon of the calendar date starting on the preceding midnight. The Moon went through perigee on 1975 Jun 14 22^h or $t_p = 2{,}442{,}578.417$. The decimal part is the number of hours since *noon*, 10, divided by 24.

---

[4] $P_{\text{anom}} = 27.5545505^d$. This period (or month) may be compared to several other periods (or months): sidereal (with respect to the stars), $P_{\text{sid}} = 27.321662^d$; synodic (with respect to the Sun), $P_{\text{syn}} = 29.530589^d$; draconic or nodal (with respect to the nodes of the orbit), $P_{\text{dra}} = 27.21221^d$; and the tropical month (with respect to the equinoxes), $P_{\text{trop}} = 27.321582^d$.

1958 April 3 21³⁰ LCT

1958 July 1 22³⁰

1958 August 4 23⁰⁵

1958 November 25 22³⁰

1959 January 21 18³⁰

1959 January 23 20¹⁵

FIGURE 10.2. Six photographic images of the Moon from which the orbital eccentricity can be found. The times are in Local Civil Time (LCT), which in this case means Central European Time. The exposures were taken with an effective focal length of 1000 mm and an f ratio (focal length/diameter) of 20. The exposure was 0.1 s on film with ASA speed 50 (DIN 18).

FIGURE 10.3. The distance of the Moon from the Earth, plotted as C/α vs. ϕ. C is an arbitrary constant to scale r from the measured value of α^{-1} to the graph paper. The Earth's center is indicated by a plus sign. The eccentricity is small enough that the orbit can be approximated by an offset circle, the center of which is marked by a small open circle.

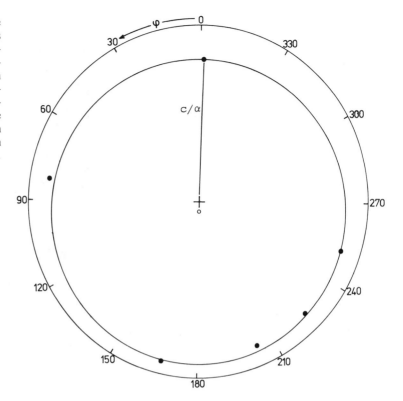

Now make a table with the following column entries: calendar date, JDN, ϕ, and α (the angular diameter of the lunar image). Calculate all values of ϕ and measure α. By taking repeated measurements and averaging, you can improve the precision of each apparent diameter. Select a suitable constant C for plot scale conversion and plot the quantity C/α against α on polar graph paper in a way similar to that in Fig. 10.3. A value of $C = 3600$ will result in the plotted value of $r \approx 60$ mm, for example.

The elliptical shape of the orbit is slight enough, and the data sufficiently noisy, for the path to be approximated by an offset circle (an "eccentric" of Hipparchus's time). In the example of Fig. 10.3, the offset circle is constructed, by compass, to pass through the data as closely as possible. The values of C/α_p and C/α_a may then be read off your graph. Recovering α_p and α_a from these values, you can then apply Eq. 10.3 to find the eccentricity. The true value of the eccentricity $e \approx 0.055$.

Not all the scatter in our plotted orbit is due to roundoff error in computing phase (i.e., in ϕ) or measuring error (i.e., in α). As you will see in the next section, the rate at which the Moon travels along its orbit varies with its orbital position. When it is closer to perigee it moves faster, and therefore it will cover a larger anomalistic phase than it will close to apogee. This contributes to the scatter by effectively smearing the phase because our phase calculation assumes a constant rate—the average rate. Moreover, the shape of the Moon's ellipse is in fact blurred by the perturbing action of the Sun. At the conjunctions (new and full Moon), the Sun tends to increase the separation of the Earth and Moon, while at the quadratures, it tends to draw them together. The perturbation on the shape of the orbit is one among many.

The differing lengths of the several different types of month (synodic, sidereal, anomalistic, and draconic) testify to the complexity of the lunar orbit. The synodic period is, of course, due to the motion of the Earth–Moon system around the Sun. The anomalistic period involves the rotation of the major axis of the Moon's orbit in its orbital plane because of the tidal action of the Sun, which speeds up or slows down the Moon in different parts of its orbit. The draconic period is due to the rotation of the orbital plane of the Moon. As noted in Chapter 3, the latter has to do with the fact that the Earth

TABLE 10.1. Areal velocity data for the Moon.

t (UT date)	Δt	$\Delta\theta$ (°)	$\dot\theta$ (°/d)	α (°)	$\dot\theta/\alpha^2$ [(°·d)$^{-1}$]
Perigee					
1975 July 10^{d}23^{h}15^m					
July 11 23 30	24^{h}15^m	14.7	14.6	0.548	49
Apogee					
1975 July 30 08 00					
July 31 09 00	25 00	13.2	12.7	0.498	51

and Sun tend to pull the Moon out of its orbital plane. The response of the Moon to the torque is the toplike precession of its orbit, *nodal regression*.

To separate out some of these effects, consider the following challenges:

1. In a diagram of the Moon's orbit, show how the motion of the Earth and Moon around the Sun causes the synodic month to be longer than the sidereal month.
2. Examine Fig. 10.1 and describe how the advancement of the line of apsides (the major axis) causes the difference between the anomalistic and sidereal months.
3. The advancement of the line of apsides is about 40.7°/y. The six exposures in Fig. 10.2 were taken over several lunations. Explain why no error in the calculated phase arises from the advancement of the line of apsides.
4. Review Fig. 3.2 and describe how the regression of the line of nodes causes the difference between the draconic and sidereal months.

10.4 Kepler's Second Law

The rate of change with time of the area swept out by the Moon at distance r from the Earth and moving through a small angle $\Delta\theta$ in a small interval of time Δt is

$$r^2 \cdot \Delta\theta / \Delta t \quad \text{or} \quad r^2 \cdot \dot\theta$$

where $\dot\theta \pm \Delta\theta / \Delta t$. The second law therefore states that

$$r^2 \cdot \dot\theta = \text{const} \qquad (10.5)$$

Now from Eq. 10.1, $r \propto 1/\alpha$, so that $r^2 \propto 1/\alpha^2$. Substituting this into Eq. 10.5:

$$\dot\theta / \alpha^2 = \text{const} \qquad (10.6)$$

The quantity $\dot\theta$, the *angular orbital velocity*, is thus proportional to the square of the Moon's angular diameter.

Given measurements of the angular diameter of the photographic images of the Moon, $\dot\theta$ can be found. From the first column of data in Table 10.1, the details of two sets of exposures made in 1975, the angular motion of the Moon has been computed. The date of apogee is not $0.5 \cdot P$ from perigee because of the perturbations. The quantity $\dot\theta$ has not been computed by using a calculation like that in §10.3 but was from the actual movement of the Moon over the daily intervals. The general formula for calculating the change in θ in terms of the change in RA and DEC is

$$\cos\Delta\theta = \sin(\delta_2) \cdot \sin(\delta_1) + \cos(\delta_2) \cdot \cos(\delta_1)$$
$$\cdot \cos(\Delta\text{RA}) \qquad (10.7)$$

Here, $\delta_{1,2}$ is the declination on successive days, and ΔRA is the corresponding difference in RA. A similar expression can be written for the dependence of $\Delta\theta$ on the change in celestial longitude, λ, and latitude, β. Since β typically is $\sim 5°$ or less, an approximation which may be used is

$$\Delta\theta \approx \sqrt{[(\Delta\lambda \cdot \cos\beta)^2 + (\Delta\beta)^2]} \qquad (10.8)$$

The angular rate per day is then computed.

A measured value for the angular diameter, α, may be calibrated in degrees by photographing a star field at the same declination as the Moon. With an exposure of length 120^s / cos δ, the star trail(s) at precisely the declination δ will closely approximate $1/2$°.

The areal velocity, $\dot\theta/\alpha^2$, appears in the last column of Table 10.1. Notice how the change in angle θ very nearly balances the change in the square of the angular size, as shown in the ratio of these quantities in the last column.

10.5 Further Activities

Should you wish to obtain and use lunar photographs of your own for this part of the exercise, you can compute the JDN for the nearest perigee, using a form of Eq. 10.4. Insert the current JDN, t in the more general formula,

$$\phi = [(t - t_p) / P_{anom}] \cdot 360 \qquad (10.9)$$

This will give you the number of cycles, n, since t_p. Then compute the JDN of the next perigee:

$$t' = t_p + (n + 1) \cdot P_{anom} \qquad (10.10)$$

To obtain the next date of apogee, compute $t'' = t' + t_p / 2$. This will be very rough, but it is sufficient to sample the orbit far from perigee. Photographs taken on two successive nights near each of these dates will provide the data needed for the perigee and apogee rates. For $\Delta\theta$ values, the Moon's equatorial or ecliptic coordinates will be needed, as described previously. The coordinates are best looked up, because of the large parallax of the Moon. Be sure to record all relevant tabular data, starting with the dates and times of the exposures. If the observational work is not feasible, use the data of Table 10.1 and work through all the steps.

This has been a lengthy exercise but some of the concepts that were introduced here will find application in other chapters.

11

The Galilean Moons of Jupiter

11.1 Introduction

In the previous chapter, the three Keplerian laws were described and the first two were demonstrated qualitatively in the context of the orbit of Earth's Moon. Kepler discovered the basic orbital descriptions by studying the planets, particularly the planet Mars. Kepler's laws are applicable, however, to the satellites of the planets (as well as to binary stars, to which they will be applied in later chapters). Because we observe the motions of the other planets from an unfavorable perch on our own moving planet, their movements in our sky are somewhat complicated (see Chapters 7 and 8). Consequently, the brightest of Jupiter's moons are incomparably better for demonstrating the third Keplerian law quantitatively.

The Galilean satellites are named for Galileo Galilei (1564–1642), who discovered them telescopically. They present a kind of miniature solar system with the following properties:

1. Nearly circular orbits in nearly the same plane as Jupiter's equator.
2. Short periods (for observing convenience!).
3. Large range in orbital radii (factor of ~4.5).
4. Small gravitational effects on each other and on Jupiter.

Condition 1 prevents their use for demonstrating the first two Keplerian laws, however. Moreover, the mutual gravitational effects and the tidal effects of Jupiter have important consequences for the inner moons, especially Io.

Here the relative form of the third law is used:

$$a_1^3 / P_1^2 = a_2^3 / P_2^2 \qquad (11.1)$$

for two bodies of masses m_1 and m_2, orbiting about a body of large mass, M. The quantities a and P represent the semimajor axis and the period of revolution, respectively. Note that a may be in furlongs (or any other distance unit) and P in fortnights (or any other time unit) if one desires, because the types of units are the same on either side of the equation. The stipulation that M be large is an important one because the Keplerian formula is an approximation to the more exact relation in which the sum of the smaller and larger mass, $m_1 + M$, and not just M, is present. In the case of Jupiter, this approximation is justified, as we shall see.

11.2 Derivation of the Third Law

The third law for vanishingly small satellite masses in circular orbits is easy to derive. To do this, though, we must describe why such objects move in circular arcs.

What prevents the moons from simply escaping from the planet along the direction of their velocities at any given moment is the circumstance that there is a change in the velocity taking place at each instant. This change in velocity is known as the *acceleration*. It is due to the planet's force of gravity acting on the satellite. The acceleration is in the direction of the force, that is, toward the planet, and it has a magnitude (or size)

$$a = v^2 / r \qquad (11.2)$$

Figure 11.1 illustrates the proof of this equation.

The acceleration due to gravity is $g \pm G \cdot M / r^2$. This is the actual cause of the changing velocity in

FIGURE 11.1. The change in velocity Δv with time Δt in a circular orbit is related to the orbital radius r and the change in angle $\Delta\theta$. By similar triangles, $\Delta v/v = r \cdot \Delta\theta / r$. Then, dividing by Δt,

$$\Delta v/\Delta t = (v/r) \cdot (r \cdot \Delta\theta / \Delta t)$$

Defining $a \pm \Delta v / \Delta t$, and noting that $r \cdot \Delta\theta / \Delta t = v$, $a = v^2 / r$.

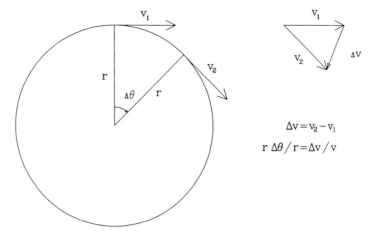

$$\Delta v = v_2 - v_1$$

$$r\,\Delta\theta / r = \Delta v / v$$

the orbit attributed to acceleration a. Therefore, $a = g$, or

$$v^2 / r = G \cdot M / r^2 \qquad (11.3)$$

Thus, $v^2 = G \cdot M / r$. Since $v = 2\pi \cdot r / P$,

$$4\pi^2 \cdot r^2 / (P^2 \cdot r) = G \cdot M / r^2$$

or

$$4\pi^2 \cdot r^3 / P^2 = G \cdot M \qquad (11.4)$$

By applying Eq. 11.4 to each of two satellites m_1 and m_2 and taking the ratio, we obtain Eq. 11.1.

11.3 The Satellite Data

A simplifying condition in the Jovian system is the dominance of the mass of Jupiter, which exceeds that of all its satellites by more than $\sim 10^4$ times. The two important quantities for our purposes are the periods and the orbital radii.

In astronomy many phenomena can be predicted for future dates, or back-calculated for past dates, even across millennia, despite the complications of multiple perturbing forces.[1] So it is with the positions of the Jovian satellites.

The periods of the moons can be obtained with high precision. Suppose for a moment that the

sidereal period of revolution were obtained from observations over only a single cycle. In this case, the error would be equal to an amount Δt. After n such cycles, however, the error would be reduced to an amount $\Delta t/\sqrt{n}$. The period of each satellite can be measured on Fig. 11.2, which portrays the apparent orbital positions of the four Galilean satellites continuously over the period of a month. The lengths of the days can be used as standards of scale, and the onset of transit or occultation — where the satellite crosses in front of or behind the disk of Jupiter — provides a sharply marked location.[2] Try measuring across as many cycles as Fig. 11.2 permits and obtain average values for each satellite.

The size of the orbit can be obtained in a similar way from Fig. 11.2. A millimeter-rule measurement can be converted into meters or kilometers by measuring the linear equatorial diameter of Jupiter ($\sim 142,800$ km). The data appearing in Fig. 11.2 and in figures like it from the *Astronomical Almanac* are designed for observational purposes only and not explicitly for comparison between observation and theory. For our purposes, however, the precision is sufficient. Alternatively, you can carry out an observing project to determine these quantities (see §11.5) for at least two satellites.

The observed periods of the satellites are not actually the sidereal periods. They are synodic

---

[1]Sometimes, as in the case of Gerald Sussman's "digital orrery," orbital modeling can be carried forward or backward hundreds of millions of years (see the comment in *Science* **240**, 986–987, 1988). Such an application requires powerful, dedicated computers and a critical time interval or "step size" for the vital reduction of error accumulation.

[2]The occultation eclipses may be disappearance or reappearance events. The satellites may also pass into Jupiter's shadow so that they become invisible to us (immersions), or suddenly reappear (emersions).

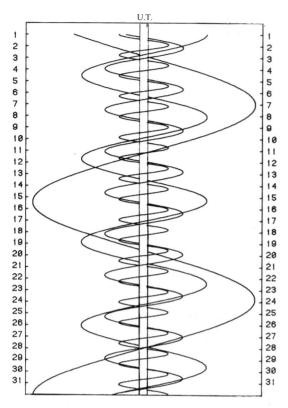

U.T.

FIGURE 11.2. The positions and motions of the Galilean moons of Jupiter as they appeared in January 1976. The tick marks indicate 0 hours UT for each date.

periods, because in the interval in which the satellite has swung around the planet, the planet has also been revolving about the Sun. The two are related by the expression

$$1/\,P_{syn} = 1 / P_{sid} - 1 / P_J \qquad (11.5)$$

where $P_J = 4332.587^d$. Figure 11.3 illustrates the difference, which is analogous to that between the synodic and sidereal months for the Earth's Moon.

There is also the effect of the Earth's motion during the interval of observation, however; the importance of this effect depends on the planet's configuration at the date of observation. It is most important at *opposition* and perhaps least near the quadratures.[3] Figure 11.4 illustrates the extreme parallactic effect across the orbit. As measured from opposition, the maximum shift is

$$\pm \arcsin(1 \text{ AU} / 5.2 \text{ AU}) \approx 11.1°$$

which can lead to an error in timing of $\pm[(11.1°/360°) \cdot P]$.

Table 11.1 contains conservative assessments of the error in period, deduced from observation, from Fig. 11.2, or from the equivalent. An error in P leads to a relative error in P^2 of $2 \cdot \Delta P / P$. Similarly, errors in the orbital radius r will lead to a relative error in r^3 of $3 \cdot \Delta r / r$. A typical measuring error in Fig. 11.2 is about ± 0.25 mm. For Io, $r = 15.8$ mm, as measured on the original diagram from which Fig. 11.2 was taken, and the uncertainty in r^3 is

$$3 \cdot 0.25 / 15.8 = 0.047, \text{ or } \sim 5\%$$

The error in P causes an error in P^2 of

$$2 \cdot 0.05 / 1.77 = 0.056, \text{ or just over } 5\%$$

As a challenge, plot P^2 against a^3 for each of the satellites and thereby deduce the third law graphically. If you provide error bars for your measured quantities, you will get a feeling for how well a straight line can be fitted to the data. Try to determine the slope for the relation. This quantity involves a mathematical constant, $4\pi^2$, a physical constant, G, and the mass of Jupiter. In Chapter 12, we use these data again.

---

[3]See Chapter 8 for definitions and discussion of these terms.

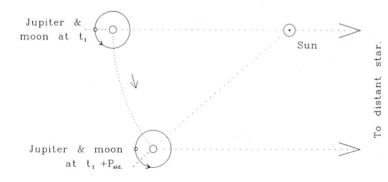

Jupiter & moon at t_1

Sun

To distant star.

Jupiter & moon at $t_1 + P_{sid.}$

FIGURE 11.3. The sidereal and synodic revolutions of a satellite around Jupiter. Sizes and distances have been exaggerated for clarity.

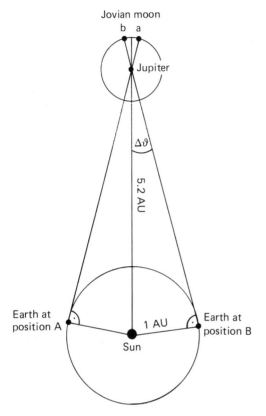

Jovian moon
b   a

Jupiter

$\Delta\vartheta$

5.2 AU

Earth at
position A

1 AU

Sun

Earth at
position B

FIGURE 11.4. The motion of the Earth introduces a parallax which changes the observed average interval between eclipses.

After you have found the Keplerian orbit relation despite the presence of error, you might like to try it again with somewhat improved data. Most of the appropriate corrections (including those for the time for light to travel across the varying distance between Earth and Jupiter) are incorporated in Table 11.2, which contains the basic data for all four Galilean satellites. Perturbations caused by the interactions of the satellites on each other and the effect of Jupiter's nonspherically distributed mass on the satellites are remaining sources of error in these data.

TABLE 11.1. Uncertainty in periods.

Moon (name)	e_p
I (Io)	0.05[d]
II (Europa)	0.11
III (Ganymede)	0.22
IV (Callisto)	0.52

TABLE 11.2. Data for the Galilean satellites.

Name	Orbit radius (km)	Diameter (km)	Brightness[a]	Sidereal period (d)
Io	422,000	3,630	5.0	1.769 137 786
Europa	671,000	3,140	5.3	3.551 181 041
Ganymede	1,070,000	5,260	4.6	7.154 552 96
Callisto	1,883,000	4,800	5.7	16.689 018 4

[a]Brightness is measured in magnitude units at mean opposition.

11.4 The Third Law

Ratios of the sidereal periods and ratios of the orbital radii may be calculated, squared, and cubed, respectively, to give

$$(P_I / P_{III})^2 = 0.06114 \quad (r_I / r_{III})^3 = 0.06135$$
$$(P_{II} / P_{IV})^2 = 0.04528 \quad (\dot{r}_{II} / r_{IV})^3 = 0.04525$$
$$(P_I / P_{IV})^2 = 0.01124 \quad (r_I / r_{IV})^3 = 0.01126$$

The satellite designations are identified in Table 11.1. To the precision with which the distances are known (±500 km), the results are in agreement. The largest discrepancy in the preceding ratios is $\sim0.3\%$. Try calculating and plotting P^2 vs. r^3 to note the improvement in the graphical fitting of the data.

11.5 Observational Activity

If you have an evening, a telescope, and clear skies, use the current edition of the *Astronomical Almanac* to predict approximately when an eclipse of one of the Jovian satellites will begin (observing an end is a bit trickier because it is harder to catch a reappearance than to note a disappearance) and plan ahead of time to observe it, carefully timing the event. At about the time of maximum separation from the planet, use a timing technique[4] to

---

[4]Through an eyepiece equipped with a crosshair, locate Jupiter and its moons, and identify the moon of interest. The crosshair should be rotated so that it is perpendicular not to the east–west diurnal motion of the Jupiter but to the line joining the planet and the moon you have selected to observe. Place the field *west* of the crosshair. Shut off the clock drive, if you have one, and time the passage of the satellite–planet distance. Also time the passage of the disk of the planet across the crosshair to provide a standard. Then

$$r_s \text{ (km)} = [t_s + \tfrac{1}{2}t_{disk}] \text{ (s)} \cdot [142,800 / t_{disk}]$$

Repeated measurement gives you both an estimate of error *and* improved precision in the resulting mean. See also §12.3 for further hints.

determine the orbital size. Note, though, that the orbital radius is measured to the *center* of the planet, so to the satellite–disk limb separation, add the radius of Jupiter. However you obtain the orbital radii and the sidereal periods of revolution about Jupiter, your work can serve a dual purpose: these data are also useful for finding the density of Jupiter, discussed in Chapter 12.

11.6 Further Challenges

If you examine the relations among the sidereal periods in Table 11.1, you will discover some interesting *resonances*: integral multiples of the angular motion. In the case of Io (satellite I), Europa (II), and Ganymede (III), we have

$$1 / P_I + 2/P_{III} - 3 / P_{II} = 0 \qquad (11.6)$$

Since the mean motion is $\dot{\theta} = 2\pi / P$, we may write Eq. 11.6 as

$$\dot{\theta}_I + 2\dot{\theta}_{III} - 3\dot{\theta}_{II} = 0$$

and therefore, integrated

$$\theta_1 + 2\theta_{II} - 3\theta_{II} = 180°$$

The satellite orbits are locked into one another and perturb one another. Io is perturbed out of its nearly circular orbit each time the other satellites are passed by it. The enormous forces of Jupiter on the innermost satellite vary strongly with distance $(1/r^3)$. As a consequence of the variable tidal friction produced by Jupiter, Io has a molten interior and is the most active volcanic site (that we know of) in the solar system. Its volcanoes spew material hundreds of kilometers above its surface, material that is quickly ionized by rapidly moving high-energy particles already caught in the intense magnetic field of Jupiter. It is a fascinating world, but no less is the ice moon, Europa. With relatively little terrain relief, this satellite appears to have a source of water—perhaps an ocean—beneath its icy crust which rapidly obliterates craters and fills in surface cracks. Huge striated areas and circular, dark-colored basins are found on Ganymede, but the chief distinguishing surface features of the larger Galilean satellites are their many craters, indicating much more ancient surfaces. The four satellites are indeed uniquely amazing worlds.

12

Planetary Packaging: Mean Densities

12.1 Introduction

An important clue to the nature of any celestial object is how it is packaged—its mass per total volume, or *mean density*, $<\delta>$.[1] This is well illustrated by the example of a pulsar, the mean density of which is a colossal 10^{17} kg/m³, or 10^{14} g/cm³. Water has a mean density of only 1 g/cm³, and the densest materials with which we are familiar have densities of up to ~20 g/cm³. The high densities found in the pulsars must be interpreted in terms of atomic nuclei densities—hence the origin of the "neutron star" designation.

Even in less exotic realms like the solar system, the mean densities of the Sun and planets provide information about the early history of the solar system. Planets can be divided into two groups on the basis of density: *terrestrial*, or *Earthlike*, planets (Mercury, Venus, Earth, and Mars) with densities ~5000 kg/m³, and *Jovian*, or *giant*, planets (Jupiter, Saturn, Uranus, and Neptune) with densities ~1000 kg/m³. The giant planets have densities similar to that of the Sun ($\delta = 1400$ kg/m³), which means that they are composed, in large part, of hydrogen, the lightest of all the elements. Except for the very low-mass planet Pluto, which is sometimes included in this category, the terrestrial planets are located in the inner part of the solar system. There they have been subject to the intense radiation of the Sun since their origins. The effect of the high solar flux has been to drive

away most of their initial hydrogen gas. The heavier elements and compounds predominate, with correspondingly higher mean densities. In this chapter, the mean densities of Jupiter and the Sun are compared to that of Earth. Earth's density is derived from the radius (obtained in Chapter 4) and the mass or alternatively the gravitational acceleration g.

12.2 Deriving the Densities

The mean density is defined as the total mass per total volume:

$$\delta = M / V$$

but it can be expressed differently. In the case of Jupiter, we can use the orbital properties of one of its satellites (see Chapter 11). The planet approximates a sphere, the volume of which is

$$V = 4/3 \cdot \pi \cdot R^3$$

where R is the planet's radius. The gravitational force on a satellite of mass m in a circular orbit of radius r may be written

$$G \cdot M \cdot m / r^2 = G \cdot 4/3 \cdot \pi \cdot R^3 \cdot \delta \cdot m / r^2 \tag{12.1}$$

The gravitational constant $G = 6.6726 \cdot 10^{-11}$ m³/(kg · s²). The corresponding centripetal force is

$$m \cdot v^2 / r = m \cdot (2\pi \cdot r / P)^2 / r \tag{12.2}$$

where v is the satellite's orbital speed and P is its sidereal period. Setting equal Eqs. 12.1 and 12.2,

$$G \cdot 4/3 \cdot \pi \cdot R^3 \cdot \delta \cdot m / r^2 = 4\pi^2 \cdot r \cdot m / P^2$$

---

[1]In this chapter we designate the mean density δ. In other contexts, however, the symbol δ usually designates the actual density that prevails as a function of position throughout the body; i.e., δ is a function of radial distance from the center and it may vary with latitude and perhaps even longitude throughout the interior.

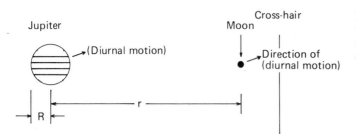

FIGURE 12.1. The ratio r/R determined from the timing of transits of the planet and satellite through the crosshair of a telescope eyepiece.

Solving this equation for δ,

$$\delta = 3\pi \cdot (r / R)^3 / (G \cdot P^2) \qquad (12.3)$$

Notice that the values of R and r need not be obtained in absolute units like meters; measured, relative units—which were used in Chapter 10—will do. By measuring the separation of one of the satellites from Jupiter's center at maximum elongation on a photograph or by timing it (both techniques were described in Chapter 10), you can measure the angular radius of the orbit, r, and, in the same units, the planet's diameter. Figure 12.1 illustrates the measurements needed, and §11.5 and §12.3 describe the telescopic technique.

A similar technique will work with the Sun and with the Earth, which is, of course, a satellite of the Sun (see Fig. 12.2).

12.3 Finding the Density of Jupiter

First, reread §11.5 and reexamine Fig. 12.1. If you elect to obtain data from telescopic observation of Jupiter and its moons, you will need a telescope with an eyepiece equipped with crosshairs. The observation is best carried out in twilight (if you have a choice—the satellite must be very close to maximum elongation or your data will be erroneous) so that the crosshair can be seen, although atmospheric conditions may cause sufficient scattering of Jupiter's light to provide background illumination (a rare benefit of a big-city or hazy sky!). Flashlight illumination can be scattered too, with a white cardboard blocking only part of an aperture. The crosshair should be rotated so that it is perpendicular not to the east–west diurnal motion of Jupiter but to the line joining the planet and the Moon you have selected to observe. Timings for the separation of the Moon from Jupiter's limb, Δt_s, and the diameter of Jupiter's disk Δt_d, can then be made. For these data, no corrections will be needed for the tilt of the orbit to the line of sight because the Moon's orbital plane is very close to

the equatorial plane of Jupiter, and we are going to use the *ratio* of the measurements, so that the sin i term cancels out:

$$r / R = [(\Delta t_s + \tfrac{1}{2} \cdot \Delta t_d) \cdot \sin i] / [\tfrac{1}{2} \cdot \Delta t_d \cdot \sin i]$$

whence

$$r / R = (\Delta t_s + \tfrac{1}{2} \cdot \Delta t_d) / (\tfrac{1}{2} \cdot \Delta t_d) \qquad (12.4)$$

The sidereal period of the appropriate satellite can be obtained from Table 11.2. From Eqs. 12.3 and 12.4 and the period, find the density of Jupiter, δ_J.

A series of timings of the Jovian satellite Callisto, made around 1980 May $16^d 19^h 30^m$ UT, illustrates the technique. Compare the resulting density of Jupiter with your own determination. The separations were measured to be 34.8^s, 35.2^s, and 34.4^s, yielding an average value of $\Delta t_s = 34.80 \pm 0.23$ s (where the error cited is the standard deviation of the mean). For the Jovian diameter in the same plane, the measurements in seconds of time were 2.6, 2.8, 2.8, 2.7, and 2.8. The mean of these is $\Delta t_d = 2.74 \pm 0.04$ s. From these quantities, the time to the middisk point is 36.17^s, and the time for the planet's apparent radius is 1.37 ± 0.02 s. Therefore, $r / R = 26.40$. From Table 11.2, $P_{IV} = 16.6890^d = 1.4419 \cdot 10^6$ s. Substituting in Eq. 12.3, we obtain

$$\delta_J = 1250 \text{ kg/m}^3$$

12.4 Finding the Density of the Sun

This part requires a sunscreen fixed firmly to the eyepiece end of the telescope. Unless your telescope is equipped with a proper solar filter, *do not* attempt to view the Sun through the eyepiece. Permanent damage can be caused to the eye even if the image is not in focus. If you have a solar filter with a crosshair but you are even slightly unsure about the filter's quality, rig up a sunscreen! Place a vertical line squarely down the center of the sunscreen, and move the telescope so that the Sun's image is

FIGURE 12.2. The ratio r/R obtained from the apparent semidiameter of the Sun, α.

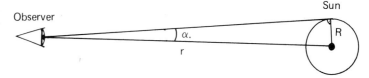

entirely east of the line. If the telescope has a clock drive, turn it off. Begin timing the Sun when its western edge touches the line and end the timing when its eastern edge reaches the line. Repeat the measurements and average them. The angular radius in arc seconds, α, may then be calculated from the formula

$$2 \cdot \alpha = 15 \cdot \cos D \cdot \Delta t \qquad (12.5)$$

where D is the declination of the Sun and Δt is the measured time interval in seconds. From Fig. 12.1, $R/r = \sin \alpha \approx \alpha$ (rad). Therefore,

$$r/R \approx 1/\alpha \qquad (12.6)$$

Given the length of the year, $P = 365.2422^d = 3.1556 \cdot 10^7$ s, the mean solar density can be calculated. The following data demonstrate the method. As a challenge, replicate the experiment. At February 27, 1978, 10^h30^m LCT, the solar diameter was timed through a high optical density solar filter equipped with a crosshair perpendicular to the Sun's motion. The timing measurements were 131.0, 131.2, and 131.0 s, giving an average interval $\Delta t = 131.1 \pm 0.1$ s. The solar declination on this date was $-8.1°$. This was read off a well-calibrated declination circle, but the solar declination is tabulated for each day in the *Astronomical Almanac*.[2] Substitution in Eq. 12.5 then yields $\alpha = 16'14''$, so that, by Eq. 12.6, $r/R = 212$, and $\delta_\odot = 1350$ kg/m³.

12.5 The Density of Earth

The gravitational acceleration at the surface of the Earth varies slightly from place to place because of the variation in height of the terrain, the rotation of

---

[2]The following formula for solar declination should suffice for present purposes, if a current almanac is not available:

$$D = 23.4 \cdot \sin (1° \cdot N + 30° \cdot M - 111°)$$

where N is the day of the month and M is the number of the month (starting with $M = 1$ for January). For days in months $M > 2$ in leap years, N should be replaced by $N - 1$. The results will be good to about $0.2°$.

the Earth, and local gravitational anomalies. A reasonably precise mean value, readily obtained in physics demonstrations, is $g = 9.81$ m/s². Since

$$g = G \cdot M / R^2 = G \cdot 4/3 \cdot \pi \cdot R^3 \cdot \delta / R^2$$
$$\delta = 3 \cdot g / (4\pi \cdot G \cdot R) \qquad (12.7)$$

The determination of the radius of the Earth was discussed in Chapter 4. If we assume the rough value $R = 6.8 \cdot 10^6$ m, we still have sufficient precision for the determination of δ: 5200 kg/m³, around four times the mean density of Jupiter and the Sun.

12.6 The Uncertainty in the Density

The computed uncertainty in the density depends on the equation used to compute the density. In the case of Jupiter and the Sun, the relation used was Eq. 12.3. Here the maximum error accruing to δ arises from the uncertainties in P and r/R:

$$\Delta\delta / \delta = |2 \cdot \Delta P / P| + |3 \cdot \Delta(r/R) / (r/R)| \qquad (12.8)$$

Of the two sources of uncertainty, the period is intrinsically better determined and its precision can be improved upon even more by averaging over many revolutions (if adequate corrections are made for parallax, synodic, and light-time effects). The precision in r/R is not improvable in this way and so it will determine the error in δ. Ignoring the error in P, we have

$$[\Delta(r/R) / (r/R)]_J \approx 3\% \quad \text{and} \quad [\Delta(r/R) / (r/R)]_\odot \approx 1\% \text{ or less}$$

For the Earth's density, $\Delta\delta / \delta \approx \Delta R / R \approx 6\%$.

12.7 Densities and Surfaces of the Moons of the Planets

The densities of the major solar system objects greatly differ, as we have seen. The densities of the moons of the planets are also interesting because many of them are largely composed of ices —water, methane, and ammonia—along with varying amounts of rocky material. The *Voyager* images of Neptune's largest moon, Triton, indicated that that body has an active surface, probably involving eruptions of nitrogen and methane from beneath an icy crust. How these objects are packed has been a matter of great interest ever since the *Voyager* spacecraft began to transmit images of the moons of Jupiter. Of particular interest are the very large contrasts seen on the surfaces of Iapetus of the Saturnian system and Miranda of the Uranian system. The latter body, especially, looks like a patchwork quilt, probably caused by its breakup through a major collision and the subsequent reaggregation of its material. Presumably the collision occurred

recently enough so that the denser rocky material and the lighter icy material have not yet had time to separate out in the newly constituted moon.

Further discussion of this topic is beyond our present scope, but as a challenge, after consulting the *Astronomical Almanac*, the *Observer's Handbook of the Royal Astronomical Society of Canada*, or some other source of satellite data, calculate the mean densities of those bodies for which masses have been determined (either through mutual perturbations or through the gravitational effects on spacecraft), and for which radii are known. For the nonspherical moons, this is a bit more of a challenge. Can you discern two classes of objects among the moons, as we have among the major planets?

13

Earth Satellites

13.1 Introduction

Artificial satellites have been circling the Earth since October 4, 1957, when the Soviet satellite *Sputnik* was launched. Their orbits are as diverse as their missions. Many scientific satellites, especially those launched to study the Earth's magnetic field, have highly elliptical orbits, so that they can sample both near and far Earth environs. Communications satellites, on the other hand, usually move in circular orbits in the plane of the equator and revolve in the direction of the Earth's rotation. Ideally, the latter should have an orbital radius $r \approx 42{,}170$ km from the Earth's center. By Kepler's third law, such an orbit has an orbital period, in this case known to very high precision. The period is $23^h56^m04\overset{s}{.}0989$ (or $0.997\ 269\ 663\ 24^d$) of mean solar time, the rotational period of the Earth.[1] A satellite with these properties is called a *synchronous satellite*; it remains stationary over a particular point on the Earth's equator, to provide continuous relay of television and telecommunications to a wide area within direct sightlines of the satellite.

Weather satellites typically have yet a third type of orbit, with an inclination so high that it is called a polar orbit. These satellites have periods of about 2 h and send back images of cloud formations to aid in weather forecasting. Usually they transmit both images in visible light (which astronomers sometimes call "optical" images) and infrared images, which reveal the moisture content of the atmosphere much more strongly. To maintain uniform image quality, the images must be obtained at the same altitude so that the satellites have almost perfectly circular orbits. All of these types of satellites have negligible mass compared to the Earth, and their motions must follow from the Keplerian laws and Newtonian mechanics.[2] In this chapter, we concentrate on the third type of satellite.

13.2 Reception of Weather Satellite Signals

Among weather satellites, the NOAA series satellites, which belong to the Automatic Picture Transmission Program of the U.S. National Oceanic and Atmospheric Administration (NOAA), are the most readily accessible. The transmission carrier frequency is 137 MHz, in the VHF region.[3] Receivers and antennae are relatively inexpensive and are commercially available from mail order electronics surplus houses. The transmitted power

---

[1]This is not quite the same as the length of the mean sidereal day, which is defined as the average time interval between two successive transits of the vernal equinox. Because of the precession of the equinoxes, the period of rotation is equal to $1.000\ 000\ 0970\ 9$ mean sidereal days (see Woolard and Clemence, 1966).

[2]The net force on a metal-skinned satellite may include electrodynamic forces also, due to its interactions with the Earth's magnetic field, the magnetosphere of charged, energetic particles, and with the solar wind (the plasma of energetic charged particles evaporated from the hot corona of the Sun).
[3]Since each satellite has a limited lifetime, and since the transmission frequencies of new satellites may well differ, it would be wise to contact the nearest government weather office before investing in a particular VHF FM receiver.

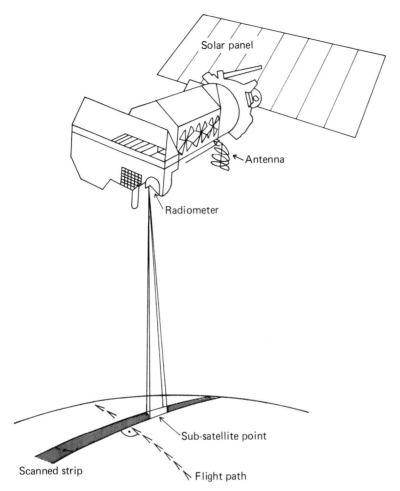

FIGURE 13.1. Schematic drawing of a weather satellite.

of these satellites is so strong that the signals can be received even with a 0-dB antenna (slide-wire, nondirectional type). A small directional antenna is required, however, for even reception over the 15-min time interval during which the satellite is within range. The satellites broadcast their visible and infrared detected signals at a line-scanning frequency of 2 Hz (i.e., twice per second). The picture content is amplitude modulated and lends itself to easy reconstruction. Figure 13.1 shows the basic structure of this type of satellite. The solar panel provides the electrical power. The detector scans the Earth's surface in a raster pattern. Every half-second a narrow strip of the Earth's surface perpendicular to the orbital motion is scanned. This scanning process produces a line of the image and can be heard in the loudspeaker of the VHF receiver as a modulation on the carrier frequency,

which has a characteristic whistle with a frequency of 2.4 kHz.

This operation is very different from that of the first weather satellites where an entire television image was first built up and then scanned and transmitted line by line to the ground. Currently, every line of the image is transmitted in real time. Each scan line (Fig. 13.2) contains the following duty cycles: calibration marks, ~150 ms; infrared data, ~175 ms; and optical data, ~175 ms. The 2.4-kHz carrier wave is modulated by a 2-Hz signal carrying this information. Picture reconstruction is easy, in principle. Besides a fax drum or rotating cylinder (like that used for barographs or seismographs but run at higher frequency), a solution to the reconstruction problem is provided by the intensity modulation (Z modulation) property of a cathode ray tube oscilloscope (see Figs. 13.3 and

FIGURE 13.2. Information content of a scanned line of an image produced by NOAA satellites. The scan cycle is marked by special information (gray scales) and other housekeeping information (shaded strips), the infrared, and the visible light scans.

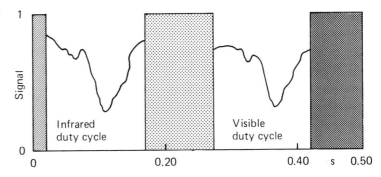

13.4). Input to the X and Y coordinates of the oscilloscope can be used for timing and stacking the intensity sweeps by means of the circuitry seen in Fig. 13.3.

The last part of this section is for the intrepid "electronicker."[4] If you are hopeless in electronics, skip to §13.3 to see what clever things can be done by just recording the output of a speaker connected to the receiver. It is important to point out that the format of this chapter was written before the use of microcomputers became widespread. A suitable challenge would be to interface the receiver to a home computer with a data acquisition software package such as Measure or Asyst[5]; graphics software packages could then be used to facilitate image reconstruction.

The circuit to provide synchronization can be constructed from commercially available units by a sufficiently motivated hobbyist. Output A provides the trigger pulse to the oscilloscope; output B furnishes an analog input voltage to the oscilloscope. The voltage at B has to be ramped—slowly made to increase—so that each successive image line lies above the last. To replicate the image geometry of the subsatellite point[6] successfully, the line pitch must be considered carefully; a change of y mm must correspond to an x-direction change of $0.3y$ cm/ms. If this is done, the screen pattern can be captured with a camera on time exposure. The quality of the image depends on the signal-to-noise ratio (S/N). However, it will always be possible to see large-scale cloud structures in strong low-pressure regions and coastlines.

If the equipment is quartz stabilized, the Doppler effect of the satellite can be demonstrated. This amounts to 3 kHz at most, so that a constancy in frequency of a few hundred hertz over the 15-min interval is required.

[4]This is an American astronomical colloquialism, attributed to Dr. Kenneth Franklin, American Museum–Hayden Planetarium, ca. 1957.
[5]*Measure* and *Asyst* are registered trade names of proprietary computer programs.

[6]This is the point just below the satellite, or the point on the Earth at which the satellite is at the zenith.

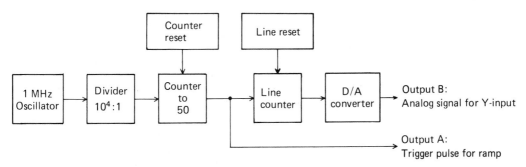

FIGURE 13.3. Block circuitry diagram for the synchronization of data lines to an oscilloscope for the reconstruction of weather satellite images.

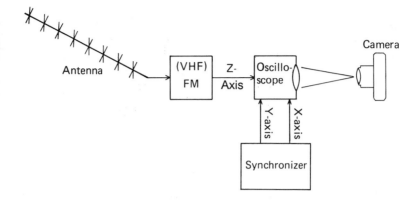

FIGURE 13.4. Construction of a satellite receiving station. The front section requires an antenna and a receiver; the second an oscilloscope and a synchronization circuit (like that in Fig. 13.3) as well as a camera.

13.3 Orbital Radii of the Satellites

The VHF signals of the satellite remain unaffected, for the most part, by the ionosphere, and propagate in a straight line. Figure 13.5 illustrates the geometry of the orbit and the limited areas of the Earth's surface which the transmissions can reach at any instant.

Given the relationships among r, R, h, and θ in Fig. 13.5, we can write

$$\cos (\theta/2) = R / r = R / (R + h) \quad (13.1)$$

The angle θ has another meaning as well. It describes the portion of the orbit visible from a fixed point on the Earth during a pass through the zenith of that point, as the shaded portion of Fig. 13.5 shows. If P is the orbital period and T is the interval over which the satellite can be observed (and received), then we have

$$\theta / 360° = T / P \quad (13.2)$$

The maximum receiving time is, of course, valid only for those satellites that pass through the zenith of the receiver (see challenge 5 in §13.5). Combining Eqs. 13.1 and 13.2, we can now derive the orbital radius of the satellite:

$$r = R + h = R / \cos(180 \cdot T/P) \quad (13.3)$$

Figure 13.6 is the volume output of a receiver tuned to the satellite frequency during two successive passes. From the diagram we obtain $P = 114.5^m$, $T = 22.5^m$. (Note that newer generations of satellites have smaller orbits, and therefore smaller values of P and T.) Setting $R = 6371$ km, we get $r = 7813$ km; therefore, $h = r - R = 1442$ km. Note, however, that a typical weather satellite is not in a geosynchronous orbit and so cannot appear at zenith for any given observer for two successive passages, even if it had an orbital inclination of exactly 90°.

From Eq. 13.3, the uncertainty in r arises from that in h (the error in the Earth's radius can be ignored), so that $e_r \approx e_h$. The latter error depends on that in T and P. Assuming the errors in T and P are the same, the relative error in P (i.e., $\Delta P / P$) is much smaller than that in T (one could argue that the uncertainty in T is actually somewhat larger because of visibility changes during the passes),

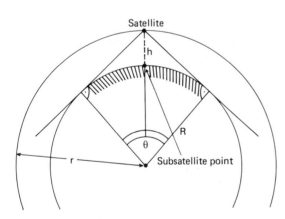

FIGURE 13.5. The line-of-sight geometry of the satellite orbit showing the connections among the Earth's radius, R, the orbital radius, r, and the subsatellite ground distance, $h = r - R$, and the central angle, θ.

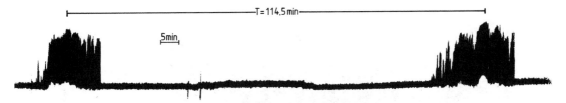

FIGURE 13.6. A recording of loudspeaker output during two successive satellite passages. The separation of maximum signal corresponds to an interval of 114.5^m. The satellite appears above the horizon a bit more than 20^m, but this interval varies because the longitude of the subsatellite point changes from pass to pass. The near-zenith passage record is at the right.

because P is about six times as large. Error theory[7] tells us that

$$e_r^2 \approx e_T^2 \cdot (\pi \cdot R / P)^2 \cdot \sin^2(\pi \cdot T/P) / \cos^4(\pi \cdot T/P) \quad (13.4)$$

where $e_T \approx 1^m$. Substituting values for r, R, P, T, and e_r into Eq. 13.4, we obtain $e_r \approx \pm 152$ km. Verify this yourself.

13.4 Reconstruction of the Satellite Image

To carry out the basic exercise in the previous section, only a receiver (with a simple undirectional antenna) and a stopwatch are essential. The actual reconstruction of the satellite weather image is a bit more complex. The discussion here assumes that you are not using a microcomputer for image reconstruction. First, you should have a small directional antenna (with gain ~ 14 dB and half-power full beam width ~ 40°) and the sychronization circuitry, like that in Fig. 13.3, must be built. This means first that you should complete the reading of §13.3 if you skipped the last part of that section. The construction of that part of the apparatus requires no more than 10 integrated circuits and a 10-bit A/D converter for uniform y deflection. Furthermore, the oscilloscope should not show anomalous, strong brightness modulation in the individual pixels. This can be tested best by expos-

ing a picture with no Z input.[8] The resulting picture should be uniformly gray. It goes without saying that one should take care to shroud the camera and oscilloscope face against external light, especially if the photography is done in the daytime.

Figure 13.7 shows a photograph from an oscilloscope face of a weather satellite map obtained with the techniques and gadgetry described here. It depicts an infrared image of northern Scandinavia as it appeared on July 16, 1976.

13.5 Further Challenges

Try your hand at some of these puzzles, and (maybe) qualify to work for a satellite communications company!

1. Verify the orbital radius of a communications satellite, given in §5.1. Why *must* this be the radius?
2. Calculate the farthest line-of-sight distance that can be reached by a communications satellite, and find out what this means in terms of the longitude and latitude range (i.e., twice the difference between the latitude of the subsatellite point and the farthest point, and the same with the longitude).
3. Nobody's perfect! Suppose the orbital eccentricity and inclination of a communications satellite

---

[7]The expression for e_r is given for its square for heuristic reasons. Here, there is no real advantage in writing it like this; the value comes if one were to attempt to include the effects of errors in P and in R. For example, to find the error in z that depends on both x and y, one writes

$$e_z^2 = e_x^2 \cdot (\partial z/\partial x)^2 + e_y^2 \cdot (\partial z/\partial y)^2$$

[8]Nearly all oscilloscopes have such an input, which, according to the voltage that is applied, changes the intensity of the beam on the phosphor. Since the Z input is, however, used mainly for beam suppression, the correlation between brightness and voltage may not be very good. The test described above checks that a constant Z voltage results in a uniform picture without ripple or other distortion. If you elect to use a microprocessor for image construction, you will need to check on the gray scale of your monitor.

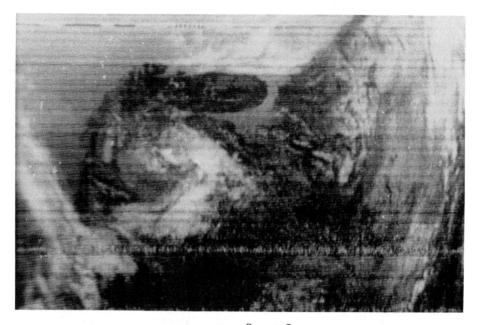

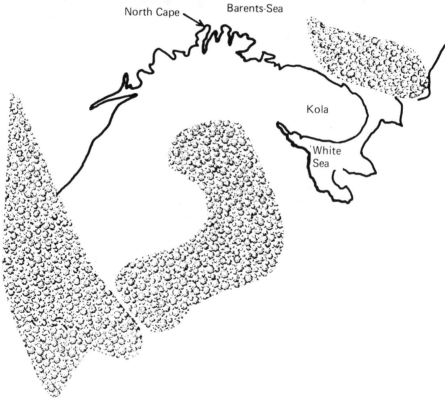

FIGURE 13.7. A weather satellite picture of northern Scandinavia in the infrared, with an identifying sketch of the area below.

were not exactly 0 and 0°, respectively. What would be the effects, separately and together, on the apparent position of the satellite in the sky?

4. Communications satellites, more generally called geostationary satellites, can actually be observed with a telescope. Taylor (1976) points out that they have typical brightnesses of magnitude 11–14. Calculate the *hour angle* for such a satellite which is located above a longitude 2 h east of your own. If that was too easy, now calculate the *azimuth* and *altitude* of the object (see Eqs. A.6 and A.7 in Appendix A).

5. Now let us return to the weather satellites. Calculate the angle θ for the case of a satellite that does not pass directly over you but over a longitude which is $\Delta\lambda = 45°$ away. How does this affect the duration of the transmission that you can receive? What change in subsatellite longitude will occur for an observer at the equator, say, during a period of 114.5^m? Finally, how many passes of that satellite can be observed consecutively by that equatorial observer?

References and Bibilography

Taylor, G. E. (1976) *Sky & Telescope* **71**, 557.

Woolard, E. W., and Clemence, G. T. (1966) *Spherical Astrononomy*. Academic Press, New York, p. 351.

14

Orbits and Space Travel

14.1 Introduction

The planets have been targets of exploration since the launching of *Sputnik* in 1957. Space probes have been sent to each of the planets known in antiquity and to all but one of the known planets. In this exercise, we explore first the properties of planetary orbits, then how to calculate their coordinates on the sky from their orbital elements, and, finally, how to design the orbit of a spacecraft to travel to a target planet.

14.2 Orbital Elements

Celestial mechanics and astrometry no longer attract the numbers of astronomers they did when they were the dominant fields of astronomy. This is unfortunate, because they are essential fields, without which a great deal of fundamental data about planets, stars, and stellar systems would have been inaccessible and space travel would be impossible.

The study of planetary motions was carried out in ancient Mesopotamia long before the Christian era. From the reign of King Ammizaduga, who ascended the throne of Babylon 146 years after Hammurabi, the "Giver of Laws," we have a record of observations of the planet Venus. Current chronology places it around 1700–1600 B.C. The Mesopotamians and later the Greeks had theories of motion of the Sun, Moon, and planets, but these were of an empirical kind, to permit prediction (or back-calculation) of the planetary positions. Modern orbital theory had to await the development of the physical laws of motion, which led from the work of Copernicus to that of Isaac Newton (1643–1727). Laplace, Lagrange, and Gauss,

among other great celestial mechanicians, made important subsequent contributions. The techniques discussed in this chapter deal with the problem of predicting the position of a planet in the sky, given the orbital elements. The detailed derivation of the equations used in this and other methods may be found in Brouwer and Clemence (1961), Danby (1988), and Moulton (1958), and elsewhere.

There are six independent parameters that define an orbit and the position of a planet in that orbit:

1. The *semimajor axis*, a, establishes the size of the orbit since the orbit of each planet is an ellipse; it is also the mean distance of the planet to the Sun. The total energy of the orbit is related to a in the following way:

$$E = -\text{const} / a$$

In Fig. 14.1, E_1E_2 is the major axis, $2a$ (also known as the *line of apsides*), and E_1C or CE_2 is the semimajor axis.

2. The *eccentricity*, e, measures the departure of the orbit from a circle; e therefore establishes the shape of the orbit. It is defined as the separation of the foci divided by the major axis:

$$e = F_1F_2/E_1E_2 = F_1F_2 / 2a = F_1C / a$$

3. The *inclination*, ι, is the angle by which the orbit is tilted to a reference plane (in the present context this is usually the ecliptic). It is illustrated in Fig. 14.2.

4. The *longitude of the ascending node*, Ω, together with the inclination, establishes the orientation of the plane of the orbit. Ω is measured along the ecliptic from the direction of the vernal equinox to the point on the ecliptic where the planet

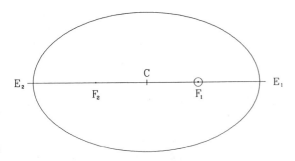

FIGURE 14.1. The orbit of a planet is an ellipse with the Sun at one focus (Kepler's first law).

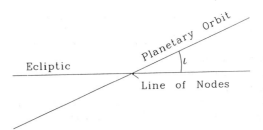

FIGURE 14.2. The inclination is measured at the line of intersection between the planes of the orbit and the ecliptic, the *line of nodes*.

crosses from south to north of the ecliptic plane. It is also the Sun-centered angle between the direction to the vernal equinox and the direction to the ascending node.

5. The *argument of the perihelion*, ω, establishes the orientation of the orbit within the orbital plane. It is measured in the orbit from the ascending node to the perihelion. It is also the Sun-centered angle between the direction to the ascending node and the direction to the perihelion. Figure 14.3 illustrates elements ι, Ω, and ω and the variables r and υ.

6. The *epoch*, T_0, is a moment in time when the planet is at perihelion. This element establishes

the location of the planet in the orbit, since the motion of the planet is periodic.

These six elements, together with the sidereal period, P, sometimes listed as a seventh element,[1] are sufficient to provide knowledge of where the

---

[1] It would appear from Kepler's third law, $P^2 \propto a^3$, that the period is redundant. It must be remembered, however, that the Newtonian form of Kepler III has both Sun *and* planet mass in it, so that Kepler III is only approximately true, but it may well be sufficient, depending on the precision with which the result is needed and the relative mass of the planet.

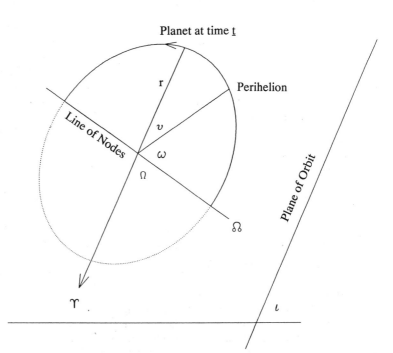

FIGURE 14.3. The orbit in the orbital plane, showing the elements ι, Ω, and ω; the line of nodes; the distance to the Sun, r, and the *true anomaly*, υ.

planet is within the orbit at any time and where the planet is with respect to the ecliptic at any time.

At this point, if you have learned all you wish or need to know about orbits, you can consult Appendix E, which gives planetary position tables and includes instructions on their use. Then return to §14.7 to complete the exercise and do some traveling! For the serious celestial mechanics student, read on and enjoy!

14.3 Locating an Object in the Orbit

The angle υ between the perihelion point and the Sun–planet line at any instant is known as the *true anomaly* (see Fig. 14.3). The Sun–planet, or *distance modulus*, r is given by

$$r = a \cdot (1 - e^2) / (1 + e \cdot \cos \upsilon) \quad (14.1)$$

The true anomaly may be calculated from the following series approximation:

$$\upsilon = M + 2e \cdot \sin M + (5/4) \cdot e^2 \cdot \sin 2M + (1/12) \cdot e^3$$
$$\cdot (13 \cdot \sin 3M - 3 \cdot \sin M) + \cdots \quad (14.2)$$

where M is the *mean anomaly*, the angle swept out by an imaginary planet moving at the average angular speed of the real one. In Eq. 14.2, M, e, and υ are in radian measure. You may need to have the arguments of the sine functions in degrees for your calculator; for those terms only, use

$$\theta° = \theta \text{ (rad)} \cdot 180 / \pi$$

M is related to the time since perihelion passage by

$$M = 2\pi \cdot (t - T_0) / P \text{ rad} \quad (14.3)$$

where t and T_0 are in Julian day numbers and decimals thereof. Note that the route to follow in calculating the orbital position with time is

$t \rightarrow M$ via Eq. 14.3,
$M \rightarrow \upsilon$ via Eq. 14.2, and
$\upsilon \rightarrow r$ via Eq. 14.1.

Once υ and r are found, the position on the sky can be computed.

14.4 Finding the Object in Ecliptic Coordinates

The coordinates x'', y'', z'' of an object specify its position uniquely in the coordinate system of the orbit. That coordinate system has the plane of the orbit as one reference plane, and a reference circle through the vernal equinox defines another. The

axes must be centered on the origin, in this case the Sun. The x'' axis is directed toward perihelion, the y'' axis toward a point in the plane 90° east of the perihelion direction, and the z'' axis is directed toward the north pole of the orbit, perpendicular to the orbital plane. In the absence of any perturbing forces, all the motion is in the orbital plane, and the coordinates x'', y'', and z'' may be written

$$x'' = r \cdot \cos \phi \cdot \cos \upsilon = r \cdot \cos \upsilon$$
$$y'' = r \cdot \cos \phi \cdot \sin \upsilon = r \cdot \sin \upsilon$$
$$z'' = r \cdot \sin \phi = 0 \quad (14.4)$$

where the angle ϕ is the coordinate above or below the orbital plane and so, in this case, is zero.

The position indicated by x'', y'', and z'' can be *transformed* into an equally unique description of the object's location in the ecliptic coordinate system, x', y', z' (see Appendix A, §A.2.4). The x' axis is toward the vernal equinox, the y' axis is 90° east of that direction in the plane of the ecliptic, and the z' axis is toward the North Ecliptic Pole. The two coordinate systems are indicated in Fig. 14.4. The transformation can be done with the help of the *direction cosines*, which provide the projections of each orthogonal axis, x'', y'', and z'', in turn, on each axis of the ecliptic system:

$$x' = \ell_1 \cdot x'' + m_1 \cdot y'' + n_1 \cdot z''$$
$$y' = \ell_2 \cdot x'' + m_2 \cdot y'' + n_2 \cdot z''$$
$$z' = \ell_3 \cdot x'' + m_3 \cdot y'' + n_3 \cdot z'' \quad (14.5)$$

The direction cosines are nothing more than the cosines of the angles between the axes involved, for example, ℓ_1 is the cosine of the angle between the x' axis and the x'' axis, m_1 is the cosine of the angle between the x' axis and the y'' axis, and n_1 is the cosine of the angle between the x' axis and the z'' axis. By means of spherical trigonometry, one may obtain the following values for the direction cosines:

$$\ell_1 = \cos \Omega \cdot \cos \omega - \sin \Omega \cdot \sin \omega \cdot \cos \iota$$
$$m_1 = - \cos \Omega \cdot \sin \omega - \sin \Omega \cdot \cos \omega \cdot \cos \iota$$
$$n_1 = \sin \Omega \cdot \sin \iota$$
$$\ell_2 = \sin \Omega \cdot \cos \omega + \cos \Omega \cdot \sin \omega \cdot \cos \iota$$
$$m_2 = - \sin \Omega \cdot \sin \omega + \cos \Omega \cdot \cos \omega \cdot \cos \iota$$
$$n_2 = - \cos \Omega \cdot \sin \iota$$
$$\ell_3 = \sin \omega \cdot \sin \iota$$
$$m_3 = \cos \omega \cdot \sin \iota$$
$$n_3 = \cos \iota \quad (14.6)$$

These equations, substituted into Eqs. 14.5, will yield the heliocentric rectangular coordinates of the planet. Since $z'' = 0$, direction cosines n need not be calculated. We will discuss how to compute

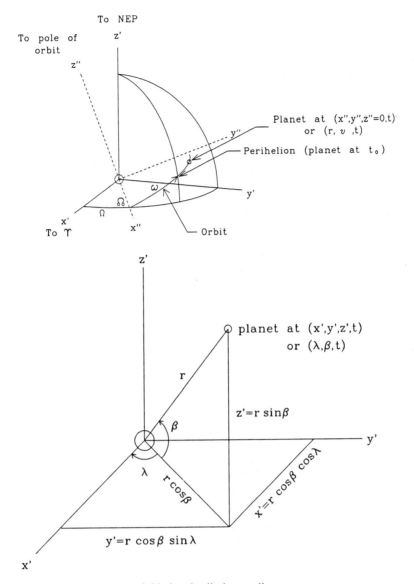

FIGURE 14.4. Orbital and ecliptic coordinate systems.

the *geocentric* ecliptic and/or equatorial system coordinates in §14.5.

The ecliptic coordinates λ and β, in the ecliptic coordinate system centered on the Sun, can now be calculated:

$$x' = r \cdot \cos \beta \cdot \cos \lambda$$
$$y' = r \cdot \cos \beta \cdot \sin \lambda$$
$$z' = r \cdot \sin \beta$$

By application of algebra, λ and β can be obtained:

$$\tan \beta = z' / \sqrt{[(x')^2 + (y')^2]}$$
$$\sin \lambda = y' / (r \cdot \cos \beta)$$
$$\cos \lambda = x' / (r \cdot \cos \beta) \qquad (14.7)$$

The latter two equations are needed to resolve the quadrant question because λ may be anywhere between $0°$ and $360°$.

Thus, the procedure for finding the heliocentric ecliptic coordinates from the orbital elements consists of the following steps:

1. Beginning with the quantities r and υ, which were obtained in §14.4, calculate x'', and y'' from Eqs. 14.4.
2. Compute the direction cosines from Eqs. 14.6.
3. Calculate the heliocentric rectangular ecliptic coordinates x', y', and z' from Eqs. 14.5.
5. Obtain the heliocentric ecliptic coordinates λ and β from Eqs. 14.7.

After you have done all this, check the results! First, the *heliocentric* longitudes and latitudes are usually tabulated for each of the planets in the *Astronomical Almanac*, for direct comparison. Second, as described later, the results can be converted into geocentric coordinates and transformed to right ascension and declination. These can then be checked both against the almanac positions and against direct observation.

14.5 Finding the Object in the Sky

To get the heliocentric x', y', and z' coordinates into a *geocentric* ecliptic system, the geocentric coordinates of the Sun must be added:

$$x_g' = x' + x_0, \quad y_g' = y' + y_0, \quad z_g' = z'$$
(14.8)

where $x_0 = r_0 \cdot \cos \lambda_0$ and $y_0 = r_0 \cdot \sin \lambda_0$, in terms of the Sun's distance in astronomical units, and its celestial longitude, λ_0, both of which are tabulated for each day in the *Astronomical Almanac*.

The geocentric distance may now be computed:

$$D^2 = (x_g')^2 + (y_g')^2 + (z_g')^2 \quad (14.9)$$

The ecliptic coodinates λ and β, in the ecliptic coordinate system centered on the Earth, can now be calculated using Eqs. 14.7.

Appendix A, §A.3.3, describes and illustrates the transformation between the ecliptic and the equatorial systems. Once the ecliptic coordinates have been determined, the transformation equations (Eqs. A.11 and A.12) can be used to compute the right ascension and declination, or one may convert from rectangular ecliptic to rectangular equatorial coordinates by means of the equations

$$\begin{aligned} x_g &= x_g' \\ y_g &= y_g' \cdot \cos \varepsilon - z_g' \cdot \sin \varepsilon \\ z_g &= y_g' \cdot \sin \varepsilon + z_g' \cdot \cos \varepsilon \end{aligned} \quad (14.10)$$

Here ε is the *obliquity of the ecliptic*, equal to about $23°439291$ (*Astronomical Almanac*, 1989).[2]

In a way similar to that by which the ecliptic coordinates λ and β were obtained, the equatorial coordinates, α and δ, can be obtained:

$$\begin{aligned} x_g &= r \cdot \cos \delta \cdot \cos \alpha \\ y_g &= r \cdot \cos \delta \cdot \sin \alpha \\ z_g &= r \cdot \sin \delta \end{aligned} \quad (14.11)$$

whence

$$\begin{aligned} \tan \delta &= z_g / \sqrt{(x_g^2 + y_g^2)} \\ \tan \alpha &= y_g / x_g \end{aligned} \quad (14.12)$$

Both $\sin \alpha$ and $\cos \alpha$ may be found to check on the quadrant of α.

In summary, beginning with the results of §14.4,

1. Look up the solar longitude and distance and compute the solar components to get geocentric rectangular ecliptic coordinates x_g', y_g', z_g' (Eqs. 14.8).
2. Recalculate geocentric coordinates λ and β from Eqs. 14.7 and transform λ and β into α and δ, using Eqs. A.11 and A.12, or use x_g', y_g', and z_g' to compute x_g, y_g, z_g from Eqs. 14.10 and convert x_g, y_g, and z_g to α and δ via Eqs. 14.11.

14.6 Carrying Out the Calculation

Try to go through the exercise described in previous sections with a planet that you know to be visible. Because of the simplifying assumptions regarding the spacecraft's orbit, avoid the planets Pluto and Mercury—unless you want to plan to rendezvous with one of these planets when it is at one of its nodes. You then have the opportunity to observe the object and compare the observed and calculated positions.

As an example, we will compute the position of the planet Mars on October 6, 1988, close to the time of nearest approach to Earth during a favorable opposition. This will show up any errors nicely, because the apparent motion of Mars will be rapid at this time and because the orbital perturbations are likely to be large. The following elements of Mars are used for the calculations:

$a = 1.52368$ au
$e = 0.093278$
$\Omega = 49.474°$
$\omega = 286.409°$
$\iota = 1.8499°$

---

[2]The value is given for the "standard epoch," A.D. 2000; for the value of the obliquity of the ecliptic of date, use

$\varepsilon = 23.439291 - 0.0130042 \cdot T - 16 \cdot 10^{-8} \cdot T^2 + 504 \cdot 10^{-9} \cdot T^3$

where $T = (t - 2000.0)/100 = (\text{JDN} - 2451545.0)/36525$, t is the desired date in years, and JDN is the current Julian day number and decimal thereof (see Appendix D).

FIGURE 14.5. The Hohmann transfer orbit between planets. Note that it is *not* a straight line between the planets' orbits at minimum separation, but a smooth, elliptical orbit, tangent to the planetary orbits at each end.

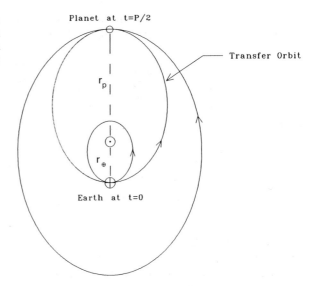

The mean anomaly is readily computed from the mean longitude, defined in the *Astronomical Almanac*, and so we need not use the time of perihelion:

$$M = L - (\omega + \Omega) = -331.4838° = 28.5162°$$
$$= 0.49770148 \text{ rad}$$

Solving for υ from Eq. 14.2, we find

$$\upsilon = 0.59666918 \text{ rad} = 34.1866°$$

Whence, from Eq. 14.1, we get

$$r = 1.402226 \text{ au}$$

Now, solving Eqs. 14.4,
$$x'' = 1.1599379$$
$$y'' = 0.7878973$$
$$z'' = 0$$

The direction cosines can be evaluated from the orbital elements (Eqs. 14.6):

$$\ell_1 = +0.9123330 \quad m_1 = +0.4087132$$
$$n_1 = +0.0245373$$
$$\ell_2 = -0.4082764 \quad m_2 = +0.9126173$$
$$n_2 = -0.0209761$$
$$\ell_3 = -0.0309664 \quad m_3 = +0.0091192$$
$$n_3 = +0.9994788$$

From these, Eqs. 14.5 are evaluated:

$$x' = +1.3802736$$
$$y' = +0.2454734$$
$$z' = -0.0287341$$

Equations 14.7 then yield the heliocentric values of β and λ:

$$\tan \beta = -0.0204961 \quad \text{giving } \beta = -1.1742°$$
$$\sin \lambda = +0.1741752 \quad \text{giving } \lambda = +10.084276°$$
$$\cos \lambda = +0.9845513 \quad \text{giving } \lambda = +10.084273°$$

The λ's are in reasonable agreement considering that we have been carrying more precision than is required given the number of significant figures in the input data. From the *Astronomical Almanac* for 1988, we find the values of λ and β:

$$\lambda = 10°05'13'' = 10.0868°$$
$$\beta = -1°10'27'' = -1.1741°$$

in satisfactory though not excellent agreement because of the perturbations of the other planets (especially Earth at this time and Jupiter), which cause departures from Keplerian motion.

The position of the Sun, to the equinox of date, was used to calculate the x and y coordinates of the Sun:

$$\lambda_0 = 192.958122° \quad \text{and} \quad \beta_0 = +0.13''$$

so that

$$x_0 = -0.9742206 \quad \text{and} \quad y_0 = -0.224166$$

Equation 14.8 then gives

$$x_g' = +0.406053$$
$$y_g' = +0.021307$$
$$z_g' = -0.0287341$$

Whence, by Eq. 14.9, $D = 0.407626$, the geocentric distance of the planet. Now, using Eqs. 14.10, we find

$$x_g = +0.406053$$
$$y_g = +0.030978$$
$$z_g = -0.017888$$

Finally, from these results and Eqs. 14.11 and 14.12, we arrive at

$$\alpha = 00^h 17^m 27^s$$
$$\delta = -2°30'54''$$

The tabulated values are $\alpha = 00^h 17^m 29^s$, and $\delta = -2°30'38''$, again in satisfactory, although not perfect, agreement.

As an instructive challenge, derive the direction cosines in Eqs. 14.6, redrawing Fig. 14.4 to suitable scale to facilitate placement of the spherical triangles that are needed.

14.7 Planning a Voyage

Now that you have succeeded in demonstrating your skill at practical celestial mechanics, you have earned passage to the planet. Begin by sketching the orbits of your planetary target and the Earth to scale on suitable polar coordinate paper. You may ignore eccentricities smaller than that of the Earth (~ 0.017) and the inclination of the other planet. We will assume that the near-space engineers will manage to get you to the Earth-orbiting space station and will provide for escape from the Earth itself. You will plan the change from an orbit at the Earth's distance from the Sun to that of the planet. What you need to do now is to plan the *Hohmann transfer orbit*, an efficient orbit using minimum energy (i.e., rocket fuel) to do the job. The transfer orbit has a major axis equal to the sum of the Earth–Sun and Sun–planet distances. It is illustrated in Fig. 14.5.

The elements of this orbit are easily obtainable, so obtain them. You know $2a$, so compute a. You also know the perihelion (and aphelion) distances of the spacecraft [$a(1 - e)$ and $a(1 + e)$, respectively], so compute e. You also know the inclination ι. The period is known because a is known and the space craft has negligible mass. The *time-of-flight* is $P/2$. Note that the *longitude of the ascending node*, Ω, is not defined in the present case. The quantity $\Omega + \omega$, the *longitude of perihelion*, may be used. It describes the "broken" angle from the vernal equinox to the perihelion. To find ω and T_0, you need to know the date of launch. This is the trickiest part, because if you do not have a computer program to find the *launch window*, you will need to find the launch date by trial and error, so that you arrive at the planet's orbit precisely when the planet arrives at that point, at a time $P/2$ after launch. You will need the appropriate annual volumes of the *Astronomical Almanac* or a multiyear computer almanac for this exercise. Good launch and bon voyage!

References and Bibliography

Brouwer, D., and Clemence, G. M. (1961) *Methods of Celestial Mechanics*. Academic Press, New York.

Danby, A. (1988) *Fundamentals of Celestial Mechanics*, 2nd ed. Willmann-Bell, Richmond, Va.

Moulton, F. R. (1958) *An Introduction to Celestial Mechanics*. Macmillan, New York.

15

Photography of the Sun and Stars

15.1 Introduction

Here we begin discussing the radiant energy from natural cosmic bodies, in particular from the Sun and stars. It was recognized that distances could be found through comparisons of brightness well before the first measurement of stellar parallax by Bessel in 1838. We begin where Johannes Kepler first explored—with the inverse square law for light.

Newton argued that stars were distant suns because they showed (as far as he knew) no discernible relative movement. Such motion would be expected if they were physically close together (by the inverse square law of gravity) and parallax would reveal if they were near to us; therefore, they were far away. He also knew that the planet Saturn intercepted approximately the same amount of sunlight as the Earth (since Saturn is around 10 times further but also around 10 times larger than Earth). But the Earth intercepts only a minute portion of the total amount of light emitted by the Sun, and Saturn appears to us like a star of magnitude 1. If the Sun were 10 times farther from us, it would appear to be 5 magnitudes fainter than it is now, or about magnitude -22. Therefore, to be comparable in apparent brightness to the relatively nearby object Saturn means that the very distant stars must be comparable to the Sun in true brightness.

We can confirm the logic of this argument in mathematical terms. First we define what we mean by brightness. The *luminosity* $\mathcal{L}$ of a star is the total amount of light energy that it emits each second. This energy can be expressed in the same units we use to quantify other kinds of energy (ergs, joules, etc.), so a star's luminosity is in units of ergs per second or, more commonly, joules per second or

watts (10^7 ergs/s)—the units of *power*. Therefore, $\mathcal{L}$ is the total radiated power of the star.

To catch all of a star's radiant energy, one would have to surround the star with a sphere, the area of which is $4\pi \cdot r^2$. It would be an interesting goal for a more advanced technological society than our own to capture all of its star's light energy. For present purposes, only the small fraction of the radiation that a telescope intercepts need be observed. At a distance r from the star, we observe its *radiant flux*, or *luminous flux*,

$$F = \mathcal{L} / (4\pi \cdot r^2) \text{ W/m}^2 \qquad (15.1)$$

The flux is the amount of light energy received per cross-sectional area per second from the star at distance r away from its center. We can increase the amount of energy received from a star by observing it longer, and with a bigger telescope.

Now we can compare the flux of the Sun to that of our star:

$$F_S / F_* = [\mathcal{L}_S / (4\pi \cdot r_S^2)] / [\mathcal{L}_* / (4\pi \cdot r_*^2)] \qquad (15.2)$$

where r_S, $\mathcal{L}_S$, and F_S are the distance, luminosity, and flux of the Sun, respectively. Expressing both r_S and r_* in astronomical units, r_S becomes 1, so Eq. 15.2 becomes

$$F_S / F_* = \mathcal{L}_S / [\mathcal{L}_* / r_*^2] \qquad (15.3)$$

As a first approximation, suppose we assume that all stars have the same luminosity, so that $\mathcal{L}_S = \mathcal{L}_*$. Then the $\mathcal{L}$'s cancel, so that

$$r_* = \sqrt{[F_S / F_*]} \qquad (15.4)$$

A bright star in our sky might have a radiant flux of about 10^{-12} that of the Sun. Consequently, the ratio

$F_S/F_* = 10^{12}$ for this case, and so $r_* = 10^6$ au, or $\sim 1.5 \cdot 10^{14}$ km. This enormous distance for our sample bright star confirms Newton's conclusion. Modern refinements do not much affect it.

These days we know that there are large differences in the luminosities of the stars, but the most brilliant stars are sparse (and none are located near the Sun) while the faintest are too faint to be seen without a telescope, even though they are so numerous that there are many in the "neighborhood" of the Sun. Moreover, by selecting a star with a spectrum like that of the the Sun we can ensure that the star has a luminosity like that of the Sun, because the spectral features indicate a star's temperature, luminosity, and the chemical makeup of its surface gases. With these qualifications, the basic result is unchanged: there are "yawning chasms" between the stars. Even the nearest star, α Centauri, is about 6000 times further than Pluto from the Sun.

In this exercise, you will be invited to compare directly the brightness of the Sun and solarlike stars.

15.2 Photographing the Sun and Stars

As with several other exercises in this book, this one requires the use of a camera. Additionally, you will need neutral density filters to dim the sun. The idea is to photograph both the Sun and the stars with the same film, taking a range of time exposures of each to ensure overlap in brightness. Since the Sun's image will be a disk, you will need to defocus the camera while photographing the star fields to obtain star images that are disks of the same diameter as the Sun's image. You can then find the Sun and star photographs that have equivalent image brightnesses and, from the transmission factor by which the neutral density filter cuts down the Sun's brightness, the f ratio which describes the degree to which the camera's aperture is stopped down, and the factor due to the time exposure ratio, compute the distance to the star in units of the Sun's distance from us—the astronomical unit. To make the proper choices about the relative exposure times and filter values, some of the properties of photographic film need to be discussed.

15.2.1 Photographic Constraints

It is important to use neutral density filters to greatly reduce the time exposure difference between the Sun and the stars, and not rely on differences in exposure times.[1] This is because of the photographic response of the film to light.

The film responds to the amount of light energy that falls on it, to the product of the *exposure time*, t, and a quantity similar to the flux, called the *illumination*, I:[2]

$$E = I \cdot t$$

Photographers call E, the *exposure*. The response of the photographic film to light falling on a certain spot is an increase of the number of silver grains deposited on that spot, with a resulting decrease in the transmission and an increase in the opacity of the film, once it has been developed, at that spot. Once the film has been developed, run through the stop bath, "fixed" (to prevent further development), washed, and dried, it can be studied by having a narrow beam of light of known brightness measured as it passes through the emulsion. The terms transmission and opacity refer to what happens to this measured light: *transmission*, T, is the ratio of the light transmitted through the film to that incident on it; the *opacity*, $O = 1/T$; thus the greater the exposure, the greater the opacity. The relation between the logarithm of the opacity, called the *optical density* D, and the logarithm of the exposure is called the "H&D," or *characteristic curve*.[3] Figure 15.1 illustrates a typical H&D curve.

Characteristic curves vary from batch to batch of the photographic emulsion. Note that the curve is not a straight line but does have a region in the middle where the density is approximately proportional to $\log E$. The slope of this linear part is called gamma, or γ. It measures the *contrast*—the rate at which the density increases as the log of the exposure increases.[4] The lower part is the *toe*, in which the emulsion is not very sensitive to changes

---

[1] At present, there is no choice because by themselves f ratios and exposure times of commercially available cameras cannot provide the necessary range in sensitivity.

[2] The unit of this quantity is the *lux*. It is equivalent to a monochromatic flux, i.e., the flux per unit wavelength interval (e.g., per Ångstrom, Å), of $1.463 \cdot 10^6$ W/(m²·Å) (see Allen, 1973).

[3] The H&D stands for Hurter and Driffield (1890), who pioneered the study of these curves.

[4] This is *not* the same thing as the speed of the emulsion, which measures essentially how far to the left (see Fig. 15.1) the characteristic curve lies. The principal scales of speed in use since the early 1960s are the ASA (American and British standards) and DIN (German standards). If we designate as E_m the point where the characteristic curve rises to a density of 0.1 above fog, then ASA $= 0.8/E_m$ and DIN $= 10 \cdot \log E_m$, assuming uniformity in processing in a standard way.

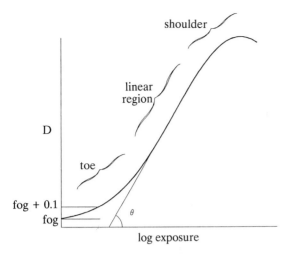

FIGURE 15.1. A stylized "characteristic curve" of a photographic emulsion. See the text for definitions of the labeled parts. The contrast $\gamma \equiv \tan \theta$.

in E, and the upper part is the *shoulder*, where the emulsion is again less sensitive. At very large values of the exposure, the sensitivity actually decreases again; this is the *solarization* part of the curve; this region should be avoided when photographing the Sun! When images differ in brightness, it is important to know if they are on the same part of the characteristic curve, and it is most desirable that they both be on the linear part.

We would like to have the exposures of the Sun and star be comparable in image brightness; if we succeed in having them so, can we be sure that we will be on the same part of the H&D curve? The answer, regrettably, is no! It is theoretically possible for a much fainter source to be exposed longer to give the same film response, but there are serious pitfalls. First, γ varies with development time (the longer the development, the larger γ up to a limit, after which it again decreases due to the effects of bringing up the faint background fog) so that we must ensure that the frames are processed in the same way. But there is a more fundamental difficulty: most photographic emulsions suffer from reciprocity failure. This means that the illumination and the time of the exposure do not behave in a symmetrical way: the density of the image of a very bright source exposed for a very short time is not the same as that of an image of a faint source exposed for a very long time, *even though they have the same value of E*. The speed decreases at both very high and very low levels of illumination and the contrast decreases at high

levels and increases at low levels. Therefore, it is advisable to minimize the differences in exposure time between the solar and the stellar exposures.

15.2.2 Selection of the Filters

A no. 1 neutral density (ND) filter has an optical transmission of about 10% in the visible region, but it is not really "neutral" or "gray." The transmission is sharply lower in the ultraviolet part of the spectrum and gradually increases to around 40% at 1 μm in the near infrared. Therefore, if we observe a blue object of the same brightness as a red object, the latter will *look* brighter through the filter. As we will explore in other exercises, stars differ in color as well as intrinsic brightness, but by choosing to observe a star that has a spectral type like that of the Sun, we can neglect the color effect difference of the transmission filter. Four no. 1 ND filters together will transmit $\sim 10^{-4}$ of the light, the equivalent of one no. 4 filter. However, other kinds of dense filters (e.g., filters with extremely thin metal coatings) may be used also, and these may actually be more gray than the carbon neutral density filters in widespread use. We have used such filters for the example in §15.2.4. The precise transmission factor of the filters can be checked with a densitometer which has a light source of similar color to sunlight. A photographer's light meter also will give an idea of the transmission factor. In the example, the transmission was determined with one of these devices. The disadvantage is a larger uncertainty, since the typical meter has a precision of only $\sim \frac{1}{2}$ stop (for a factor of $\sqrt{2}$).

15.2.3 Selecting the Star

Table 15.1 contains information about several relatively bright stars whose spectra resemble that of

TABLE 15.1. Solar type stars.

Star	Magnitude	Spectral type	α	δ
η Cassiopeiae	3.44	G0V	00^h48^m4	$+57°45'$
HR 483	4.9	G2V	01 40.6	+42 31
κ Fornacis	5.20	G1V	02 22.0	−23 52
κ Ceti	4.83	G5V	03 18.8	+ 3 20
χ^1 Orionis	4.41	G0V	05 53.7	+20 16
β Virginis	3.61	F9V	11 50.1	+01 50
β Canum Venaticorum	4.27	G0V	12 33.2	+41 25
λ Serpentis	4.43	G0V	15 45.9	+ 7 23
δ Equulei	4.49	G1V	21 13.9	+ 9 58
HR 8501	5.37	G3V	22 17.5	−53 41

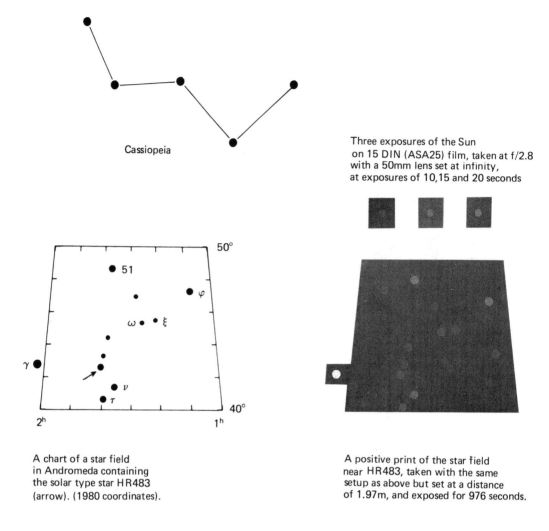

Cassiopeia

Three exposures of the Sun
on 15 DIN (ASA25) film, taken at f/2.8
with a 50mm lens set at infinity,
at exposures of 10,15 and 20 seconds

A chart of a star field
in Andromeda containing
the solar type star HR483
(arrow). (1980 coordinates).

A positive print of the star field
near HR483, taken with the same
setup as above but set at a distance
of 1.97m, and exposed for 976 seconds.

FIGURE 15.2. The field chart of the solar-type star HR 483 (*left*) and the exposures of the Sun and the star field (*right*). HR 483 can be identified on the chart at left by the arrow. The film speed was DIN 15, which is equivalent to ASA 25.

the Sun. The spectral type is given by a letter–number combination, such as G2, and a luminosity class, also derived from the spectrum, such as V for a dwarf star. The Sun's spectral classification is G2V. Select a star that is convenient for the time of year and your latitude, and prepare your observing program.

15.2.4 Making the Exposures

The idea behind the extrafocus image of the star is to obtain images that have about the same brightness per unit area.[5] If the star areas are exactly the same,

we will then be comparing the correct brightness level of the two objects. Some experimentation will be required to get the star image to be the same size

[5]The brightness per unit area is referred to in several ways. For an illuminated surface, the present context, it is called the *illuminance* and is in units of *lux* (lumens/m2) or, sometimes, in units of *phot* (10^4 lux). For a luminous surface, like the surface of the Sun or a star, it is called *luminous emittance* (lm/m²), or simply *radiant flux*, *luminous flux*, or sometimes *surface brightness* [in W/m² or ergs/(cm²·s)].

as the Sun image, but the following imperfect example can be used as a guide.

Figure 15.2 shows the field chart used to find the selected star, HR 483, a star from the *Bright Star Catalogue* (Hoffleit, 1982). It also compares three exposures of the Sun with an out-of-focus exposure of the field containing the star. The exposure times are given as 10^s, 15^s, and 20^s for the Sun and 976^s for the star. Four gray filters, each having a transmission of 1/400 or $2.5 \cdot 10^{-3}$, were used. The estimate is that a 12.5^s exposure for the Sun would have been about right. The resulting relationship between the Sun and stellar exposures is

$$I_S / I_* = F_S / F_* = (400)^4 \cdot 976 / 12.5 = 2.0 \cdot 10^{12}$$

From Eq. 15.4, we find that

$$r_* = \sqrt{(2.0 \cdot 10^{12})} = 1.4 \cdot 10^6 \text{ au}$$

15.3 Distances to the Stars

The resulting distance to HR 483 is a large number, as expected. A more convenient way to state this number is in terms of the unit known as the parsec (pc). The Earth's orbit is the baseline for measurements of a star's trigonometric parallax, and a star that undergoes a parallactic shift of 1 arc sec across the radius of the orbit (i.e., the astronomical unit) has a distance of 1 pc. To find the distance in astronomical units and in kilometers, consider that any distance laid off on the periphery of a circle at that distance will subtend a radian angle equal to $180°$ / $\pi = 180 \cdot 60 \cdot 60'' / \pi = 206265''$. One *arc second* is therefore 1 / 206265 radian. Any arc s at a distance r subtends an angle θ, and the following relationship holds among them:

$$r \cdot \theta = s \tag{15.5}$$

When r is the distance of a star and s is the arc closely approximating the chord drawn from the Earth to the Sun, then θ is p, the parallax. Figure 15.3 illustrates the geometry. Thus, we have

$$r \cdot p'' / 206254 = 1 \text{ au}$$

or

$$r \cdot p'' = 206265 \text{ au}$$

or, in parsecs,

$$r \text{ (pc)} = 1 / p'' \tag{15.6}$$

In our example, the distance of HR 483 is 6.9 pc. The accepted distance of this star, for which parallax measurements are available, is 11.6 pc. The error arises from a variety of factors, most of which

Distant Stellar Background

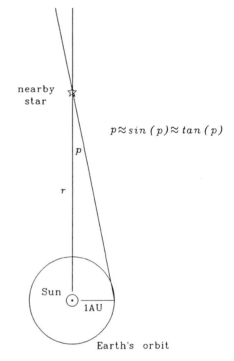

nearby star

$$p \approx \sin(p) \approx \tan(p)$$

Sun
1 AU

Earth's orbit

FIGURE 15.3. The definition of trigonometric stellar parallax. The distance is so great that $\sin p'' \approx \tan p'' \approx p$ (rad).

we have discussed, but mainly from the uncertainty in the filter transmissions. From Eq. 15.4, and again taking

$$I_S / I_* = F_S / F_*$$

we have

$$r^2 = (I_S / I_*) = (I_* \cdot f_F \cdot f_T / I_*) = (f_F \cdot f_T) \tag{15.7}$$

where f_F is the combined filter factor and f_T is the exposure time ratio. Other factors, if known, could be folded into this equation as well [e.g., the ratio of (f ratio)^2 or the ratio of atmospheric transmission]. The relative error in r as computed with Eq. 15.7 is

$$e_r / r = \tfrac{1}{2} \cdot [(e_f/f_F)^2 + (e_T/f_T)^2] \tag{15.8}$$

where e_F is the error in f_F and e_T is the error in f_T. In §15.2.2, we noted a factor $\sqrt{2}$ uncertainty range in each no. 1 ND filter. For the four filters, then,

$$e_F / f_F \approx \pm \; ^{1}/_{2} \cdot [(\sqrt{2} \cdot f_{1F})^4 / (f_{1F})^4] \approx 2 \quad (15.9)$$

Assuming that the error in f_T is negligible, the error in r is about $\pm \; r/2$, so that r is predicted to lie in the range 3–11 pc. This result could be obtained to much higher precision by densitometry measurements of the filters. It could also benefit from having the exposure times closer, and with emulsion tests it would be possible to assess the effects of reciprocity failure. For "Press" plates, familiar to scientists of an older generation, a factor of only 2 in exposure times can cause a density difference of 0.3 (corresponding to a linear factor of 2) in some parts of the H&D curve. Astronomical emulsions can be much better. Even with the best precision, however, photographic photometry is unlikely to yield an error in distance smaller than about 10%. It has been supplanted by photoelectric photometry and electronic imaging techniques almost completely.

Parallax measurements yield very high precision for distances to the nearest stars, but beyond 200 pc, results are typically as uncertain as the present example or more so. For such distances, special techniques and photometry must be employed.

For suggestions of other photographic targets, see Appendix F.

References and Bibliography

Allen, C. W. (1973) *Astrophysical Quantities*. Athlone Press, London.

Hoffleit, D. (1982) *The Bright Star Catalogue*, 4th ed. Yale University Observatory, New Haven, Conn.

Hurter, F., and Driffield, V. C. (1890) *Photo-Chemical Investigations and a New Method of Determination of the Sensitiveness of Photographic Plates*. Journal of the Society of Chemical Industry, Liverpool Section, p. 455.

16

The Speed and Color of Light

16.1 Introduction

The nature of light has intrigued and puzzled philosophers and scientists since ancient times. In Book IV of *On the Nature of Things*, Lucretius, living in Rome of the first century B.C., contends that we are able to see objects around us because they emit replicas of themselves, just as snakes or cicadas shed their skins at certain times, although these replicas would be much thinner and would be emitted continuously. In his view, the passage of these images is instantaneous; thus when a smooth-surfaced body of water is exposed to a starlit sky, images of stars and sky immediately descend and are seen in the water. Even in Lucretius's time, the stars were at the extreme of the heavens.

Isaac Newton, convinced that light behaved like a stream of particles because it propagated in straight lines, devised a "ballistic" theory of light. After Newton's time, the wavelike characteristics of light were fully explored climaxing with the electromagnetic equations of James Clerk Maxwell (1831–1879). The realization that light does not require a medium in which to propagate, and that it does indeed sometimes act like particles, as Newton had proposed, revolutionized physics at the beginning of the twentieth century. The quantum theory of Max Planck (1858–1947) demonstrated a different character of light and matter—a discrete, noncontinuous character not fully appreciated before, despite the musings of Lucretius and the brilliant intuition of Newton. In the twentieth century, light *and* atomic particles were seen to possess "wave–particle duality," exhibiting the properties of both waves and particles. In 1905, Albert Einstein (1879–1955) provided an explanation for the phenomenon known as the *photoelec-*

tric effect,[1] by treating light as a collection of particles (photons), each possessing a discrete amount of energy, $E_\nu = h \cdot \nu$, where ν is the frequency of the light and h is a constant. Finally, in the 1920s Louis de Broglie (1892–1987) predicted, and C. J. Davisson and L. H. Germer experimentally confirmed, the *wave properties* of electrons. In the context of quantum theory, the interaction of light and matter can finally be understood. Modern scientists use the vocabulary of wave theory interchangeably with that of quantum theory in discussing the properties of light. The quantum, it has been realized, is a wave packet of energy.

Here and in later chapters, we use both of the official languages of light as they apply in particular contexts, beginning with a few definitions and examples.

The question of the finiteness of the speed of light may have been raised for the first time by Francis Bacon (1561–1626). In his principal work, *Novum Organum Scientiarum* (1620), he conjectures that the "twinkling stars" may no longer exist even though we continue to see their light. Galileo Galilei (1564–1642) seems to have attempted the first measurement of the speed of light. He had two assistants bearing shuttered lanterns attempt to signal each other from adjacent hilltops. The attempt was not successful because of human reaction time, which is measured in tenths of seconds. Shutter action in fractions of microseconds is required for such an experiment to succeed. Light does indeed have a finite speed, and much of this chapter is

---

[1]The ejection of electrons from a metal surface illuminated with light of wavelength shorter than a certain critical wavelength. The energy of the emitted electrons increases as the wavelength is decreased.

TABLE 16.1. The velocity of light from Jovian occultations of Io.

Opposition, T_{opp}	Last eclipse before opposition T_0	P_{eff}	First eclipse after conjunction Observed T_n	Calculated	δt	$\delta\Delta$ (10^8 km)	c_{eff} (10^3 km/s)
1900+			1900+				
$60^y06^M20^d$	$06^M13^d23^h23^m$						
		1.769856^d	$61^y03^M08^d5^h31^m$	21^m	10^m	2.43	405
61 07 25	07 18 23 05						
		1.769903	62 05 21 3 49	43	6	1.62	450
62 08 31	08 30 00 57						
		1.769937	63 06 18 1 59	54	5	1.82	608
63 10 08	10 05 19 34						
		1.769933	64 06 29 1 56	48	8	2.36	491
64 11 13	11 08 19 41						
		1.769887	65 08 03 1 58	48	10	2.39	399
65 12 18	12 13 19 36						
		1.769856	66 08 22 3 31	16	15	2.56	285
67 01 20	01 17 19 18						
		1.769824	67 09 10 4 53	34	19	2.82	247
68 02 20	02 20 00 21						
		1.769807	68 10 14 4 24	03	21	2.78	220
69 03 22	03 18 21 53						
		1.769802	69 11 02 5 34	12	22	2.53	192
70 04 21	04 21 02 49						
		1.769810	70 12 14 6 51	32	19	2.78	244
71 05 23	05 19 00 22						
		1.769832	72 02 10 6 29	14	15	2.39	266
72 06 24	06 21 23 56						
		<1.769859 ± 15>	Arithmetic mean				<346 ± 40>
			Harmonic mean				<305 ± 33>

devoted to reviving, with the help of hindsight, some astronomical history—the measurement of the speed of light using a technique developed in the seventeenth century. It was not fully appreciated until the turn of the twentieth century that the speed of light is independent of the speed of the source or of the receiver.

16.2 The Velocity of Light from Jovian Eclipses

The first evidence for a finite speed of light was provided by the Danish astronomer Ole Rømer (1644–1710), who used observations of eclipse phenomena of Jupiter's moons. His well-received presentation to the French Academy of Sciences on November 22, 1675, is a distinguished example of the importance of astronomy to the development of physics. Rømer's result for the speed of light was $c = 227,000$ km/s, compared to the modern result, $c = 299,792.5 \pm 0.1$ km/s. The errors in his determination were considerable. The difficulties

which Rømer and his successors faced can be appreciated by examining the satellite data.

Our study of these data builds on the knowledge of the Jovian system, particularly on the knowledge of the repeatability of the *transit* or *occultation*, *immersion* or *emersion* events[2] involving Jupiter's moons, described in Chapter 11. The sources of error in these data are several:

First, the edge of Jupiter's shadow is not sharp and the satellites have finite size so they do not disappear or reappear instantaneously, resulting in limited precision for the timings;

---

[2]These phenomena occur when the satellite crosses in front of or behind the disk of Jupiter and enters or leaves the shadow, respectively. The immersion and emersion events are analogous to our lunar eclipses. Shadow immersions are easier to see than satellite disappearances behind the Jovian disk because of the lack of contrast in the latter case. Emersion events are difficult to catch because even the most experienced observer cannot fully anticipate the event; even when they are caught, human reaction time will make timing of the events inevitably late compared to immersion observations.

second, it is difficult to follow the events across the full diameter of Earth's orbit, especially when Jupiter is near superior conjunction.[3]

third, the eclipses vary in onset and duration because the orbits, although close to Jupiter's equatorial plane, do not lie in the ecliptic plane. Only when Jupiter crosses the ecliptic, at the ascending or descending nodes of its orbit, can the eclipses be central;

and fourth, as mentioned in Chapter 11, Jupiter's oblateness and mutual satellite perturbations cause departures from Keplerian orbits.

Given this litany of difficulties, it must seem rather miraculous that reasonable results ensued. Careful, systematic work is needed to obtain good results from noisy data, but results Rømer certainly did obtain. We will show here that the data are, nevertheless, good enough to obtain a finite velocity for light.

The strategy is to average out some sources of error by taking data around the Jovian orbit—that is, over a 12-y interval. Table 16.1 lists the data in this interval for the immersion of the satellite Io into the Jovian shadow. The term "opposition" refers to the moment when the planet is opposite the Sun in our sky (more precisely, when the celestial longitudes of Sun and planet differ by 180°).

First, we list the time and date of the last Io eclipse before Jovian opposition for each year (column 2). The type of eclipse selected is an immersion into the Jovian shadow. From the interval between successive last occultations, and the number of revolutions of Io, we calculate the "effective" synodic period of Io, P_{eff}, over each interval (column 3). The marked scatter in the 11 individual values of P_{eff} indicates the presence of the sources of error discussed earlier, but the mean of those 11 values is, indeed, very close to the true synodic period. P_{eff} is then used to compute the instant of another eclipse, close to conjunction. The geometry of the orbits of Io, Jupiter, and Earth is such that an immersion into Jupiter's shadow can be studied only during half of Earth's orbit: after a conjunction, and not before, and before an

opposition, not after (see challenge 1 in §16.4). P_{eff} is then used to calculate a time of immersion, T_n, for the first eclipse just after conjunction, using the formula

$$T_n = T_0 + n \cdot P_{\text{eff}} \qquad (16.1)$$

where T_0 is the instant of the eclipse near the previous opposition and n is an integer. Column 4 of Table 16.1 lists the observed instant of eclipse and the calculated one (T_n). The difference, δt, is tabulated in column 5. This difference is basically due to the extra distance $\delta \Delta$ tabulated in column 6, which the light must travel across Earth's orbit as the planets move from opposition to conjunction. The distance Δ is calculated from the geometry of Fig. 16.1. The cosine law of plane trigonometry can be written

$$\Delta^2 = r^2 + 1^2 - 2 \cdot r \cdot \cos \alpha \qquad (16.2)$$

where r is the Jupiter–Sun distance in astronomical units (au). Setting $r = 5.2$ au, $\Delta = \sqrt{(28.04 - 10.4 \cdot \cos \alpha)}$ au. The heliocentric angle, α, increases uniformly with time, and runs from 0° to 360° during a synodic period of Jupiter, 398.87^{d}. From the time of opposition, T_{opp}, α increases by time T to a value

$$\alpha = 360° \cdot (T - T_{\text{opp}}) / 398.87^{\text{d}} = 0.9025°/\text{d}$$
$$\cdot (T - T_{\text{opp}}) \qquad (16.3)$$

The distance Δ can then be calculated. The opposition distance[4] is then seen to be 4.2 AU and the difference, $\delta \Delta$ computed from the expression

$$\delta \Delta = (\Delta - 4.2) \text{ AU}$$

Finally, the "effective" speed of light is calculated for each Jovian synodic period by dividing column 6 by column 5. The result appears in the last column. The true velocity of light does not vary around the orbit—the variation arises from effects already discussed—but the arithmetic mean is only about 1 standard deviation from the accepted value. A better result may be expected from the harmonic

---

[3]That is, when Jupiter has the same celestial longitude (or sometimes right ascension) as the Sun. "Superior" means that it is on the other side of the Sun—the only type of conjunction that a superior planet can have with the Sun as seen from Earth. "Inferior" planets (Mercury or Venus) may undergo either superior *or* inferior conjunctions.

[4]Our treatment assumes circular orbits for Jupiter and Earth. Higher precision can be gained by using the correct expression for the distance of each (see Chapter 14): $r = a(1 - e^2)/(1 + e \cos \upsilon)$, but the work to get this extra precision is not justified here because of the error in the timings. Another, relatively slight source of error is the use of the moment of opposition, rather than the instant of the last eclipse before opposition, in computing α.

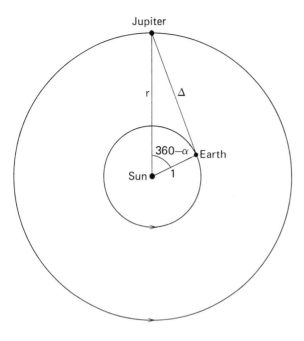

FIGURE 16.1. The Earth–Jupiter–Sun triangle. The Earth–Jupiter distance, Δ, is a function of the heliocentric angle α between the planets, and the distance between each planet and the Sun. Note that as viewed from the North Ecliptic Pole direction, looking toward the SEP, α increases in the CCW direction so that the angle shown is $360° - \alpha$ or just $-\alpha$.

mean,[5] because the time intervals involved in the averaging appear in the denominator of the ratio $\delta\Delta / \delta t$. The error in the harmonic mean is given by error theory as

$$e_h = e_z / z^2$$

where z is arithmetic mean of the reciprocals and e_z is the standard deviation of z.

The following conclusions can also be drawn from Table 16.1:

1. The eclipses may be delayed 5 to 22 min, and the speed of light can be found from the data.
2. An interval of 12 y is a useful interval over which some of the systematic errors cancel out in the averaging process.
3. The harmonic mean even more than the arithmetic mean provides reasonable results for the speed of light using this method.

16.3 Light and Color

The wave nature of light permits us to talk about its *wavelength*, λ, measured in nanometers (or Å,

---

[5]The harmonic mean is the reciprocal of the mean of the reciprocals. For example, let $1/y = (1/x_1 + 1/x_2 + 1/x_3)/3$. Then y is the harmonic mean. For definition of the arithmetic mean, see Appendix B or Chapter 1.

µm, cm, m, etc.) and *frequency*, ν, measured in hertz (Hz), the number of cycles per second of the undulating wave. Red light has a longer wavelength and lower frequency than does blue light. The product of the two quantities has units of velocity or speed, and is the speed of light:

$$\lambda \cdot \nu = c \qquad (16.4)$$

The particle nature of light enables us to talk of the *energy* in a single photon:

$$E = h \cdot \nu = h \cdot c / \lambda \qquad (16.5)$$

where $h \pm 6.6261 \cdot 10^{-27}$ erg·s $= 6.6261 \cdot 10^{-34}$ J·s, or *Planck's constant*. Finally, the *momentum* of a photon can be expressed as

$$p = h / \lambda \qquad (16.6)$$

The wavelength of the light emitted by a hydrogen atom as it moves from the first excited state to the lowest, or ground, state is 121.5 nm or 1215 Å (1 Å $= 10^{-8}$ cm $= 10^{-10}$ m). By Eq. 16.4, the frequency associated with this radiation is $\nu = 3 \cdot 10^8 / 1215 \cdot 10^{-10}$ s^{-1} $= 2.469 \cdot 10^{15}$ Hz. The unit of frequency is the hertz (cycles/second). Its quantum energy, by Eq. 16.5, is $E_\nu = 6.6261 \cdot 10^{-27} \cdot 2.469 \cdot 10^{15} = 1.636 \cdot 10^{-11}$ erg $= 1.636 \cdot 10^{-18}$ joule $= 10.2$ eV (1 electron volt $= 1.602192 \cdot 10^{-12}$ erg). This is also the energy required to raise the electron from the ground level to the first excited level in the hydrogen atom. The momentum of the photon is $p = 6.6261 \cdot 10^{-27} / 1.215 \cdot 10^{-7} = 5.454 \cdot 10^{-20}$ dyn·s or $5.454 \cdot 10^{-25}$ N·s.

16.4 Further Challenges

1. Draw diagrams of the Io–Jupiter system as seen before and after Jupiter's conjunction and opposition in the skies of Earth, and explain why the technique described in this chapter can be used only over the part of the orbit between conjunction and opposition.
2. Test the truthfulness of our statements regarding the relatively small improvements to be gained by using the correct orbital distances and the correct value T_0 in computing Δ and $\delta\Delta$.
3. How could you improve on the speed of light results obtained during this experiment? Outline a plan of action that would increase the precision of the speed of light using the timing of astronomical events like the Io eclipse.
4. Calculate the frequency, energy, and momentum for the light at the following wavelengths: 1 Å (x ray), 100 Å (soft x ray), 912 Å (far ultraviolet), 365 nm (near ultraviolet), 550 nm (visual), 656.3 nm (red), 1.65 μm (near infrared), 60 μm (far infrared), 1.35 cm (microwave radio), 8.1 m (ultra shortwave radio), 462 m (radio).

Reference and Bibliography

Lucretius. (1952) *The Nature of the Universe*. Trans. R. Latham. Penguin Books, Baltimore.

17

Planetary Surface Temperatures

17.1 Introduction

It is no news that the major source of heat energy in the solar system is the Sun. Its energy comes from nuclear reactions that take place at its core. The energy of a particular photon created in these reactions is absorbed and reradiated many times before reaching the surface, a process that takes on average about a million years. The photon also undergoes a great degradation in energy in the course of repeated interactions with matter and with other photons throughout the Sun, emerging, most likely, as a photon of yellow-green light. A great number of photons of this light, along with fewer γ ray, x ray, ultraviolet, and other photons, interact with the atmospheres or surfaces of the planets. Almost all of the visual light of the planets is reflected sunlight. The fraction of the sunlight that is not reflected back into space is absorbed by each planet, warming its atmosphere and/or surface, which then radiates infrared radiation. The temperatures of the planets are much lower than those of the Sun because of the enormous energy output of the Sun's nuclear furnace. It may surprise you to learn, however, that the planets have their own internal sources of heat, although these sources are much weaker in some planets than in others. In general, the planets emit more radiation than can be accounted for by current levels of solar radiation alone.

The surface temperature of a planet helps in a major way to determine the atmospheric composition of the planet. Other factors are the chemical makeup of the bulk of the planet and its mass. The mass and radius of the planet determine whether a particular atmospheric gas will be retained over long periods of time. This important topic is the subject of the next chapter.

Here we calculate a planet's temperature on the assumption that the Sun provides all the energy to a planet, and we compare the calculated temperature to that deduced from planetary radiation measurements. Noting the dependence of the luminosity of the Sun on the photospheric temperature, we will attempt an experiment to measure the Sun's temperature. The assumption that the Sun's radiation alone is the sole source of a planet's heat energy is not strictly true for any planet and is clearly in error for three of the Jovian planets; but it is, a priori, a reasonable assumption.

17.2 Predicting a Planetary Temperature

The *luminosity*[1] of the Sun is $\mathcal{L}$. At a planetary distance from the Sun, r, the *luminous flux*, F, may be defined as

$$F = \mathcal{L} / (4\pi \cdot r^2) \qquad (17.1)$$

This is the radiated solar power per unit area at r. The flux is intercepted over a small portion of the sphere of radius r by the cross-sectional area of the planet. If R is the planet's radius, the intercepted solar power is

$$\pi R^2 \cdot F = \pi R^2 \cdot \mathcal{L} / (4\pi \cdot r^2) = (\mathcal{L}/4) \cdot (R/r)^2 \qquad (17.2)$$

According to thermodynamics, heat cannot continue to be absorbed indefinitely by a body, or the

---

[1] The radiation emitted by the Sun across the entire electromagnetic spectrum. See Chapter 15 for the definitions of most of the radiation-related quantities mentioned here and Chapter 20 for a fuller discussion.

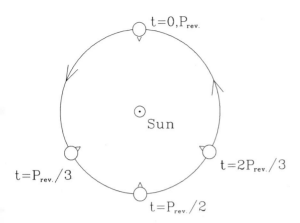
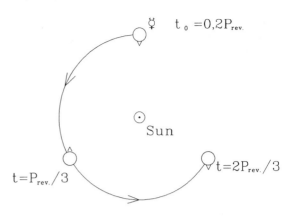

FIGURE 17.1. *Left*: A hypothetical planet tidally locked to the Sun so that $P_{rotn} = P_{rev}$. *Right*: The planet Mercury is locked into a different mode, $P_{rotn} = \frac{2}{3} P_{rev}$, so that it rotates three times for every two revolutions and the same point faces the Sun every 176^d. The well-named Caloris Basin, for example, faces the Sun at every other perihelion passage.

planets would eventually become hotter than the Sun! An equilibrium becomes established whereby as much radiation is emitted as is absorbed by the body receiving the heat energy. The amount of the emitted radiation depends on the temperature of the source. If the emission depends exclusively on the temperature, the source is called a "black body." A black body absorber will absorb *all* of the radiation falling on it, thus appearing black. The planets are not true black bodies, but they are similar to black bodies that absorb just the amount of radiation absorbed by the planet itself. The fraction of solar flux which is not absorbed is called the *albedo*, A. Equation 17.2 must be modified to include only that portion of the solar flux which is absorbed, $1 - A$. The absorbed power thus becomes

$$P_a = (\mathcal{L}/ 4) \cdot (R / r)^2 \cdot (1 - A) \quad (17.3)$$

To calculate the emitted power, we must consider the two factors involved. One is the luminous flux—the amount of energy per unit time emitted by each unit area of the surface. The emitted flux is σT^4 W/m², where T is the surface temperature and $\sigma = 5.671 \cdot 10^{-8}$ J/(m² · s · deg⁴) is the Stefan-Boltzmann constant. The second factor is the area. In order to write down the area of emission we must know if the planet is rotating, and if so, how the rotation axis is oriented in space. Consider the case of a planet that rotates with a period of rotation equal to its sidereal period of revolution around the Sun (left-hand side of Fig. 17.1).

This condition would cause one side to face the Sun perpetually and the other to suffer eternal, frigid night. At one time Mercury was thought to be locked into this configuration.[2] Effectively, the sunlit area alone would emit radiation because it alone would be able to absorb the solar radiation; this area is half that of a sphere or, $2\pi \cdot R^2$. In the case of an infinitely long solar day, therefore, the emitted power becomes

$$P_e = 2\pi \cdot R^2 \cdot \sigma T^4 \quad (17.4)$$

[2]The Italian astronomer G. V. Schiaparelli and others near the end of the nineteenth century had derived a rotation period of 88^d (the same as the sidereal period of revolution) from a study of faint markings seen with a telescope. Since Mercury was not expected to have an atmosphere, the only way heat energy could get to the dark side was by conduction through the planet—an inefficient process because of the pulverized nature of the surface debris from eons of meteoritic bombardment. By the 1960s, flux measurements showed that the temperature of the "dark side" was higher than could be reasonably accounted for by conduction. Subsequently, Doppler radar measurements showed that its sidereal period of rotation is 58.65^3, exactly two-thirds of the orbital period. The older optical data were also shown to fit the 59^d period. The figure of Mercury is so oriented that the same longitude faces the Sun every other perihelion; at the alternate perihelia, the opposite longitude faces the Sun (see right-hand side of Fig. 17.1).

For a rapidly rotating planet, with equator close to that of the orbit, the entire planet will emit, and the emitted power will be

$$P_e = 4\pi \cdot R^2 \cdot \sigma T^4 \qquad (17.5)$$

These two cases are called the slow and rapid rotation cases, respectively.

The condition of equilibrium requires that Eq. 17.3 be set equal to either 17.4 or 17.5, depending on which rotation case applies. Considering first the slow rotation case, we have:

$$(\mathcal{L}/4) \cdot (R/r)^2 \cdot (1 - A) = 2\pi \cdot R^2 \cdot \sigma T^4$$

Solving for T,

$$T_s = \sqrt[4]{[(\mathcal{L}/ 8\pi\sigma) \cdot (1 - A) / r^2]} \quad (17.6)$$

Considering next the rapid rotation case, we have

$$(\mathcal{L}/4) \cdot (R/r)^2 \cdot (1 - A) = 4\pi \cdot R^2 \cdot \sigma T^4$$

Here

$$T_r = \sqrt[4]{[(\mathcal{L}/ 16\pi\sigma) \cdot (1 - A) / r^2]} \qquad (17.7)$$

Notice also that

$$T_s = \sqrt[4]{2} \cdot T_r = 1.189 \cdot T_r \qquad (17.8)$$

so that the temperature of a very slow planet can be expected to be about 20% greater on its heated side but very much cooler on its unheated side.[3] Note that the radius of the planet does not directly affect the temperature (at least not in this relatively simple treatment). The quantity in the first parenthesis in each of Eqs. 17.6 and 17.7 is independent of the planet; it reminds us, however, that as the solar luminosity changes, the temperature of a planet—the Earth included—will change too. In the next section, you will be asked to attempt to measure the value of the solar luminosity. Normally this is done by measuring the solar flux at the Earth, a quantity known as the solar constant, and then using Eq. 17.1 to find $\mathcal{L}$.

17.3 Measuring the Solar Luminosity

The surface temperature of a planet is determined basically by the Sun's luminosity together with the distance to the Sun and the albedo. The solar luminosity is determined by two main factors: its "surface" temperature (the temperature of a region we call the photosphere) and its radius. The luminosity or radiated power of the Sun is thus

$$\mathcal{L} = 4\pi R_0^2 \cdot \sigma T^4 \qquad (17.9)$$

The solar temperature can be derived from this expression. Because the Sun and stars are not really black bodies—their spectral distributions are cut into by myriads of spectral features, from the discrete and continuous absorptions and emissions of atoms and ions in the outer regions of the solar atmosphere—the temperature derived this way is called the *effective temperature*, T_{eff}. It indicates the temperature of a black body that would give the same radiant flux as the Sun.

The surface temperature of the Sun is so high that it is above the vaporization points of all the elements. While atoms may briefly combine to form radicals and compounds, their liaisons are short lived, and many of the atoms, especially those of the metallic elements, are ionized to form positive ions.[4] Thus the Sun and the other stars are gaseous spheres.

The basic experiment is to compare the flux of the Sun to that of another light source, such as a light bulb. While not of high precision, the result will be sufficient to prove some important points. The incandescent light bulb radiates a good deal of its light—perhaps 90%—as infrared light, which we perceive as heat. The Sun radiates most strongly in the visual region of the spectrum, a condition that reflects its higher temperature; it radiates x rays to radio waves, but relatively weakly in the areas outside the optical range, and most radiation of shorter wavelength than ~ 330 nm fails to penetrate the atmosphere. The proper instrument to measure the total flux at *all* wavelengths has not been invented, but for present purposes, a sufficient approximation to it is a *bolometer*, a device that measures infrared radiation through a change in electrical conductivity or resistance. Here, your skin will function as a bolometer, since the skin responds to all the solar radiation falling on it.

First, in a setting in which you can see the unobscured Sun directly (i.e., *not* through a window

---

[3]In the case of slow (but not infinitely slow!) solar motion in the sky of Mercury, data from the *Mariner* 10 space probe showed a midday surface temperature ~ 700 K, a sunset temperature ~ 150 K, and a midnight temperature ~ 100 K. A slightly lower temperature is expected just before sunrise.

[4]One or more of the electrons bound to the nucleus by electrostatic forces have been wrenched away in interatomic collisions or by encounters with high-energy photons. There is an important exception, however. The H^- ion is a negative ion caused by the temporary association of a hydrogen atom with a second electron. The dissociation of this combination is a major contribution to the opacity of the solar atmosphere to visible light.

pane), set up a glowing light bulb of at least 100 W (the bigger the better) about face level and in the general direction (within 180°) of the Sun. Next, rig up a meter or yardstick on a stand so that it extends horizontally away from the lightbulb, with origin next to the center of the light bulb. Now, close your eyes and position yourself toward or away from the lamp until the warmth of the bulb on your face is equal to that of the sunlight. (Be careful not to move too quickly or you may touch the light-bulb and burn yourself; also, it would be unwise to stare either at the Sun or at the light bulb, even for relatively short times.) Have someone measure the distance of your skin from the light bulb at that point. Averaging several measurements will increase the precision of the results.

By taking the mean distance, $<d>$, from the lamp, and assuming a light bulb power p and a distance to the Sun r, the equality of the illumination means that

$$\mathcal{L} \, / \, (4\pi \cdot r^2) = p \, / \, (4\pi \cdot <d>^2)$$

so that

$$\mathcal{L} = p \cdot (r \, / \, <d>)^2$$

As an example, a certain experiment gave $p = 100$ W, $<d> = 0.1$ m, resulting in $\mathcal{L} = 2.3 \cdot 10^{26}$ W. This result is not particularly precise or accurate; the true value is $3.83 \cdot 10^{26}$ W. But it does show the enormous power emitted by the Sun, and to within a factor 2.

Improvement in precision can be gained by increasing the wattage of the bulb, because the distance d will then increase and, measured to the same precision, the relative error will be smaller. In the present case the error in d is ~ 20 to 30%. A slight improvement in accuracy can be realized if corrections are made for the extinction of the visible sunlight as it passes through the Earth's atmosphere (a typical effect is ~ 10 to 20%). This is discussed in Chapter 19. A small factor due to the fact that some of the solar radiation (γ rays, x rays, far UV, microwaves, etc.) does not penetrate the atmosphere at all should be included as a correction (~ 5%). Finally, another source of uncertainty is the efficiency factor of the skin in converting energy of wavelengths shorter than the infrared into heat.

The traditional way to compute the solar luminosity is through the *solar constant*, which is the flux, f, of solar radiation received outside the Earth's atmosphere at mean Earth–Sun distance.[5] Several sources yield a value of $f = 1360$ W/m². Therefore, within round-off error,

TABLE 17.1. Solar system temperatures.[a]

Object	r (AU)	e	A	$T_{r/s}$	T_{ss}	T_{obs}
Mercury	0.387	0.206	0.106	517s	615	700[b]
Venus	0.723	0.007	0.65	299s	356	740[b]
Earth	1	0.017	0.367	248r	351	288
Moon	1	0.017	0.12	320s	381	384[b]
Mars	1.524	0.093	0.15	216r	306	210
Phobos	1.524	0.093	0.06	222r	314	296 ± 15[b]
Jupiter	5.203	0.048	0.52	102r	144	134 ± 4
Io	5.203	0.048	0.61	96r	136	141 ± 3[b]
Europa	5.203	0.048	0.64	112s	134	134 ± 3[b]
Ganymede	5.203	0.048	0.42	126s	151	145 ± 3[b]
Callisto	5.203	0.048	0.20	137s	163	160 ± 3[b]
Saturn	9.528	0.054	0.47	77r	109	97 ± 4
Titan	9.528	0.054	0.21	101s	120	130 ± 5[b]
Uranus	19.18	0.046	0.51	53r	75	59 ± 2
Neptune	30.1	0.006	0.41	44r	63	57 ± 2
Triton	30.1	0.006	0.4	53s	63	55:
Pluto	39.3	0.246	0.3	48s	57	45:
Charon	39.3	0.246	0.3	48s	57	45:
Ceres	2.766	0.078	0.035	166r	235	135 ± 6
Pallas	2.771	0.234	0.05	165r	233	159 ± 4
Vesta	2.361	0.091	0.27	167r	237	143 ± 3
Eros	1.458	0.223	0.23	216r	305	

[a]The albedo and other data have been taken from the *Astronomical Almanac* (1989) and from Abell, Morrison, and Wolf (1987). Asteroid temperatures are weighted means from Webster et al. (1988) and Johnston et al. (1989).
[b]Maximum temperature measured.

$$\mathcal{L} = 4\pi \cdot r^2 \cdot f = 4\pi \cdot (1.496 \cdot 10^{11})^2 \cdot 1360 \text{ W}$$
$$= 3.82 \cdot 10^{26} \text{ W}$$

17.4 Comparing Calculated and Observed Temperatures

The effect of the luminosity of the Sun on the planets is summarized in Table 17.1, where the temperatures, $T_{r/s}$, for both rapid (r) and slow (s) planetary rotation cases have been computed from Eqs. 17.6–17.8. The subsolar temperature, T_{ss}, the equilibrium temperature of a square meter of area which is perpendicular to the incident radiation, is also provided; but this is meaningful only for the intermediate and slower rotators, where local equilibrium conditions can occur. Local temperatures may be higher by a factor of $\sqrt[4]{2} = 1.19$, or more, than T_s. Where infrared flux observations are available, the temperature they imply has been

---

[5]The determination of this quantity was done before the space age, by means of determining the extinction of sunlight through the Earth's atmosphere. See Chapter 19 for the basic ideas behind the technique.

given. For Mercury, the highest temperatures are achieved near perihelion, while the lowest are on the night side just before dawn (~ 90 K); this planet undergoes the largest temperature variation in the solar system. Completely black absorbers will have a subsolar temperature 1.028 times greater than that computed. The very large observed surface temperature for Venus is a result of a runaway "greenhouse effect." In a greenhouse, visual light enters, is absorbed by plants and the interior, and is reradiated at a characteristic temperature. The radiation is mainly in the infrared, because the peak of the emitted spectrum of a black body is given by

$$\lambda_{max} \ (cm) = 0.2898 \ / \ T \ (K) \qquad (17.10)$$

the relation known as *Wien's law*. At $T = 5500$ K, the radiation curve peaks at $\lambda = 5.27 \cdot 10^{-5}$ cm $\approx$ 530 nm; at $T = 300$ K, it peaks at $\lambda = 9.66 \cdot 10^{-4}$ cm $\approx 9.7 \ \mu$m. Thus the interior of the greenhouse, heated to 300 K, glows in the infrared. The glass is opaque to this infrared radiation, however, and the subsequent reflection and reradiation cause further warming of the interior. The analogy has its limits but the atmosphere of Venus has massive amounts of carbon dioxide (CO_2) and some water vapor; both absorb infrared radiation strongly. A similar but much more modest explanation holds for Earth's atmosphere. Earth has a small, variable amount of water vapor, and an even smaller amount of CO_2. The greenhouse effect is apparently increasing for Earth, as more and more fossil fuels are consumed resulting in a buildup of CO_2; unchecked, it can be expected to cause widespread climatic changes in the near future.

The Moon undergoes wide temperature variations because of its slow rotation and lack of atmosphere; on the night side the temperature dips to 100 K. Just as the local temperature varies widely on Earth, it does so on Mars. Mars has almost the same length of day and the same tilt with respect to its orbital plane. The maximum T_{ss} seen is ~ 300 K (+20° C) and the minimum, ~ 120 K (−150° C).

At the outset it was noted that all the planets radiated more energy than they received from the Sun. There are two reasons for this. First, each planet possesses an initial amount of heat energy left over from its formation when material was accreted from the solar system nebula nearly 6 billion years ago. The heat of formation produced during the collisions heated the planets to a sufficiently fluid state to permit the denser materials — iron, nickel, and so on — to settle toward the planetary cores, a process known as *differentiation*. Second, radio-

active material — naturally occurring radioactive isotopes of potassium, uranium, thorium, rubidium, and others — emit high-energy particles during decomposition which heat up the surrounding area.

In the terrestrial planets these sources of heat are not major, compared to the solar energy contribution to their heat budgets. In three of the Jovian planets, the effect of internal heat sources is much greater. The rapid temperature is the relevant predicted quantity in Table 17.1 to compare with the observed temperatures for the giant planets. They are all rapid rotators with extensive atmospheres. Jupiter radiates 2.5 times more energy than it receives from the Sun; some of the excess may be due to slow contraction of Jupiter's extensive gaseous bulk under its own weight.

The selected satellites of the giant planets are tidally locked to their planets, so that the rotation and revolution periods are the same. Only Io with $P_{sid} = 1.77^d$ period could be considered a rapid rotator, and the question is moot as to which rotation case is more appropriate. It is also speckled with hot spots on its surface — active volcanic vents — which are 400 K hotter than the surrounding areas.

Finally, the selected asteroids all have periods of rotation of 10^h or less, and so qualify as rapid rotators (this assumes that the polar axis is not in the plane of the orbit — a reasonable assumption in most cases — but see challenge 4 in the next section for the consequences if this is not true).

17.5 Further Challenges

1. We have discussed planet-wide consequences of solar illumination of planetary surfaces and subsequent reradiation by the surface. Common sense, and our own knowledge of the effect of low Sun altitudes, however, tells us that the temperature varies widely during the day at any particular place on the Earth. See if you can derive an expression for the temperature, T_{ss}, of 1 m² of ground at a place where the Sun is overhead, that is, at the sub-solar point.

2. Now try to derive the temperature at zenith angle[6] z:

$$T_{local}(z) = T_{ss} \cdot \sqrt[4]{(\cos z)}$$

where, again, T_{ss} is the temperature at the sub-solar point, (at $z = 0$).

---

[6] $z = 90° - h$, where h is the altitude.

3. Now you are in a position to verify the subsolar temperatures (for a square meter of surface) for each of the objects listed in Table 17.1.

4. The planet Uranus has it polar axis pointed nearly into the plane of its orbit. Its revolution around the Sun causes one of the poles to face the Sun every half-revolution, or 42^y. Its rotation period, however, is less than a day. When does it belong to the slow and when to the rapid rotation case? How does the equilibrium temperature vary with Uranian seasons?

5. How should equilibrium temperatures vary for a planet with large eccentricity—for example, Mercury, Pluto, and even Mars? Recall from earlier chapters how the distance varies from aphelion to perihelion and calculate the ratio of the temperature extrema in these cases.

6. One interesting satellite we have not discussed here is Iapetus in the Saturnian system, with orbital period 79.3^d and rotation synchronous with this period. The albedo of this moon varies from 0.5 on its bright side to 0.05 on its faint side. Calculate the effect of this albedo range on the subsolar temperatures in those areas.

7. The constancy of the luminosity of the Sun is a matter of lively debate and widespread interest. Attempts thus far to find evidence of variation of the Sun's size from the study of past eclipse records have failed to produce statistically significant evidence of size variation. There is even less evidence over geological time regarding the variation of the Sun's radius. That there has been variation in its luminosity is sug-

gested, however, by evidence of vegetation in the high arctic, and even the presence of dinosaurs in Antarctica in the Mesozoic Era.[7] Calculate the change in solar temperature required (a) to cause Earth's equilibrium temperature to decrease to 200 K, creating the kind of deep ice age that Mars is currently experiencing; and (b) to cause Mars's equilibrium temperature to reach 270 K.

References and Bibliography

Abell, G. O., Morrison, D., and Wolff, S. C. (1987) *Exploration of the Universe*. Saunders, Philadelphia.

Johnston, K. J., Lamphear, E. J., Webster, W. J., Lowman, P. D., Seidelmann, P. K., Kaplan, G. H., Wade, C. M., and Hobbs, R. W. (1989) *Astronomical Journal* **98**, 335–340.

Webster, W. J., Johnston, K. J., Hobbs, R. W., Lamphear, E. S., Wade, C. M., Lowman, P. D., Kaplan, G. H., and Seidelmann, P. K. (1988) *Astronomical Journal* **95**, 1263–1268.

---

[7]The change of latitude of these sites because of continental drift may not have been great enough to explain the effects. An alternative explanation is that conditions on Earth, such as a more active period of volcanic outgassing, caused an enhanced greenhouse effect and that this coincided with long-term periodic variations in the Earth's orbital properties to trigger warmer and cooler episodes. The latter effect, the Milankovitch Theory (after Milutin Milankovitch, 1879–1958), works reasonably well for cycles of ~ 10^5 y; its applicability to longer periods (10^7 to 10^8 y) is unknown.

18

Planetary Atmospheres

18.1 Introduction

In Chapter 17, you had an opportunity to consider what temperature to expect on a given planet and the calculated values were compared to the observed values. It was noted that the atmospheric composition depended on the temperature in an important way and that the gravity would also help to determine the chemical makeup of the atmosphere over eons of time. In this chapter, those relationships are discussed; in the next chapter, the effects of our own planet's atmosphere on starlight is explored.

18.2 Sources and Sinks of Atmospheric Gases

The formation of the planets undoubtedly occurred at high temperatures. We know this because the Earth underwent *differentiation* – the separation of components of different densities. The presence of a molten mass of material will cause iron and other heavy metals to sink toward the center while the lighter materials float on the surface. The evidence that this occurred is the presence of high-density materials toward the center of the Earth, light silicaceous materials in the crust of the Earth, and the evidence for a still extant melt zone in the outer core of the Earth. The evidence for a separation of materials is seen not only in the Earth, but also in the differentiated materials of meteorites. Iron meteorites, mostly consisting of iron and nickel, are thought to have originated from the cores of parent bodies, whereas the stony-irons are thought to have come from the mantles of these bodies. It is also possible, however, that bodies like the Moon

were never completely melted all at once but underwent melting in zonal regions. The presence of the undifferentiated material of the chondrites (the bulk of the stony meteorite class) and the comets, from the outer fringes of the solar system, by contrast, provides still more evidence of the changes undergone by materials in molten conditions in the larger bodies. Much of the original heat that caused the melting must have come from the kinetic energy of impacting bodies whose accretion resulted in the planets and main asteroidal bodies. The presence in the early solar system of short-lived radioactive isotopes, like aluminum 26, also contributed.

For the less massive planets nearer the Sun – the terrestrial planets – the very light hydrogen envelope was quickly swept away by the effects of the solar wind and the high temperatures, for reasons we will explore shortly. The cooling process in a molten primordial planet must have begun with the outer regions, which became the crust. The conduction of heat from the interior must have been impaired with the buildup of *regolith*, the pulverized debris produced by many impacts of still accreting asteroidal fragments. Any gases retained on the surface from the primordial stage were joined by that which was trapped in the interior and released from the molten rock through volcanism, to become the atmosphere.

Atmospheric components can be lost by evaporation from the planet, by chemical reactions, and by photodissociation and collisional dissociation followed by evaporation or chemical reaction. The outgassing of water vapor, for example, could be followed by dissociation of the molecule by ultraviolet solar radiation in the high atmosphere, with the subsequent loss of hydrogen; oxygen may then

TABLE 18.1. Principal constituents of the Earth's atmosphere.

Gas	μ	Percent (by weight)	v (m/s)
Nitrogen (N_2)	28.013	75.52	567
Oxygen (O_2)	31.999	23.14	482
Water vapor (H_2O)	18.015	<0.02 var.	642
Argon (Ar)	39.948	0.01	431
Carbon dioxide (CO_2)	44.010	$5 \cdot 10^{-4}$	411

chemically combine with surface material to produce oxides, like the iron oxides of the Martian crust.

18.3 The Evaporation of Gases from a Planet

The kinetic energy of a molecule is given by

$$E = \tfrac{1}{2} \cdot mv^2$$

where m is the mass of the molecule and v is its speed. The energy of a molecule derives from that of its surroundings, a measure of which is the temperature, T, in Kelvins (K) on the absolute scale. The energy of a molecule, acquired through collision with its surroundings, is partitioned according to the ways in which the molecule can move. For a monatomic molecule, which consists of a single atom, like a helium molecule, the number of degrees of freedom, n, is 3; for a diatomic molecule, like hydrogen, oxygen, or nitrogen, which can rotate (about either of two axes perpendicular to the molecule's axis) as well as translate in any of the three spatial coordinates, n is 5. For more complicated molecules, the situation is more complicated; vibrational motion becomes important, and in nonlinear molecules, rotation may contribute a third degree of freedom. The energy in the translational motion alone can be expressed as $\tfrac{3}{2} \cdot kT$, where $k = 1.38062 \cdot 10^{-23}$ J/deg, known as the Boltzmann constant. Thus we have $\tfrac{1}{2} \cdot mv^2 = \tfrac{3}{2} \cdot kT$, so that

$$v = \sqrt{[3kT/m]} \qquad (18.1)$$

The mass of the hydrogen atom, $m_H = 1.673 \cdot 10^{-27}$ kg, so we can write $m = \mu m_H$, where μ is the mean molecular weight, in units of the hydrogen atomic mass. The speed of the molecule defined in this way is called the root mean square (rms) velocity. Inserting the constants into Eq. 18.1, we obtain

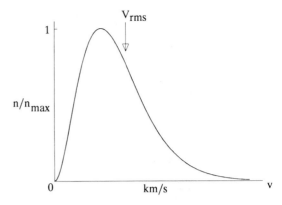

FIGURE 18.1. The velocity distribution for a typical molecule. Note the high energy tail of the distribution.

$$v = 157.3 \cdot \sqrt{[T/\mu]} \ \mathrm{ms}^{-1} \qquad (18.2)$$

Table 18.1. lists the principal atmospheric constituents of the Earth and the rms speed at the nominal surface temperature of 300 K. For a molecule of hydrogen (H_2), the rms speed at 300 K is 1927 m/s, and for helium, it is 1362 m/s. The Earth's atmosphere has only a trace of these gases.

The question of whether a particular constituent can be retained by the atmosphere for a long time can be answered by considering the *escape velocity*. In order to escape from a planet's surface, an object must acquire escape velocity,

$$v_{esc} = \sqrt{[2 \cdot GM/R]} \qquad (18.3)$$

where $G = 6.672 \cdot 10^{-11}$ N·m²/kg² is the gravitational constant, and M and R are the mass and radius of the planet, respectively. With $M = 5.976 \cdot 10^{24}$ kg, $R = 6.378 \cdot 10^6$ m, $v_{esc} = 11,180$ m/s; the escape velocity is much larger than any of the speeds in Table 18.1, but it is also larger than the rms speeds of the hydrogen and helium atoms. Thus the critical question is not *whether* v_{esc} exceeds v but *by how much*. The reason is that there is a distribution of speeds, with a small percentage of the molecules having speeds many times the rms value (Fig. 18.1); these high-speed molecules form what is known as the "high-energy tail" of the distribution. The gas evaporates away through this high-energy tail. If $v / v_{esc} \approx \tfrac{1}{3}$, the gas will be lost in weeks; if $v / v_{esc} \approx \tfrac{1}{4}$, in ~ 10^4 years, if $v / v_{esc} \approx \tfrac{1}{5}$, in $\approx 10^8$ y; and if $v / v_{esc} \approx \tfrac{1}{6}$, it should, barring temperature increases, be around for the lifetime of the solar system. For Earth, $\tfrac{1}{6} \cdot v_{esc} \approx 1860$ m/s. Therefore, we might expect that

TABLE 18.2. Planetary atmosphere parameters.[a]

Planet	v_{esc}	$\frac{1}{6} v_{esc}$	T	v_{H2}	v_{He}	v_{N2}	v_{O2}	v_{H2O}	v_{CO2}
Mercury	4,250	710	517	2,520	1,790	680	630	840	540
Venus	10,360	1,730	299	1,920	1,360	510	480	640	410
Earth	11,180	1,860	248	1,745	1,240	470	440	580	370
Moon	2,375	320	320	1,980	1,410	530	500	660	420
Mars	5,020	840	216	1,630	1,155	440	410	545	350
Ceres	590	100	166	1,430	1,010	380	360	480	305
Jupiter	59,570	9,930	102	1,120	790	300	280	370	240
Saturn	35,560	5,930	77	970	690	260	240	325	210
Uranus	21,330	3,560	53	810	570	220	200	270	170
Neptune	23,800	3,960	44	735	520	200	180	245	160
Pluto	1,330	220	48	770	545	210	190	260	160

[a]All velocities are in meters per second, T in Kelvins.

hydrogen would not be retained, as is the case, but that helium could be. It appears likely, therefore, that Earth's primordial helium was lost when the Earth had a surface temperature much greater than 300 K. Solving Eq. 18.1 for T,

$$T = 4.041 \cdot 10^{-5} \cdot v^2 \cdot \mu \qquad (18.4)$$

For $v = \frac{1}{5} \cdot v_{esc}$, or 2236 m/s, $T \approx 800$ K, and for $v = \frac{1}{6} \cdot v_{esc}$, or 1864 m/s, $T \approx 560$ K. Since the oceans would have boiled away at ~ 370 K, and the geologic record does not support such an event, the high-temperature era must have been close to the time of formation, billions of years ago. Table 18.2 lists the adopted planetary parameters and the rms speeds (in m/s) of their principal atmospheric constituents. The temperatures are taken from Table 17.2. They are computed equilibrium temperatures for the rapid or slow rotation conditions, whichever seems appropriate. These may not be the most appropriate temperatures for every case, but they provide a more or less consistent base to investigate conditions further. At first glance, it might seem incorrect to use the equilibrium tem-

TABLE 18.3. Data for some interesting satellites.

Moon	M (10²² kg)	R (km)
Io	8.89	1815
Europa	4.78	1569
Ganymede	14.8	2631
Callisto	10.7	2400
Titan	13.5	2575
Triton	11.3	1900
Charon	0.33	750

perature for Venus, but temperatures at the cloud tops (~ 250 K) are much lower than they are on the ground, so that T_s may be more typical of the upper atmosphere from which the evaporation into space takes place. In any case, the situation on Venus is complicated.

18.4 Further Challenges

1. The large satellites of the solar system have inspired much interest since the *Voyager* missions. The Galilean satellites, Saturn's Titan, Neptune's Triton, and Charon, the only known moon of Pluto, have excited our imaginations. Investigate the potential of each of these satellites to hold an atmosphere over short and long intervals of time. Table 18.3 and Table 17.2 provide the fundamental data you will need for the calculations.

2. In Chapter 19, the detailed effects of the Earth's atmosphere on starlight as well as sunlight are discussed. In the course of that discussion, the pressure scale height and the optical depth of the Earth's atmosphere are examined. The *pressure scale height* is the distance through the atmosphere required for the pressure to drop by a factor $1/e$, where $e \approx 2.71828$. It can be defined as $H = kT/mg$. The *optical depth* effectively measures the depth into which light can penetrate a medium; for an optical depth of 1, the light will drop by a factor $1/e$. The mean molecular weight of Jupiter's atmosphere was determined originally from a ground-based observation of the occultation of a star by Jupiter. From the distance over which the star's light diminished by

$\sim 1/e$, the data became available to compute $\mu = m / m_H$. Try it: If $H \approx 8$ km, find μ.

3. The possibility of life on other planets is one of the more interesting topics of speculation. Now that you can appreciate why a planet may have a particular gas in its atmosphere, you are in a position to investigate the "habitable zone" of the solar system, the range of distance from the Sun over which water vapor may be retained by an Earth-like planet. Examine Table 18.2 and comment about the possibilities for Mars and Venus. Carl Sagan has suggested that Venus's atmosphere could be "seeded" with microorganisms that would thrive on the massive CO_2 content of the atmosphere. How would this help to make the planet more habitable?

19

Atmospheric Effects on Starlight and Sunlight

19.1 Introduction

Starlight is affected in many ways by its passage through the Earth's atmosphere. The light is refracted or bent, scattered and dispersed, and dimmed and reddened by the atmosphere. The Sun makes an excellent probe for studies of atmospheric effects; it is especially useful for studying refraction, extinction, and color effects that occur near the horizon and the polarizing property of the atmosphere.

Atmospheric refraction was mentioned as early as the first century A.D. by Cleomedes and independently by Ptolemy (discussed in his *Optics*), ca. A.D. 150. The phenomenon can be demonstrated by putting a stick partly into water. To St. Augustine and, later, to others in the Middle Ages, this confirmed the untrustworthiness of the senses, a conclusion that led Western society away from empirical science for more than a thousand years.

Atmospheric dispersion effects are seen in the rainbow, in sundogs, and in the "green flash" phenomena, all of which involve the Sun. They are also seen in the colorful scintillation of starlight, which is visible to the naked eye, and in wedge refraction, which requires a telescope.

Scattering of starlight by the molecules of the atmosphere is the principal source of extinction and reddening; we are most familiar with this phenomenon when we observe a red sunset against a blue, not black, sky.

19.2 Atmospheric Refraction and Dispersion

A stick put partway into water appears to be bent, the water acting to bend the light reflected from the stick. The phenomenon is called *refraction*. The atmosphere acts like this too, and the effect is to bend a ray of starlight so that it is closer to the zenith than if there were no atmosphere. Figure 19.1 illustrates the effect.

Refraction is the action of a translucent substance to change the direction of a ray of light. Snell's law, or the law of refraction, states that as light moves from one medium, characterized by an *index of refraction n*, into another, characterized by an index n', it will undergo a change in direction from an angle z with respect to the normal of the interface to an angle z':

$$n \cdot \sin z = n' \cdot \sin z' \qquad (19.1)$$

The index of refraction measures the speed of light in a substance compared to that in vacuum; therefore, in the vacuum of space it is 1 and in air has a value around 1.000277 (for green light, ~550 nm).[1] Atmospheric refraction can be defined as the difference between the original direction of the star and the refracted direction:

$$R(z') = z - z' \qquad (19.2)$$

Substituting Eq. 19.2 in Eq. 19.1, we get $\sin(z' + R) = \sin z'$. Applying the sine law for plane triangles to this,

$$\sin z' \cdot \cos R + \cos z' \cdot \sin R = n' \cdot \sin z'$$

Now R is relatively small, typically ≤ 1 arc min for $z' < 45°$ (but larger for bigger angles, as discussed later), so that $\sin R \approx R$ (expressed in radian measure), and $\cos R \approx 1$. With these approximations,

---

[1]The value of n_{air} depends on wavelength, temperature, and pressure as well. The value quoted for dry air is for the conditions $T = 15°C$ and pressure $P = 1.013 \cdot 10^5$ Pa.

$$\sin z' + R \cdot \cos z' = n' \cdot \sin z'$$

or

$$R = (n' - 1) \cdot \tan z' \qquad (19.3)$$

The quantity $(n' - 1)$ is the refraction correction in radians at $z' = 45°$, when $\tan z' = 1$. Multiplying it by $206265''$ (the number of arc seconds in a radian angle) gives $R(45°) \approx 57.3''$. A more precise formula, useful to $z \approx 60°$, is given by Woolard and Clemence (1966, p. 84):

$$R = 60.29'' \cdot \tan z' - 0.06688'' \cdot \tan^3 z' \qquad (19.4)$$

The atmospheric refraction is zero at the zenith and increases to a maximum value near the horizon, where it amounts to about $0.5°$, the angular diameter of the Sun or Moon. This means that when these or any other objects appear on the horizon, they are actually still below the horizon but have been bent upward just enough to be visible. Beyond about $z' \approx 60°$, Eq. 19.4 is no longer very precise, and the Pulkovo Observatory Tables are often used. Tricker (1970, p. 16) gives an approximation to these tables; for $z' = 75°$, $R \approx 3.5'$; $80°$, $5.2'$; $85°$, $9.6'$; and for $z' = 90°$, $R \approx 33.8'$. The actual amount of refraction at any one time, however, depends a great deal on atmospheric conditions along the line of sight. Woolard and Clemence discuss several ways to determine the amount of refraction at very low angles. These include measuring the observed distances between the brighter stars of the Pleiades and measuring the apparent separations of selected points on the Moon's disk. Try doing this experiment at different times of the year, recording carefully the temperature, humidity, barometric pressure, date, and so on, to see if you can verify Tricker's (1970, p. 17) finding that a barometric pressure increase of 25 mm causes a rise of 3.5% and a rise of 9°C causes it to decrease by 1%.

The bending process is slightly different for rays of different color; the blue rays are bent more than the red. If you view very low stars through a telescope and the atmosphere is very calm, you can see the effect of this dispersive action of the atmosphere. It is called *wedge refraction*, and the result is to produce tiny spectra of the stars. The length of the spectra depends on atmospheric conditions, but a typical length is about 2 to 3″ at a zenith angle of 75°.

The *green flash* is an example of variable atmospheric dispersion. This phenomenon occurs when patches of air of different densities refract the (setting or rising) Sun so as to momentarily isolate the bluer part of the spectral image. The effect,

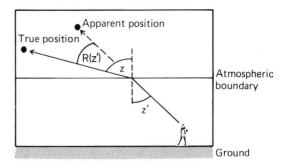

FIGURE 19.1. The refraction of starlight.

best seen from a very high vantage point, when the Sun sets over a low, flat horizon, preferably a marine horizon, usually requires some optical aid to see it. An interesting—but by no means easy—project would be to catch the green flash on a sunscreen with a movie camera or videocasette recorder. As we have cautioned elsewhere, *do not* view the Sun directly.

Dispersive refraction effects also play a role in the colors of rainbows, which involve refraction in water droplets, and of sundogs, which involve ice crystal refraction.

19.3 Twinkling Stars

To the naked eye stars sometimes seem to brighten and dim rapidly, sometimes accompanied by sharp changes in color. Astronomers regularly get telephone calls from the public regarding such phenomena, especially if there are scudding clouds in the sky, because a bright star will appear to vary, change color, and "move" around the sky—a classical UFO!

The explanation for the twinkling—astronomers prefer the term *scintillation*—has to do with variations in the refractive properties of air over the line of sight to the star. Any such sightline will have in it many parcels of air at differing pressures and temperatures. The refractive index will vary from parcel to parcel. The parcels are not stationary either, moving in and out of the sightline. The net effect is variable refraction, leading to displacement, brightness, and color variation. The size of the parcels of air is important in determining the effect on star images.

Amateur astronomers are familiar with the term "seeing," which characterizes the quality of the night—really the steadiness of the images permitted by the sky. A small telescope equipped with a high-power eyepiece can discern the displacement

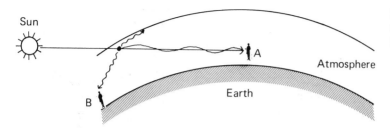

FIGURE 19.2. The cause of the blue sky and red sunset (and sunrise) phenomena.

of the images readily. With a larger telescope, which gathers a larger bundle of starlight rays, the effect is usually an image that swells and shrinks as the refractive effects of the cumulative parcels vary. The seeing is always poorer for higher zenith angles, but on a poor night, it will be bad even for objects at zenith. One way to estimate seeing is to note a star's apparent diameter. The stars are too far away to reveal solarlike disks, so the apparent size has to do with the effects of refraction, which are summarized in the word seeing. This can be measured by means of a reticle scale set in the eyepiece, behind which a focused star image can be placed. The reticle can be calibrated with a double star of known separation (rotating the eyepiece as needed), or with the timing technique described in earlier chapters. At the Dominion Astrophysical Observatory, in Victoria, British Columbia, typical seeing is 3 to 5 arc sec; at the Rothney Astrophysical Observatory in Calgary, Alberta, it may be as low as $1/2$ arc sec, or, during a chinook,[2] as large as 10 arc sec or more.

Because limited sizes of air cells are involved, planets normally do not twinkle. We can use this circumstance to compute an upper limit on the size of the cells. On a night in December 1976 in Bochum, West Germany, Venus had an angular diameter of 16 arc sec and was not twinkling. As a rough approximation to the cell size during observation, consider that the atmosphere has a scale height of ~ 8 km. At this distance, 16 arc sec is subtended by an arc,

$$d \leq 8\,\text{km} \cdot 16'' / 206{,}265 = 6.2 \cdot 10^{-4}\,\text{km} \approx 0.6\,\text{m}$$

Therefore, the cell size was of the order of 1 dm or less. The interplanetary and interstellar media cause radio sources to scintillate also. Decametric radio waves from the environment of the planet

Jupiter undergo interplanetary scintillation, and pulsars undergo scintillation produced in the interstellar medium. This "radio twinkling" allows us to probe the properties of the interstellar medium.

19.4 The Extinction of Starlight

Cloud, smoke, and haze frequently obscure our view of the universe, and the scattered light of the Sun and Moon make it difficult to see even the brightest stars. Even when the Sun and Moon are not in the sky, the background may be brighter because of the presence of airglow and aurora, or scattered light. In this section we talk about the extinction of starlight.

The scattering process which produces Earth's glorious red sunsets and beautiful blue skies (Fig. 19.2) is a consequence of *Rayleigh scattering* by atmospheric molecules, the efficiency of which varies as $1 / \lambda^4$. The blue color of distant mountain peaks (the "purple mountains' majesties" of "America the Beautiful") is also caused by scattering. The most important difference between the blue light of distant scenery and the light of the sky is the high component of dust scattering, which contributes to a light path close to the horizon. The dust scattering has little spectral dependence but causes a bright haziness. In dust-free regions of the Earth, the blue coloration of terrestrial objects that are 8 km away or more corresponds, in fact, to the color of a cloudless sky at zenith. The 8-km distance is the order of magnitude of the *scale height* of the atmosphere. The scale height is a characteristic vertical distance over which the atmospheric pressure decreases to some fraction, say, $1/e$ ($\sim 37\%$), of the surface value. (The quantity e is the base of the natural logarithm system; it has a value of $2.7182818\ldots$). On a clear day, try to verify, and maybe improve on, Table 19.1. With these data, you can obtain the order of magnitude of the scale height of the Earth's atmosphere.

---

[2]A wind phenomenon on the leeward slopes of mountains, called the *sirocco* in the Mediterranean area and the *Föhn* in Switzerland, Austria, and Germany.

The molecules of the air affect starlight as well as sunlight. They scatter blue light much more efficiently than they do red, resulting in a blue sky and a reddened celestial object. Collectively we do not receive enough luminous flux from the stars, compared to the Sun, for the sky to appear blue at night, but that is the net result: reddened starlight and a slight "blueing" of the sky. The scattering is a major component of the dimming of starlight as it passes through the atmosphere, especially in the visual region of the spectrum ($\gamma \approx 500$ to 600 nm), where atomic and molecular absorption play relatively small roles. The scattered light contributes to the general background.

The extinction follows an exponential law: the light lost is proportional to the incoming intensity[3] in the following way:

$$dI = - k \cdot I \cdot dx \qquad (19.5)$$

where dI is the light intensity lost through a path length dx, I is initial intensity of the radiation, and k is a constant of the medium, in this case, the atmosphere. Through the mathematical process of integration,[4] the solution is

$$I = I_0 \cdot e^{-kx} \qquad (19.6)$$

TABLE 19.1. Purple mountain effect.

Object distance (m)	Blue coloration level
100	Hardly noticeable
1,000	Observable
10,000	Pronounced

This equation is also used in nuclear physics in neutron diffusion studies as well as in the study of radiation transfer in stellar atmospheres. In atmospheric extinction work, the quantity x is defined as *air mass*, the relative distance traversed through a plane-parallel atmosphere. Figure 19.3 shows that compared to the zenith, the air mass,

$$X = 1 \ / \cos z = \sec z \qquad (19.7)$$

For large values of X, the plane-parallel approximation breaks down and a correction in the form of a polynomial approximation is usually applied. An approximation good to $z \approx 80°$ is given by Bemporad and reproduced by Hardie (1962):

$$X = \sec z - a(\sec z - 1) - b(\sec z - 1)^2 - c(\sec z - 1)^3 \qquad (19.8)$$

where $a = 0.0018167$, $b = 0.002875$, and $c = 0.000808$. The decrease in intensity in starlight as

[3]As we explain elsewhere in this book, when starlight is being considered, usually the luminous or radiant flux is intended, not the intensity, which is usually defined as the flux per unit solid angle. For point sources, like stars, there is no difference in these quantities, however, and traditionally the intensity is used in the equation of transfer (Eq. 19.5).

[4]This process can be sidestepped somewhat, for those not familiar with integration, by assuming that I has the form $I = A \cdot e^{Bx}$. After differentiating, this equation gives $dI = AB \cdot e^{Bx} \cdot dx$. Setting this equal to Eq. 19.5 it follows that $B = - k$ $A = I (x = 0)$ or I_0.

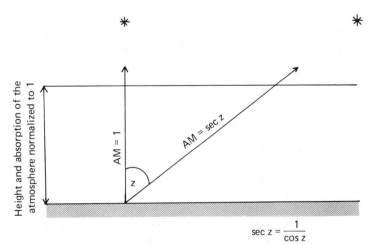

FIGURE 19.3. The air mass defined as $1/\cos z$ or $\sec z$. The extinction of starlight grows with zenith angle, z. At large values of air mass, the curvature of the atmosphere becomes important, and a correction must be applied.

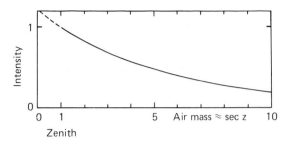

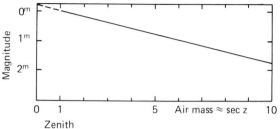

FIGURE 19.4. The extinction of starlight, in units of intensity, as a function of air mass.

FIGURE 19.5. Extinction plotted as magnitudes vs. air mass. Note that the magnitude outside the atmosphere can be obtained by extrapolating the linear relation to zero air mass. Although there is little evidence of strong departures from linearity for visible light in the region between 0 and 1 airmass, infrared extinction undergoes some important changes there.[5]

a function of air mass is indicated in Fig. 19.4. Note that the curve is not really defined between 0 and 1 air mass.

Since the days of Hipparchus, astronomers have used the *magnitude* measure, instead of flux or intensity, to indicate stellar brightness. Taking the base 10 logarithm of both sides of Eq. 19.6, we obtain

$$\log I = \log I_0 - k \cdot X \cdot \log e$$

Multiplying both sides by the factor -2.5 and absorbing the constant factors $(-2.5 \cdot k \cdot \log e)$ into a new constant, k', we obtain

$$m = m_o + k' \cdot X \qquad (19.9)$$

where

$$m = -2.5 \cdot \log I + \text{const}$$

and

$$m_o \pm -2.5 \cdot \log I_0 + \text{const}$$

the observed and outside atmosphere magnitudes, respectively. The evaluation of the const term is unnecessary here, since it cancels out. The zero-point of the magnitude system depends on the properties of the particular photometric system involved. (We discuss this topic in Chapter 23.) Magnitudes would be better called indexes of faintness, not brightness, because the lower the intensity, the larger the magnitude. Thus an object of magnitude 3.7 will be brighter than one of magnitude 6.2 by a factor 10 in flux.

Figure 19.5 shows the extinction in magnitudes as a function of air mass. Note the curve has become a straight line. The slope of that line is the *extinction coefficient*, k', and the zero point (where the line reaches zero air mass) is m_0. A rough value for the extinction coefficient can be found in the following example: In the early evening hours of

December 29, 1976, the third-magnitude star ι Orionis was first seen as it was rising at an altitude $h = 4°$ ($z = 86°$). The Pleiades, which were already high in the sky, revealed that the naked eye sky limit was about sixth magnitude. Very roughly,

$$X \approx 1 / \cos 86° = 14.3, \quad \text{so } k' \approx (6 - 3) / (14.3 - 1) = 0.2^{\text{m}}/\text{A.M.}$$

This crude example gives approximately the right extinction coefficient. Much more precise values can be determined from many observations of the same star (the Bouguer method) or observations of several standard stars around the sky (Hardie method), or some combination. The precision of the results can exceed 0.001^{m} (or 0.1%).

The extinction coefficient varies with wavelength of the light and the photometric system, so if one is photographing the stars, the extinction will be different from that seen by the eye. The cofficient k' is larger toward the blue than toward the red. If $k'_V = 0.2$, k'_B may be 0.3 and k'_U, 0.5 (for the visual, blue, and ultraviolet regions, respectively). It should be noted, however, that the coefficients can vary from night to night, and from one site to the next. The coefficients are usually smaller and undergo less variation at mountaintop observatories than at lower sites.

---

[5]In fact, infrared extinction is a more complicated phenomenon in several other ways. First, it is heavily dependent on the water vapor content at the time of observations; second, refraction requires recalculation; and third, the air mass correction is different, intermediate to that indicated by Eqs. 19.7 and 19.8. For further details see Milone (1989).

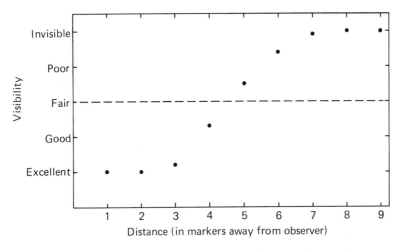

FIGURE 19.6. The visibility of evenly spaced signposts through fog.

Space telescopes are a valuable addition to the arsenal of astronomy because they will be able to measure starlight without the bright background produced by light pollution of terrestrial cities, the scattered sunlight, moonlight, and starlight, aurorae and airglow, and the effects of scintillation and extinction. They will be few in number, however, and they will be dedicated primarily to problems on the frontiers of astronomy involving faint and faraway objects or those requiring the specialized instrumentation. Therefore, ground-based astronomy is and will remain indispensable into the foreseeable future.

19.5 Further Challenges

1. The quality of a night can be gauged in several ways. In §19.3, we discussed how to provide an estimate of "seeing." Another estimate is the naked eye "sky limit," a measure of the faintest star that can be seen with the naked eye. It is a gauge, as the example in §19.4 showed, of the extinction and sky background brightness quality of the night. A star atlas will help to provide suitable stars. The chosen stars should be at least 30° from the horizon. Try to determine these figures of merit over a period of time from your own site and keep a record of these observations; they will help you to become better acquainted with your night sky and inform you of any deterioration of conditions.
2. Now for some foggy night astronomy. At first glance, you might think this challenge is about

theory, rather than observation. Well, in part it is, but it also involves observations. If you live in a location that is given to fog and mist, this challenge is for you. First you need to determine some standards of distance so that the depth of penetration through the fog can be gauged. From a selected vantage point that you can always recognize, select several signposts, markers whose distances from each other are the same as the distance between you and the nearest one. Fenceposts in a field, telephone poles, or lampposts will do nicely.

On reflection, you will recognize that things do not "simply disappear" in fog or mist, but that various gradations can be assigned. For each of your signposts, you can assign a mark of visibility, like 1 for good, 2 for slight impairment, 3 for intermediate, 4 for greatly reduced, and 5 for completely invisible. Figure 19.6 illustrates the idea.

What does this have to do with astronomy? It shows what happens when light passes through a scattering medium—like the atmosphere or interstellar dust. In Figure 19.6, the halfway visibility point is about $4.6 \pm \sim 0.6$ markers away. Over a distance r, the light will be cut down by the factor e^{-kr}, where k is a constant. In each volume element of fog, the amount of scattered light will be $k \cdot e^{-kr}$. By adding up all these volume elements (i.e., by integrating them from 0 to the distance of a marker R), we get $\int k \cdot e^{-kr} \, dr = 1 - e^{-kR}$, the total scattered light between you and the first marker. Figures 19.7 and 19.8 show the dimming of the transmitted light and the increase in the

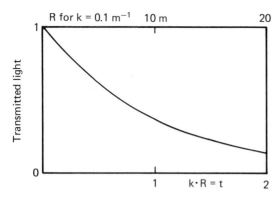

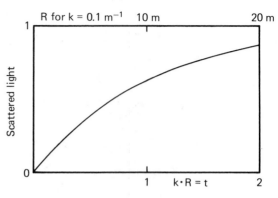

FIGURE 19.7. The extinction of light through fog. Transmitted light is plotted against the optical depth (lower scale) and distance in meters (upper scale).

FIGURE 19.8. The increase in scattered light produced by fog. Foreground brightness is plotted against optical depth and distance.

scattered light, respectively, as a function of the product of k and R. This product is called the *optical depth*, τ. When $\tau \ll 1$, an optically thin case, both extinction and stray light are minimal; when $\tau \gg 1$, an optically thick case ($\tau > 2$ more or less suffices), the transmission is strongly diminished and the light scattered is practically everything.

Therefore, $\tau \approx 1$ is an important quantity, because it provides an indication of the depth of penetration into the fog. So, what is *your* value of R for an optical depth of 1? Should the quantity k vary with distance (in a patchy way—like many fogs and the interstellar medium do!), then the optical depth is defined as

$$\tau = \int k(r)\, dr$$

In a similar way, the gaseous photosphere of the Sun appears fairly sharp. The depth of penetration along the line of sight (i.e., $\tau \approx 1$) is only about 400 km. At the distance of the Sun 400 km *on the sky* subtends less than 1 arc sec:

$$\alpha = [400 / (1.5 \cdot 10^8)] \cdot 206265''/\text{rad} = 0.55''$$

3. Light as electromagnetic radiation can be thought of as a combination of electric and magnetic fields each of which undergoes an oscillation or vibration (at the frequency of the light) in a plane perpendicular to the direction of propagation of the light. In the course of scattering light from the Sun, the atmosphere also *polarizes* it, that is, preferentially scatters light which is vibrating in a particular plane. Normally we see unpolarized light, that is, all the planes are represented. A pair of polaroid sunglasses acts as a filter because it restricts to a particular plane the light that passes through it. Now it turns out that the eye itself has a weak but real ability to detect Polarized light because of a slightly *dichroic* retina (i.e., the eye is sensitive to two colors along two different axes). To use this ability, you need to develop your "Haidinger

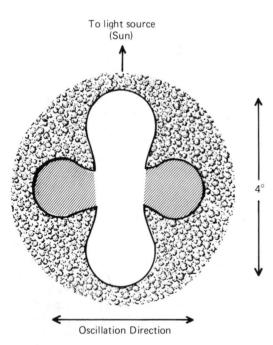

FIGURE 19.9. The Haidinger bundle—the human eye's polarization detector. The white part of the cruciform figure represents the yellow region, the crosshatched, the blue.

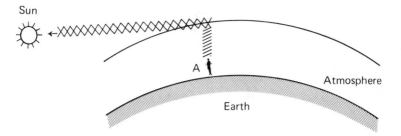

FIGURE 19.10. Atmospheric molecules strongly polarize the light they scatter into directions 90° from the source. A Polaroid filter will reveal this easily; rotating the filter to maximum gives you the direction of the vibration of the polarized light. The unaided eye too can detect this effect with the Haidinger bundle.

bundle."[6] This is a blue and yellow cross-shaped figure, with very little contrast, which lies in the field of direct vision. The figure differs from person to person and measures about 4° in the long direction, which is in the direction of the light source and of the magnetic vector. This direction should appear slightly yellow, whereas the perpendicular direction, the vibration direction, should appear slightly blue (Fig. 19.9) Practice your ability to see this by looking through a Polaroid filter at a white piece of paper at arms' length or at a uniformly gray sky. It may help to tilt your head, but the Haidinger bundle should be immediately apparent. If you are not afraid of acquiring a reputation doing this, you can now try it on a clear sky just after sunset. Look at the zenith, tilting your head as needed. The yellow axis points toward the Sun. The light scattered from the sky in a direction 90° from the sun is strongly polarized (Fig. 19.10). With some practice, you can detect a degree of polarization down to 25% with your eyes alone.

References and Bibliography

Hardie, R. H. (1962) Photoelectric Reductions. In W. A. Hiltner (ed.), *Astronomical Techniques*. University of Chicago Press, Chicago, p. 180.

Milone, E. F. (ed.) (1989) *Infrared Extinction and Standardization*. Springer-Verlag, New York.

Minnaert, M. (1954) *The Nature of Light and Colour in the Open Air*. Trans. H. M. Kremer-Priest; rev. K. E. B. Jay. Dover, New York.

Tricker, R. A. R. (1970) *Introduction to Meteorological Optics*. American Elsevier, New York.

Woolard, E. W., and Clemence, G. (1966) *Spherical Astronomy*. Academic Press, New York.

---

[6]In German, *Haidingeres Büschel*, named for W. Haidinger (1795–1871). See Minnaert (1954) for further discussion.

20

Sources of Cosmic Radiation

20.1 Introduction

Our only source of information about most of the universe is electromagnetic radiation. The development of our knowledge about the nature of this radiation and the terminolgy used to describe it were treated in Chapter 16. In this chapter, we describe the radiation laws, which are basic to our understanding of how electromagnetic radiation is produced and how it can be detected.

In Chapter 17, you saw applications of the Stefan–Boltzmann law ($F = \sigma T^4$) and Wien's law ($\lambda_{max} = \text{const} / T$). Here we discuss applications of black body or thermal radiation laws and the conditions giving rise to nonthermal radiation, and we examine the spectral distributions of astronomical objects. First we summarize the black body radiation laws.

20.2 Black Body Radiation Laws

A black body is an idealized, perfect radiator. By definition, it absorbs all the radiant energy that falls on it, reflecting nothing. It will reach an equilibrium temperature quickly and radiate at the same rate as it receives the radiation. Because its radiated power, spectral distribution, and color are completely described by one quantity—its temperature—black body radiation is often called thermal radiation, and black bodies are called thermal sources. A black body radiator can be constructed simply: it is a well-insulated box with completely black, opaque interior walls, an internal heat/light source, and a tiny opening through which a minute portion of the radiation can escape for measurement purposes. For this reason, black body radia-

tion is sometimes called cavity radiation. The correct energy distribution of a black body was derived first by Max Planck (1858–1947) in 1900. Most of the other laws follow from it. The Planck function (sometimes called the Kirchhoff–Planck function) is illustrated for a variety of temperature sources in Fig. 20.1. The function may be expressed as in terms of λ or of ν, and the corresponding curves are drawn for both. Differentiating it with respect to wavelength and setting the result equal to zero gives Wien's law relating the wavelength of the peak of the curve to the temperature. Integrating the Planck distribution over all wavelengths yields the Stefan–Boltzmann law, relating the total radiant flux to the fourth power of the temperature. The Wien approximation and the Rayleigh–Jeans law were earlier approximations to the black body energy distribution and are sometimes still used for convenience to represent the distribution over limited wavelength regimes, but they must be used carefully. They are suitable for opposite sides of the radiation curve. Wien's law is an adequate representation over short wavelengths; the Rayleigh–Jeans approximation works adequately for long wavelengths. The latter predicts a continuous increase in flux with decreasing wavelength. This incorrect prediction is called the "ultraviolet catastrophe."

Note that the black body radiation curves of Fig. 20.1 are completely nested: the higher temperature curves envelop the lower. The area under each curve thus increases with temperature, and the calculation of this area gives the Stefan–Boltzmann law. Notice also that the peaks of the curves shift to shorter wavelengths and higher frequencies with increasing temperature, in accordance with Wien's law.

The black body radiation laws are summarized in Table 20.1. In these relations, wavelength is in

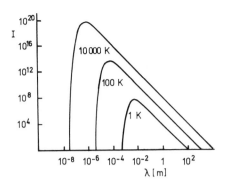
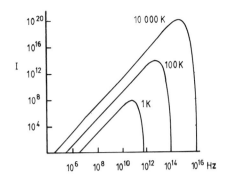

FIGURE 20.1. The spectrum of black body radiation for black bodies at three different temperatures. The dependent variables are λ (on the left) and ν (on the right). Note the nested behavior of the curves and the shift of the peak to the blue (shorter wavelengths and higher frequencies) with increasing temperature.

meters, frequency in reciprocal seconds or hertz, and temperature in Kelvins. The following constants are used in the equations:

Planck's constant:
$$h = 6.6261 \cdot 10^{-34} \text{ J·s}$$
Boltzmann's constant:
$$k = 1.3807 \cdot 10^{-23} \text{ J/deg}$$
Velocity of light:
$$c = 2.9979 \cdot 10^{8} \text{ m/s}$$

The radiation curves of astronomical sources that resemble black bodies are seen in Figs. 20.2 to 20.5. In each case, we note the peak wavelength and the "Wien temperature" obtained from Wien's law. The curves are based on observations taken over a broad bandwidth, that is, over an extended range of wavelength (the size of the bandwidth varies with the wavelength regime: it may be ≤ 1 μm in the near infrared and the optical region, but much wider in the millimeter and microwave regions). Detailed spectral information is lost when this is done, but more flux is obtained and the precision of the measurement is greatly increased.

Figure 20.2 is the radiation curve of the "big bang," the so-called three-degree background radiation.[1] Its discovery by Arno Penzias and Robert W. Wilson of Bell Labs in Holmdel, New Jersey, in the mid-1960s, earned them the Nobel Prize in 1978. The radiation represents the remnant glow from a certain stage of the cosmic fireball in which all known matter in the universe is thought to have been concentrated more than 10 billion years ago. That stage was critical in the development of the universe: the temperature of the fireball had cooled to 10^4 K, not high enough for atoms to be continuously ionized; thus light and matter became "decoupled" and the universe became "transparent" to the radiation. From that time to this, the universe

[1]The microwave measurements of the background radiation obtained from radiotelescopes actually lie on a 2.7-K black body curve but the millimeter-wave measurements obtained by rocket and balloon-borne detectors (because this radiation does not penetrate the Earth's atmosphere), where the curve should peak, suggest 2.96 K. The millimeter observations are difficult to obtain and to reduce so the question remains open.

TABLE 20.1. Black body radiation laws.

Relation	λ Scale	ν Scale
Planck function	$(2\pi hc^2 / \lambda^5)/(e^{hc/k\lambda T} - 1)$	$(2\pi h\nu^3/c^2)/(e^{h\nu/kT} - 1)$
Wien approximation	$(2\pi hc^2/ \lambda^5)/(e^{hc/k\lambda T})$	$(2\pi h\nu^3/c^2)/(e^{h\nu/kT})$
Rayleigh–Jeans approximation	$(2\pi c/\lambda^4) \cdot kT$	$(2\pi\nu^2/c^2) \cdot kT$
Wien's law	$\lambda_{max} \cdot T = 0.002898 \text{ m·K}$	$\nu_{max}/T = 0.5879 \cdot 10^{11} \text{ Hz/K}$
Stefan–Boltzmann law	$F = 5.6705 \cdot 10^{-8} \cdot T^4 \text{ W/(m}^2\text{·deg}^4)$	

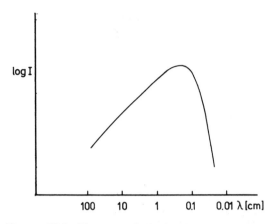

FIGURE 20.2. The spectral distribution of the three-degree black body radiation of the cosmic background. The peak wavelength of the distribution lies at ~1 mm.

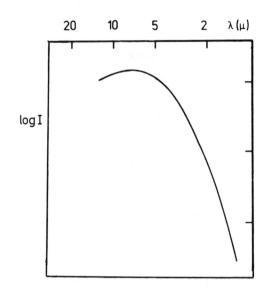

FIGURE 20.4. The spectral energy curve of an infrared object bearing the Infrared Catalog designation IRC+10011. The peak is ~7μm, so the Wien temperature is ~410 K.

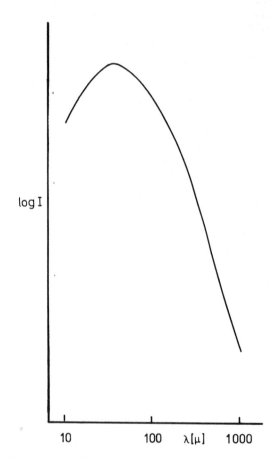

FIGURE 20.3. The spectral energy distribution of an infrared source in the Orion nebula. Note the peak of the curve near 36 μm implying a Wien temperature of ~80 K.

has continued to expand and to cool while the image of the fireball fades and reddens, since, as Boltzmann showed, expanding black body radiation remains black body radiation but characterized by ever decreasing temperature.

Figure 20.3 shows the radiation curve of an infrared source in the Orion nebula (M42). The curve represents a fit to observations from disparate sources: 22 μm ground-based observations; 25–300 μm data from a detector on board a jet aircraft flying observatory; and data ~1 mm from a millimeter-wave radio telescope. Figure 20.4 shows the radiation curve of the infrared object IRC+10011, from ground-based observations.

The radiation curves of two stars, β Peg (top) and α Aur (bottom) seen in Fig. 20.5 are based on broad-band ground-based observations. This latter can be compared to the slightly more detailed continuum observations of another G-spectral class star, the Sun (Fig. 20.6). Notice that while the overall shape conforms to that of a black body, there are definite departures.

Figure 20.7 depicts the radiation curve of the hot star, γ Orionis (spectral class B).

The connection between the peak of the color curve and temperature is apparent when we turn on the electric range. The heating elements begin to glow with a deep red and gradually become bluer and brighter as they become hotter. When the

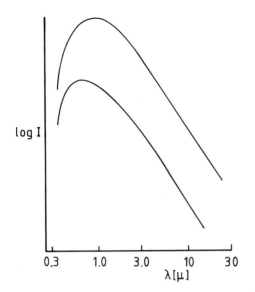

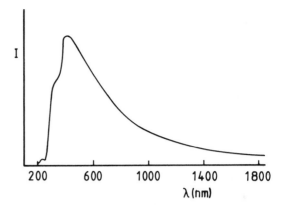

FIGURE 20.6. The radiation curve of the Sun from narrow-band observations. Such features as spectral lines are not included, but the effects of atomic ionization can be seen in the ultraviolet region of the spectrum. The peak of the curve at ~0.46 μm implies a Wien temperature of ~6300 K.

FIGURE 20.5. The spectral distribution of the cool star β Pegasi (spectral class M), shown at the top of the figure, peaks at ~1 μm. Its Wien temperature is therefore ~2900 K. The lower curve is that of yellow-white giant star, α Aurigae (Capella), of spectral class G. It peaks at ~0.8 μm implying a Wien temperature of ~3600 K.

range is turned off and they become cooler, they become redder and dimmer. Physics laboratory exercises that involve the direct or electrical heating of wires can illustrate the quantitative aspects of the black body radiation laws; we leave them for the imaginative (but careful!) reader who has access to a well-equipped physics lab.

It is important to realize that higher temperature sources do not necessarily have higher luminosity. The radiant flux will certainly be greater for a hotter source, but the flux is the radiated power *per unit area* of the surface; therefore an enormous, cooler body can provide as much power as a very much smaller and hotter source.

20.3 Non-Thermal Sources

Radiation of a different kind is produced by charged particles moving in electrical and magnetic fields. When electrons are forced by alternating voltages to move back and forth rapidly (say at several hundred kiloherz) within a wire or rod, they generate radiation at the oscillation frequency. This is what a broadcast antenna does. In the Earth's ionosphere, electrons can absorb radio waves from an astronomical object by oscillating at the fre-

quency of those waves (plasma oscillations); because they radiate in all directions but absorb from only one direction, they will be seen by a radio astronomer to absorb the radiation from the distant object.

From a particle accelerator, nuclear physicists sometimes see a ghostly blue light. As electrons are forced by changing magnetic fields to accelerate in growing spiral paths within a containment area (a cyclotron), they radiate at a frequency that depends on their speed and on the strength of the magnetic field. This is called *cyclotron radiation*. When the electrons are accelerated to speeds approaching that of light, requiring a larger ringlike structure known as a synchrotron, the

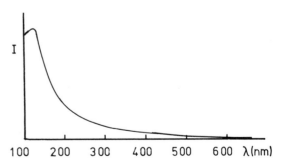

FIGURE 20.7. The spectral distribution of the blue star γ Orionis. The effects of many species of ions have caused departures from the black body curve. The peak occurs ~0.12 μm implying a Wien temperature of 24,000 K.

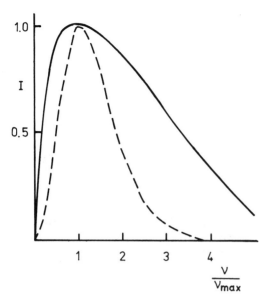

FIGURE 20.8. A comparison between the radiation curves of a black body or thermal source (dashed) and a synchrotron source (solid), set to the same frequency of maximum emission. Note the greater breadth of the synchrotron spectrum, especially its extension to higher frequencies.

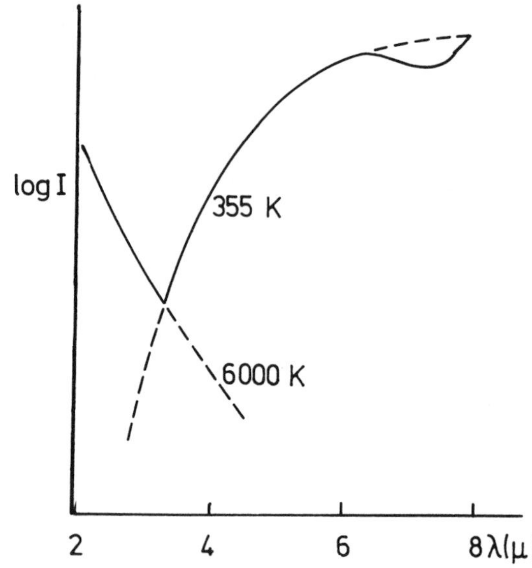

FIGURE 20.10. The spectrum of the Moon showing contributions from reflected sunlight (~6000-K black body) and thermal emission (from the heating of the lunar surface by the Sun; see Chapter 17). The latter is characteristic of a radiator at a mean temperature of ~82°C.

radiation is known as *synchrotron radiation*. As the speed increases, the radiation is directed more and more in the direction of motion of the electron, creating a kind of headlight effect. As each electron's cone of light sweeps past the direction to the observer's eye, it contributes a range of Doppler-shifted light, centered on a frequency related to that of the electron's circling motion. The collective effect of a large number of emitters is the observed synchrotron spectrum.

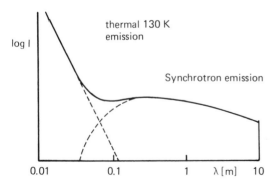

FIGURE 20.9. Radiation from the planet Jupiter is a mixture of thermal and synchrotron emission.

A characteristic synchrotron emission curve is compared to that of a black body with the same peak frequency in Fig. 20.8. The curves are plotted against multiples of the peak frequency. The main difference between these curves is the breadth of the synchrotron curve compared to the thermal radiation, especially at high frequencies (short wavelengths). Many cosmic sources produce this kind of radiation. The universe is full of (natural) particle accelerators.

Both synchrotron and black body radiation are produced by the accelerations of charged particles. In the case of the synchrotron radiation, the acceleration is caused by a magnetic field; in the thermal case, the accelerations are caused by the motions of the electrons past other charged particles. Both types are sometimes called *bremsstrahlung*, or *braking radiation*; thus we have magnetic bremsstrahlung and thermal bremsstrahlung. Both types of radiation are emitted by the Sun, where the nonthermal radiation varies greatly with the sunspot cycle. During radio noise storms, the brightness temperature (see challenge 1) may reach millions of degrees. Both types of radiation are seen in Jupiter also, as charged particles are accelerated by Jupiter's very strong magnetic field

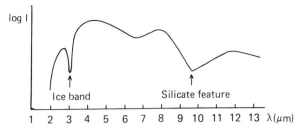

FIGURE 20.11. The spectrum of the Becklin–Neugebauer object in the Orion nebula. The star's spectrum is affected by an extensive amount of circumstellar and interstellar material; two of the strongest features – a water ice band near 3 μm and a broad silicate feature ~ 9 to 10 μm have been identified.

and create radio noise bursts. Figure 20.9 illustrates the combined spectrum.

The continua of cosmic sources may become very complicated as contributions from different types of emission are combined. Figure 20.10 shows the spectrum of the Moon at very long wavelengths. It shows the reflected light from the Sun, characterized by an approximately 6000-K black body curve, but shows an increase at much longer wavelengths characteristic of a thermal source at ~ 354 K (this is a rough average over the entire disk of the full Moon and will be substantially lower than the highest subsolar temperatures).

Discrete emission and absorption features in spectra depend on the temperature but also vary with other conditions and therefore qualify as nonthermal effects. The spectral lines, as they are called, are created when each electron absorbs a photon of a particular energy and uses the energy to shuttle to a higher orbit about the atomic nucleus. The photon is reemitted when the electron falls back to the lower orbit. An ionization "edge" can be seen in stellar spectra too, in addition to discrete spectral features. A particular type of atom will have a particular ionization energy, and if that species of atom is abundant in the atmosphere, photons having that energy or more will be selectively absorbed. This shows up as a decrease in the stellar flux at wavelengths to the violet of the ionization limit. Molecules create very interesting fluted absorption patterns in spectra, and complicated grain structures can produce quite broad features. Figure 20.11 shows the spectrum of the infrared Becklin–Neugebauer object in the Orion nebula. This complicated spectrum is due to a very young star heavily embedded in a thick shroud of gas and dust. Two strong absorption features have been identified as due to water ice and to silicates, two types of material abundant in our own solar system.

20.4 Further Challenges

1. Temperatures of an astronomical source may be derived from the Planck, Stefan–Boltzmann, and the Wien functions, if the source is a black body or an approximation to one. The temperatures are found as follows: the Planck function temperature is that which gives the best fit of the Planck function to the observed spectral distribution of the source (it is called the *color temperature*); the Stefan–Boltzmann relation temperature (*effective temperature*) is that which gives the same total flux as that seen from the source; and the Wien relation temperature is that of a black body that peaks at the same wavelength as the source. Often observations are made at only a single wavelength, however. The temperature of a black body that has the same flux as that which is observed is the *brightness temperature*. This is common in radio astronomy where the Rayleigh–Jeans law is easily applied. If the source were truly a black body, these temperatures would all be the same. Stars are not really black bodies, and these temperatures will differ from each other and from other measures of temperatures like those derived from atomic-level populations, from spectral classification, and from color photometry (all of these topics are considered in later chapters). The spectral distribution of the Sun peaks at ~ 0.46 μm, and the total radiant flux emitted by the Sun is $F = 6.284 \cdot 10^7$ W/m². Compare the temperatures derived from these two quantities.

2. Compare the radiation curves of a 20-W light bulb with that of a 100-W light bulb. Which gives a closer approximation to the spectral distribution of sunlight, five 20-W bulbs or one 100-W bulb?

3. Moonlight is very romantic and can even help us to see, but do we get any warmth from it? Calculate as best you can the lunar flux at the Earth and the sublunar equilibrium temperature that this implies.

4. In §20.3, the origin of synchrotron radiation was described. Discuss why the Doppler-shifted light is *not* centered on the actual frequency of the electron's circling motion but rather on a much higher frequency. How are the two related?

21

Optical Aids and Their Limitations

21.1 Introduction

What we know about the universe is gleaned mostly from our local surroundings. The detection of radiation from space is needed in order to test the universality of locally derived descriptions of the physical behavior of matter and energy. But, quite apart from any utilitarian motive, we also want to more fully comprehend the physical nature of the universe. Therefore, we have to be serious about doing it correctly: it is important to ensure that astronomical detectors are indeed measuring the radiation from a distant target and not from the telescope,[1] the dome, or the sky around the source. Moreover, it should be done in the most efficient way; there is no point in using an apparatus that requires a hundred hours of observation time when another technique or device will provide the same information in one hour. Practical considerations in the real world sometimes require us to use what is available instead of what we would like to use, but this does not remove the scientific imperative to use the best apparatus for the purpose. In this exercise, we examine the purposes and capabilities of our basic instruments and the uses to which they can be put.

21.2 Binoculars and Telescopes

The invention of the telescope (ca. 1608, by H. Lippershey and also possibly J. Metius)[2] and its use on the heavens (ca. 1610, by Galileo and, accord-

ing to TS-K, independently by S. Marius)[3] marked an upturn in the history of astronomy and the history of mankind. Discoveries now possible would bury forever the medieval world view and those philosophical traditions that could not come to terms with the telescope's revelations. The telescope would launch an investigation of the universe that continues to this day.

The earliest telescopes were *refractors*, that is, they depended on the bending action of a convex or positive objective lens. Around 1670, Newton produced a second type, the *reflector*, whose principal element is a concave mirror. Since their invention, both types of telescopes have undergone many improvements. For the refractor, one of the most important improvements was the invention of the achromat by C.M. Hall in 1729. The index of refraction depends on the wavelength of the light, so that a simple lens will bring red and blue light to slightly different foci. The achromat—a combination of lenses of differing indexes of refraction[4] — brings both blue and visual light to approximately the same focus. For the reflector, the introduction of a secondary mirror made observing more convenient and improved the definition of the images off the optic axis. But the early speculum mirrors

---

[1] You may recall the interesting scene from Ingmar Bergman's 1957 film *Wild Strawberries* in which a physician dreams he can see nothing through his microscope but the reflection of his own eye.

[2] When Metius's claim went unrecognized and a grant to produce telescopes was not forthcoming, he allegedly destroyed all his notes and his working model, thereby making impossible any subsequent verification of his claims. This is a good example of how not to conduct one's research!

[3] There is an interesting controversy surrounding the latter, who provided detailed and careful observations but whose claim for priority over Galileo was treated with derision by Galileo and his supporters.

[4] More precisely, lenses with different changes of refractive index over wavelength.

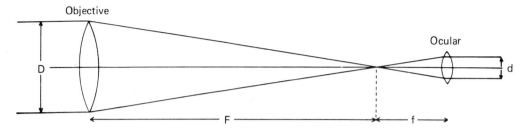

FIGURE 21.1. The basic optics of the telescope.

suffered from poor reflectivity even later when glass was coated with silver, the easy tarnishing of silver limited the popularity of reflectors until the late nineteenth century. At that point, the ease in mounting large mirrors, compared to large lenses that had to be supported on their edges, became apparent.

Terrestrial telescopes have prisms to create an upright image. Binoculars have these for the same purpose but also to shorten their length; otherwise they resemble refractors in form. The erecting and compacting elements do not affect the basic principles.

Even research-grade telescopes, which feed light from the main optics directly into instruments placed at the effective focal plane of the system, make use of eyepieces, or oculars, to examine the images that the telescope produces. A single lens may serve as an eyepiece, as in Galileo's own telescopes, or the eyepiece can be a multiplet (with several elements—to correct for chromatic and spherical aberrations, for example). If the ocular is placed at a distance equal to its own focal length behind the focal plane of the objective, the resulting rays will emerge parallel and will not form an image. The lens of the eye then focuses the light, which falls, in focus, on the retina.

21.2.1 Light-Gathering Power

The distinguishing characteristics of a pair of binoculars are summarized as "magnification objective diameter." As far as light-gathering power is concerned, the objective diameter is the relevant quantity. A telescope is usually specified only by the diameter of its primary mirror or objective lens.[5]

The diameter, D, of a telescope determines the light-gathering power, which is in fact proportional to D^2. Assuming that a dark-adapted eye has a diameter of 6 mm, a 6-cm telescope will have 10^2 or 100 times the light-gathering power; of course this does not take into account differences in the efficiency of transfer through either set of optics.

The magnification of an image, on the other hand, is the quotient of the focal length of the objective lens (or mirror), F, to that of the eyepiece, f: $M = F / f$. This can be related to image size through the geometry of Fig. 21.1: $d / f = D / F$, where D is the diameter of the objective and d is the diameter of the exit pupil of the eyepiece. The light, refracted and focused by the objective, enters the eyepiece, here represented as a single lens. In Fig. 21.1, the lens is shown a focal length f behind the primary focal plane so that light emerging from the exit pupil (diameter d) is parallel. The focusing action of the eye is sketched in Fig. 21.2. The diameter of the pupil of the eye is variable; it is in fact negatively phototropic—the brighter the ambient light, the smaller the diameter becomes, as in a good single-lens reflex (SLR) camera. The dark adapted eye may be as wide as 7 mm. The eye lens focal dis-

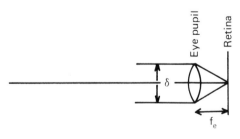

FIGURE 21.2. The basic optics of the human eye, with diameter δ and focal length f_e.

---

[5]Some manufacturers of small telescopes describe their wares only by stating some "power" of magnification; such a specification is meaningless, however. Huge magnifications are possible, in principle, for any telescope; one may achieve a magnification of 1000 with a 1-cm telescope and see nothing but a blur.

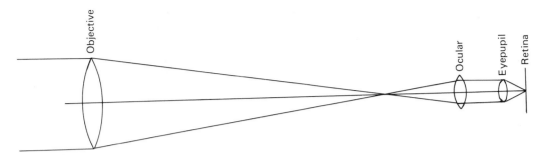

FIGURE 21.3. The combination telescope–binocular and eye.

tance is about 16 mm (in air), so that the aperture ratio[6] $\delta/f_e \approx 1{:}2$.

The sensitivity of the human eye may be tested by trying to detect the faintest stars on a star chart. The brightness scale is usually indicated in size of image against ranges of magnitude (see Chapter 15). The stars used for the test should be at least 30° from the horizon to minimize the effects of extinction (see Chapter 19).

21.2.2 Optical Images

Light leaves the exit pupil of the telescope and falls through the entrance pupil—the iris—of the eye and onto the retina (Fig. 21.3). An important fact to note is that if the exit pupil diameter, d, is greater than that of the iris, δ, some light will be lost to the retina by vignetting. On the other hand, if $d \ll \delta$, the *eye relief* is said to be poor; the image is being "undersampled" and the eyepiece may be uncomfortable to look through.

The *speed*, S, of a telescope–binocular depends on whether the source is a point source or an extended source. A *point source* is a source that, even under magnification, remains unresolved. Its luminous energy therefore falls on the smallest possible area on the retina, the area roughly corresponding to ~1 arc min. If $d < \delta$, all the light will be received (not considering reflective and absorptive losses), so the light-gathering power of the telescope compared to that of the eye alone is

$$\pi \cdot (D/2)^2 / [\pi \cdot (\delta/2)^2] = D^2 / \delta^2 \quad (21.1)$$

Since $D \gg \delta$ in most cases, there is a real gain in light received by using the telescope. With

extended sources, the light-gathering power is the same, but the situation is a bit different.

Light energy gathered from an *extended source* by the telescope compared to that gathered by the eye alone is again increased by a factor $(D/\delta)^2$, but it is spread out on the retina by a factor M^2, where M is the magnification. Each resolution element (in detectors this is known as a *pixel*) receives an amount of light proportional to $(D/\delta)^2/M^2$. This is the factor by which the telescope brightness per pixel exceeds that of the eye alone; this is a kind of measure of the *gain* of the telescope. But since D/M = d, this factor is $(d/\delta)^2$, and with $d \le \delta$, it is unity or less for extended sources. This means that the surface brightness of the image as seen through a telescope is not, in general, greater than that with the unaided eye.

An "8 × 30" pair of binoculars thus has an objective diameter of 30 mm and a magnification of 8×, and so $d^2 = (30/8)^2 \approx 14$. Compared to that of the eye alone, however, $(d/\delta)^2 \approx 0.3$. So, what do we gain? We gain detail in the case of an extended object (we have already seen that for a point source we clearly gain more flux, because the light per unit area on the retina is not changed). To discuss the perception of detail, we must discuss resolution, or resolving power.

21.2.3 Resolving Power

The acuity of your own eyes can be tested on the surface features of the Moon. With a lunar chart, such as those found in star atlases, examine the face of the Moon during twilight, when the glare of the disk is not too great. The diameter of the Moon, remember, is about 30 arc min. A further test of the acuity is the eye's ability to separate a close visual binary (see Table 2.1).

---

[6]This is the inverse of the focal ratio, which is the focal length over the diameter and is written, in this case, $f/2$.

Were the acuity of the human eye even only a bit better, people living in the Stone Age could have discovered lunar mountains, the crescent of Venus, and the existence and movement of Jupiter's Galilean moons. The consequences are interesting to speculate about. Quite possibly our scientific age would have begun not just a few centuries ago but several thousand years prior to this.

We can characterize the resolving power of the eye by examining the pattern of light produced when light is made to pass through a circular aperture and brought to focus. The central disk of this *diffraction* pattern has a size equal to $1.22 \cdot \lambda / D$ (rad), or

$$R = 1.22 \cdot 206265 \cdot \lambda / D \text{ (arc sec) (21.2)}$$

In the focal plane this corresponds to a linear separation,

$$a = (1.22 \cdot \lambda / D) \cdot F \qquad (21.3)$$

Two objects that are at least this far apart in either angular measure (Eq. 21.2) or linear measure (Eq. 21.3) may be considered to be resolved. This is known as "Rayleigh's criterion" and it is somewhat arbitrary because under superb seeing conditions, duplicity can be detected through a distortion of a stellar image equal to only one-third or so of this limit. However, even this criterion can be realized in practice only if the magnification is sufficient to match the resolution of the eye. Allowing 2 arc minutes for comfortable viewing, we have:

$$120'' = M \cdot 1.22 \cdot 206,265 \cdot \lambda / D$$

so that at $\lambda = 500$ nm (5000 Å), the ratio $M / D = 954$ m^{-1} or ~1 mm^{-1}, we obtain an "optimal" magnification to achieve the theoretical resolving power of the telescope. That magnification is

seen to be equal to the diameter of the objective expressed in millimeters. Thus, in the case of a 4-in. telescope objective, $D \approx 100$ mm and $M \approx 100$. If the telescope had an effective focal ratio $f/10$, then the objective's effective focal length is 40 in. and the focal length eyepiece needed to give a magnification of 100 is 0.4 inches or 10 mm.

There is an empirical test of the Rayleigh criterion that is fairly easy to do if you have a small portrait camera, that is, one equipped with a ground-glass screen to permit direct measurement of scale of your image. This will let you photograph the pattern shown in Fig. 21.4 from a distance of about 1 m, adjusted so that the scale marked 1 cm is in fact 1 cm long on the ground-glass screen. The black disks will appear as clear dots on the processed negative with diameters of 0.5, 1.0, and 2.0 mm. In a pinch, you can of course create such apertures by puncture but the precise sizes take a lot of practice and some luck. These will constitute an array of apertures for the test.

To proceed, open this book to the page containing the "Siemens star," a radial grating (Fig. 21.5). Illuminate it well, take your negative, and look through the smallest (0.5 mm) aperture from a distance of about 1.5 m. At this distance, the sectors will appear to blend together into a gray mass at some fraction of the radius of the figure, in this example, ~0.4. This point can be marked and its placement checked through successive trials. At that radius, the sectors have a thickness of ~2 mm.

Table 21.1 gives the results of three trials. For one trial, no filter was used [this means a passband centered on the eye's most sensitive region (~ 500 nm)]; the others were carried out with a narrowband filter centered on 557.7 nm and with a broadband filter centered on 600 nm, respectively. With these data, you can evaluate the constant, K, in the expression

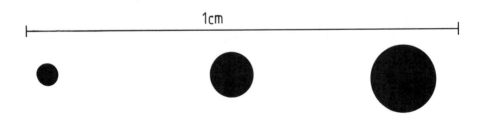

1cm

FIGURE 21.4. A pattern for creating apertures in a photographic negative.

TABLE 21.1. Resolution experiment results.

Wavelength (m)	Distance (m)	Aperture (m)	Sector thickness (m)	K
$557.7 \cdot 10^{-7}$	1.4	$5 \cdot 10^{-4}$	$1.95 \cdot 10^{-3}$	1.25
$500 \quad \cdot 10^{-7}$	1.4	$5 \cdot 10^{-4}$	$1.75 \cdot 10^{-3}$	1.25
$600 \quad \cdot 10^{-7}$	1.4	$5 \cdot 10^{-4}$	$2.05 \cdot 10^{-3}$	1.22

$$a = K \cdot l \cdot \lambda / D \qquad (21.4)$$

where a is the thickness of the sectors which are just resolved, l is the distance between the eye and Fig. 21.5, and D is the aperture diameter. From these data, a mean value $<K> = 1.24 \pm 0.01$ is obtained, a value not significantly different (see Appendix B) from the predicted value $K = 1.22$.

21.3 The Heisenberg Uncertainty Principle and Optical Resolution

A basic difference between classical and modern physics is the limitation the latter places on the precision with which the timing and position of events may be determined. For a particle like an electron, for example, the product of the uncertainty in its position, say x, and the uncertainty in a property of its motion, the *momentum*,[7] is

$$\Delta x \cdot \Delta p \approx h \qquad (21.5)$$

where $h = 6.6261 \cdot 10^{-34}$ J·s is *Planck's constant*. The same limit of precision characterizes the

---

[7]The product of its mass and its velocity, $P = m \cdot v$.

FIGURE 21.5. "Siemens star" to test the resolution of the eye.

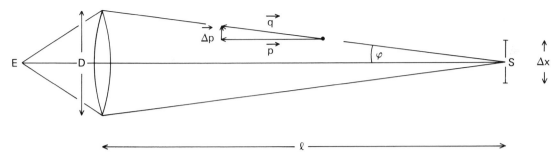

FIGURE 21.6. The uncertainty principle in an optical system with diameter D examining a point source S at a distance ℓ away. E marks a receiving element. The angle ϕ = arc tan $D / (2 \cdot \ell)$ is therefore assumed to be of the order of an arc minute. This means that $\phi \approx \tan \phi \approx \sin \phi$, and that Δp is perpendicular to p for all practical purposes.

product of the uncertainties in the total energy and the instant of observation of the particle:

$$\Delta E \cdot \Delta t \approx h \qquad (21.6)$$

Werner Heisenberg uncovered these relations in 1927.

In general, the existence of resolution limits of optical systems is treated in terms of diffraction and interference phenomena; in other words, it is tied to the wave nature of light (see §16.1). Here we are going to show that resolution limits can be understood in terms of the particle nature of light because they depend on the uncertainty principle.

A lens, even one manufactured to the highest optical precision (i.e., diffraction limited), images a point source, S (Fig. 21.6), not as a point but as a small disk, called an *Airy disk*, with much smaller amounts of light spread in diffraction rings beyond. The diameter of the Airy disk is just $1.22 \cdot \lambda / D$. Here we are interested only in the order of magnitude of the phenomenon, not in the details of the diffraction ring distribution.

The source, say S, of a photon that falls onto an element of the detector (say a receptor cell on the human retina at point E) cannot be identified unambiguously; at best we know only that the photon came from somewhere within a region $\pm \Delta x / 2$ about S. In the wave theory of light, we can connect the position difference, Δx, with the diffraction relation:

$$\Delta x = 1.22 \cdot \ell \cdot \lambda / D \qquad (21.7)$$

From the particle theory of light, we know that the energy of a photon is $E = h \cdot \nu$, and its momentum, p, is

$$p = E / c = h \cdot \nu / c = h / \lambda$$

But the photon traveling through the center of the lens aperture has a slightly different *vector* direction than does, say, a photon traveling at angle ϕ with respect to the axis defined by point S and the lens midpoint. That difference, designated Δp in Fig. 21.6, is

$$\Delta p \approx p \cdot \tan \phi \approx (h / \lambda) \cdot (D/2) / \ell$$

This difference in momentum may be considered an uncertainty in that quantity, and the position difference, Δx, can be considered an uncertainty in that coordinate, so that their product should be of the order of h:

$$\Delta p \cdot \Delta x = (h / \lambda) \cdot [D / (2 \cdot \ell)] \cdot 1.22 \cdot \lambda \\ \cdot \ell / D \approx 0.6 \cdot h$$

a result that confirms the order of magnitude of the Heisenberg uncertainty principle expressed in Eq. 21.5.

22

Laser Light and Speckle Interferometry

22.1 Introduction

Thanks to the invention of the laser, especially the continuous laser by T.H. Maiman in 1960, and the development of the principle of holography by D. Gabor in 1947, we are able to provide an extremely interesting experiment on the wave nature of light and on *granulation*.[1] Without these developments, the experiment could certainly have been done, but not nearly as convincingly.

The basis of the laser, *stimulated emission*,[2] has long been known. So has the existence of long-lived "metastable" states of the atom, a requirement for directional stimulation to work. The 21-cm line of atomic hydrogen, which is detected by radio astronomers from vast interstellar regions of the Milky Way and other galaxies, arises from a metastable configuration the lifetime of which is millions of years. Observation of the emission from any single atom in such a metastable state is exceedingly unlikely. Yet the numbers of atoms are so great that radio astronomers were able to observe it within a few years of its prediction.[3] Stimulated emission affects the visible region too. Without the occurrence of stimulated emission in

stellar atmospheres, the blue stars in the constellation Orion, for example, would appear only half as bright. Even examples of the laser itself can be found in the sky. Interstellar clouds of the hydroxyl radical (OH) can act as "masers" (microwave lasers) to produce microwave radiation.

In this exercise, we explore the ideas behind speckle imagery and interferometry and describe an experiment that you can carry out with a low-power laser.

22.2 Speckle Imagery

If you direct a laser beam to a white surface, like the wall of a room, you can immediately recognize a pattern of dark and light granulations. This characteristic phenomenon of laser light is closely related to a recently developed observational technique in astronomy, speckle interferometry. First, however, we need to explore the nature of the granulation.

If you examine the laser light reflecting off the wall, which acts a diffusing surface, or a "diffuser," at different distances from it (e.g., try to direct your eyes to their near point; if that causes eye-strain, use a short focal length lens), you will notice that even when the wall is out of focus, the granulation pattern is still seen! This suggests that the pattern appears on the retina of the eye. In fact it occurs on any surface that intercepts the scattered laser light. This means that a photographic film or plate, a ground-glass screen, or the retina of the eye itself can manifest the effect. The focusing action that takes place in the lens of the human eye is not really relevant. What is the source of the granulation? A simple experimental setup can help to answer the question in both qualitative and

---

[1]This should not be confused with solar granulation, a very large-scale phenomenon in the solar atmosphere caused by convection.

[2]This term describes the situation when an excited atom is induced by a passing photon of a certain energy to emit a photon of the same energy, in the same direction. "Laser" is an acronym for light amplification by stimulated emission of radiation.

[3]H. C. van de Hulst's attempt to make the first detection was cut short by a disastrous accident in his laboratory. Today joint credit for the discovery is given to him and to E. Purcell and E. M. Ewen, who detected it at Harvard in 1951.

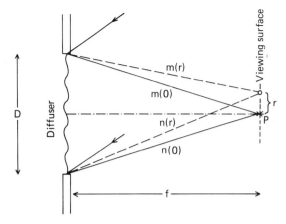

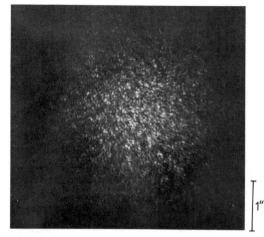

FIGURE 22.1. An experimental setup to investigate the origin of the granulation pattern in laser light. The light is directed to a diffusing surface, the illuminated spot with diameter D. Each tiny element of D can be thought of as the source of a circular radiating wavelet (like that seen in a pond). Because the diffusing surface is not smooth (by definition), the phases of the wavelets as seen on any parallel but smooth surface differ at any given time. Their interference on any surface to the right of the diffuser causes the characteristic granulation. The point P marks a bright spot, where (by chance) a majority of wavelets interfere constructively. Suppose wavelets originating at the edges of the diffuser are in phase at P. Their corresponding rays are marked $m(0)$ and $n(0)$. Suppose further that at distance r from P the phase difference between the wavelets originating from the edges of the diffuser [whose rays are marked $m(r)$ and $n(r)$] is $\lambda/2$. Then r describes the radius of a granule.

FIGURE 22.2. A speckle image of the star Bellatrix (γ Orionis). The angular diameter of the star is immeasurably small, even for the high-resolution speckle technique. Consequently, the image shows an "undistorted" appearance composed of statistically randomized granules, or speckles.

quantitative ways. Figure 22.1 illustrates the setup. The arrows signify the initial direction of the laser light, which strikes the diffusing surface and reflects into the path of an observing surface, say a ground-glass screen. We will refer to the spot on the diffusing surface of the wall simply as the diffuser and to the diameter of the spot on the wall as D. From each point on the surface we can envisage a tiny circular Huygens wavelet[4] spreading out. These wavelets will not be in phase as viewed in any cross section to the right of the diffusing surface, because of the relatively rough surface of the wall. The irregular surface of the diffuser can be said to destroy the coherence of the light that was present in the original beam. At any intersecting surface, to the right of the diffuser in Fig. 22.1, each of the Huygens's wavelets will

interfere with each other, causing patterns of constructive and destructive interference, which result in the observed granulation.

The geometry seen in Fig. 22.1 permits the determination of a characteristic size of the granulation, an average diameter of a granule. To do this, we can assume that point P, along a central normal to a flat plane through the laser spot on the wall, marks a bright spot. Such a bright spot occurs in a purely statistical way, an accidental coincidence of a larger-than-average number of constructive compared to destructive interferences. Suppose that the wavelets arising from the edges of the diffuser are among those interfering constructively at P. The corresponding rays (or normals to the spreading Huygens's wavelets) of these wavelets are designated $m(0)$ and $n(0)$ in Fig. 22.1. At a small distance r from point P on the observing screen, the brightness will fade away and at that point, the wavelets will arrive out of phase, so that the path difference between the rays $m(r)$ and $n(r)$ will be

$$| m(r) - n(r) | = \lambda/2 \qquad (22.1)$$

The minimum marks the edge of the granule, thus r marks its radius and $d = 2r$, its diameter. From Fig. 22.1, we can describe the path lengths of the rays:

---

[4]After Christian Huygens (1629–1695).

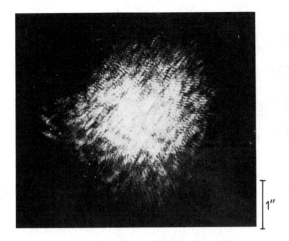

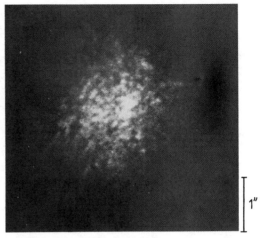

FIGURE 22.3. The speckle image of the double star Capella (α Aurigae). The pattern of double granules permits the determination of the separation of two components (0.05 arc second) as well as the orientation of the two stars to each other (upper right–lower left).

FIGURE 22.4. The speckle image of the red supergiant star Betelgeuse (α Orionis). This star, although approximately 1400 light years from us, is so large that its angular size (0.06 arc sec) is in fact twice the resolving power (∼0.03 arc sec) of the technique discussed here. The granules are noticeably larger as a consequence.

$$m(r) = \sqrt{[f^2 + (D/2 - r)^2]} = f \cdot \sqrt{\{1 + [(D/2 - r)/f]^2\}}$$

$$n(r) = \sqrt{[f^2 + (D/2 + r)^2]} = f \cdot \sqrt{\{1 + [(D/2 + r)/f]^2\}} \qquad (22.2)$$

Now both D and r are small compared to f, so that the ratio $(D/2 \pm r)/f \ll 1$. This condition validates application of the approximation

$$\sqrt{(1 + \varepsilon)} \approx 1 + \varepsilon/2, \qquad |\varepsilon| \ll 1 \qquad (22.3)$$

where $\varepsilon = [(D/2 \pm r)/f]^2$. Therefore, the difference between Eqs. 22.2 becomes

$$m(r) - n(r) \approx$$
$$(f/2) \cdot \{[(D/2 - r)/f]^2 - [(D/2 + r)/f]^2\}$$

$$\approx (1/2f) \cdot \{[(D/2)^2 + r^2 - r \cdot D] - [(D/2)^2 + r^2 + r \cdot D]\}$$

or

$$|m(r) - n(r)| \approx r \cdot D/f \qquad (22.4)$$

Setting Eqs. 22.1 and 22.4 equal, we arrive at

$$r = (\lambda/2) \cdot f/D \qquad \text{or} \qquad d = \lambda \cdot f/D \qquad (22.5)$$

It is interesting to note that the granule size is governed by the same relation as that which determines the resolving power for optical devices (see Chapter 21).

The isolated bright spots, like that at point P, are images of the almost point source cross section of the original laser beam. Many statistical fluctuations appear in the image, however, so that

(1) there is a rapid succession of different images; and (2) these individual images show strong distortions. Moreover, it is plausible that (3) from the average characteristics of these images, the original image can be reconstructed; and (4) in especially favorable cases, a single granule, or "speckle," suffices to give the original image.

An interesting question that can be asked at this point is what happens if two close and parallel laser beams, instead of one, were to illuminate the wall. A quick answer is that the granular pattern would double; that is, we would see pairs of granules. If you can obtain a second laser (or have access to a beamsplitter), try to verify this. With this, we can understand the principle of *speckle interferometry*.

Let us imagine two very distant, but very bright, sources of light illuminating a "spot" 4 m in diameter, that is, $D = 4$ m. Suppose that the separation between the two beams is comparable with that of the components of a very close double star, say Capella, with angular separation of 0.05". The "angular resolution" of the diffuser can be said to be $\alpha = d/f$, and, by Eq. 22.4,

$$\alpha = \lambda/D \qquad (22.6)$$

where $\lambda = 500$ nm $= 5 \cdot 10^{-7}$ m and $D = 4$ m, so that

$$\alpha \approx 1.3 \cdot 10^{-7} \text{ rad} = 0.03 \text{ arc sec}$$

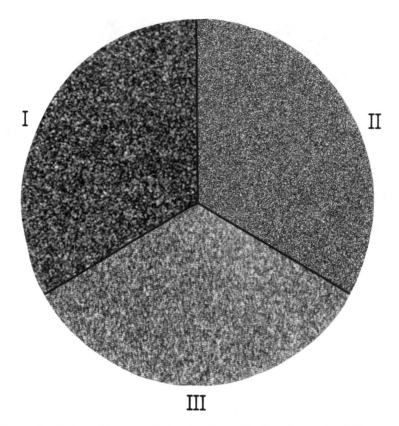

FIGURE 22.5. Composite photographic record (reduced by about a factor of 2 from the originals) of the granulation pattern of a small helium–neon gas laser ($\lambda = 632.8$ nm) under the following conditions:

Sector I: $D = 1.5$ mm, $f = 1.52$ m

Sector II: $D = 6$ mm, $f = 1.52$ m
Sector III: a simulation of the Capella case

The photographs reproduced in sectors I and II permitted the determination of d. The film speed was 50 ASA, and the exposure time a few minutes.

thus our diffuser has about the same "resolving power" as a 4-m telescope. In fact, as we have already noted, the focusing action of intervening lenses (or mirrors) is not really relevant. Our binary star is the analog of two close, parallel beams from an immensely great distance, and our screen would seem to be a 4-m diameter telescope mirror (presumably, however, much smoother than our wall diffuser). Of fundamental importance to speckle interferometry is the circumstance that the atmosphere has no influence on the results—it acts merely as a further diffusing element without any effect on the resolution.

The analogy is not perfect, however. A star image at the effective focal plane of the telescope is the diffraction pattern of the mirror, and the energy is largely concentrated in the central Airy disk of diameter $\sim 1.22\ \lambda/D$ (see Chapter 21).

This is unlike the case of our diffusing wall where the entire hemisphere opposite the wall is illuminated. Moreover, the diffusing "surface" is not really the mirror (unless the mirror is very dusty); it is in fact the atmosphere. The atmosphere, however, is a seething diffuser—its "elements" constantly in motion. None of these differences touch the essence of the analogy, but to recover the information, two special procedures are required: the photographic (or electronic) exposures must be very short (~ 0.01 s); and the sharpest granular images require monochromatic light, which can be provided by narrow-band interference filters. These two procedures incur a heavy cost. They so weaken the light that image enhancing techniques are then required to improve the signal to noise ratio. An image intensifier may help.

Figures 22.2 to 22.4 show the enhanced speckle images of three stars.

22.3 An Experimental Challenge

Widen a beam from a helium–neon gas laser to produce a spot a few millimeters wide on a white screen. Opposite the screen, about 1 m away, set up a ground-glass screen,[5] the image on which can be examined. A magnifier of 40 mm focal length or so (magnification $\geq 6\times$) should reveal the granulation immediately. With a millimeter rule and your magnifier, verify Eq. 22.5 for a range of values of D and f. Figure 22.5 shows the patterns observed in the experimental setup described earlier. Sectors I and II are similar but with different values for D. Sector III shows the characteristic pattern of two sources of illumination (a simulated double star). As in Fig. 22.3, there is a doubling of the granulation pattern, which makes it possible to find the separation and the position angle of the simulated double.

----

[5]Single-lens reflex camera, from which the lens assembly has been detached, can be erected in place of the ground-glass screen on a sturdy tripod, to let the light from the diffuser fall directly on the film plane during exposures. We recommend that the visual procedure be followed first, so you can perceive the scale of the effects, and then repeated with the camera to carry out the photography.

23

Photometry

23.1 Introduction

In earlier chapters we discussed such topics as the radiant flux emitted by the Sun, planets, and the stars; the luminosity, surface temperatures, and observed flux from these objects at the Earth; the photographic detection of starlight and sunlight; and the surface brightness and intensity of extended objects. These topics all involve photometry to some degree. Now we describe how the light of the stars is detected by photoelectric photometry, a high-precision method of measuring the flux and color of starlight, and photometry by area detectors such as charge-coupled devices.

Photoelectric photometry has been used in astronomy since the early decades of the twentieth century when P. Guthnick in Germany and Joel Stebbins and colleagues at the Washburn Observatory performed experiments and obtained light curves of bright variable stars. The systematic work of Harold Johnson and colleagues in defining and employing his UBV system[1] and A.W.J. Cousins's careful work in similar passbands in the Southern Hemisphere, beginning in the 1950s, were pivotal in establishing the system. With the UBV system, the temperatures and luminosities of the stars could be obtained from the magnitudes and from the differences among magnitudes in different passbands, the color indices. Moreover, a correction for interstellar reddening could be obtained (at least for early type stars) from UBV photometry alone. Other systems, such as the multicolor systems of Kron and Stebbins, the four-color system of Strömgren, and the DDO (for David Dunlap Observatory) system, were devised for particular astrophysical purposes and are still in use.

In this chapter, we describe the techniques required to carry out precision photometry with a pulse-counting photoelectric photometer, a device that is finding widespread use in the amateur astronomy community with the availability of low-cost commercial photometers. Their use in professional astronomy is now essentially limited to observations of single objects like variable stars, while photometry of star clusters is now largely carried out with area detectors like charge-coupled device (CCD) arrays. We also discuss the advantages and disadvantages of these two types of detectors and the optimal programs for each.

In a later chapter we discuss the analysis of the light curve of an eclipsing binary star; here we assume only that the desired target is a star that undergoes brightness changes with time. The goal is to obtain a *light curve*, a registration of the stars' brightness, sometimes in measures of magnitude, sometimes in relative flux units (relative to, say, the peak brightness) against time or phase. The phase is discussed later. In the case of a nonperiodic star, the time must be used. Two examples of light curves are seen in Fig. 23.1.

23.2 The Basic Photoelectric Photometer

The photometer consists of two main segments: the photometer head and the detector module.

The photometer head connects the detector to the telescope. The photometer head contains the

---

[1]Acronym for ultraviolet, blue, and visual passbands, respectively (Johnson and Morgan, 1953). The system was later enlarged to UBVRI (with red and infrared), and then extended further to the infrared (J, K, L, M, N, Q; H was later added between J and K).

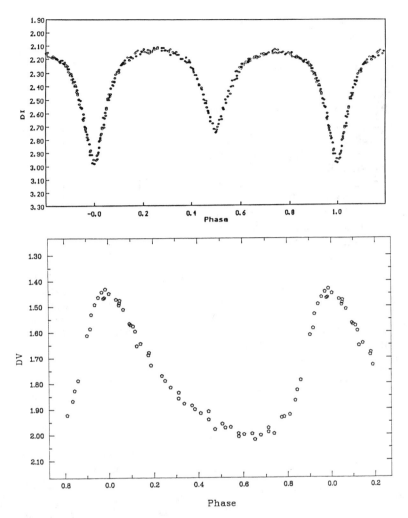

FIGURE 23.1. Light curves of two variable stars obtained by student observers (principally James van Leeuwen) at the Rothney Astrophysical Observatory of the University of Calgary. *Upper*: The near infrared (I) light curve of the eclipsing variable star IR Cassiopeiae. *Lower*: The V light curve of of the pulsating variable star DY Pegasi.

viewing optics, filters to further define the wavelength passbands, diaphragms to limit the region of background around the target, a dark shutter, and an imaging lens, the purpose of which is to image the telescope mirror on the surface of the photomultiplier tube (PMT). Figure 23.2 illustrates the arrangement for the photometer at the Cassegrain focus of a reflecting telescope.

The lens, called a *Fabry lens* (after its inventor), is important because the surface of the photomultiplier is not uniformly sensitive and if a star moving about in the diaphragm were imaged directly, it would move about on the surface of the PMT also. The Fabry lens images the mirror illuminated by the object, however, and provided that the diaphragm aperture does not vignette the object, the image of the mirror stays fixed in position on the PMT face.

The heart of the photoelectric photometer is the photomultiplier tube, which converts photons of light into electrons. The light impinges on the photocathode of the tube. By the photoelectric effect, photons which are sufficiently energetic (i.e., have a short enough wavelength) can ionize atoms and release electrons from the photocathode material. The material is specially selected for the wavelength region of interest but usually this is as wide a region as possible, with reasonable sensitivity. Photocathodes of gallium arsenide are currently favored because they are sensitive from the ultraviolet to the near infrared. As its name implies, the photocathode is sensitive to the light impinging on it and is at a large negative potential, so that the electrons which are released are impelled away from the photocathode. They are accelerated by a potential difference toward another surface, called a *dynode*, where their impact can cause the release of still more electrons by collisional ionization. Because of the arrangement of the many dynode stages, a cascading occurs, producing as many as 10^7 electrons per photon at the last dynode stage, the anode. Figure 23.3 shows a photomultiplier tube; the actual configuration varies from tube to tube.

The high voltage direct current across the tube may reach 1500 to 2000 V, although some tube types (like the famous RCA 1P21, which has been in use for several decades) operate at a much lower voltage (~ 600 to 900 V). The voltage is usually selected to give the best signal to noise (S/N) ratio, because the higher the voltage, the more sensitive the tube, and also the more noise it has. By noise

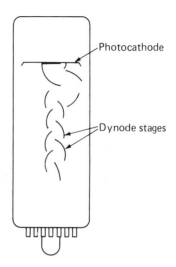

FIGURE 23.3. Basic photomultiplier tube geometry. The electric potential between each successive dynode stage accelerates the liberated electrons, causing collisional ionization and increased emission at each stage. The actual arrangement of the dynodes varies with tube type.

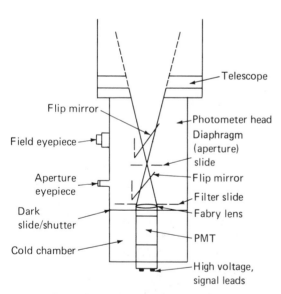

FIGURE 23.2. The basic photoelectric photometer. Light passes from the telescope, is inspected in the field eyepiece, is allowed to pass through the diaphragm, where the object is centered and focused, then through a filter, and finally to impinge on the detector, here considered to be a PMT. The cold chamber surrounds the detector. Notice the Fabry lens, which plays a critical role in reducing noise in the signal due to sensitivity variation across the face of the PMT.

here we mean the *dark current*, the output when no light falls on the tube. Some of this noise is produced by cosmic rays, but much of it is caused by the random motions of electrons and the few remaining ions in the evacuated glass tube. Because this random motion decreases with temperature, cooling the tube to the temperature of dry ice ($-76\,^\circ$C) substantially reduces the dark current and improves the S/N ratio. Therefore, most photomultipliers are surrounded by a light- and water-tight jacket around which is a container for the pulverized and packed dry ice.[2] The cooling of the PMT chamber requires the isolation of the photomultiplier tube from the atmosphere so that condensation does not occur on the face of the PMT or on the electronic components at its base. The isolation takes the form of an evacuated cylinder with end plates of thin-fused quartz,[3] connected to the chamber. The outer plate can be heated by a low-power wire heating element to prevent window frosting.

---

[2]Sometimes this is mixed in an alcohol slurry to ensure good thermal contact with the walls adjoining the PMT chamber. This is not easy to implement unless the photometer can be located at a fixed focus (the Coudé or Springfield mounting, for example), because the slurry can spill out; a water-tight plug cannot be used because the sublimed carbon dioxide must be permitted to escape.
[3]Chosen for its high transmission, compared to other lens material, in the near ultraviolet.

The base of the PMT usually houses the electronics, which can treat the signal in various ways. The current may be measured directly with a very sensitive ammeter, or the current may be integrated using large capacitors, which are read out after preselected intervals; alternatively, the current may be fed to large precision load resistors and the voltage drop across these resistors can be measured, or, finally, the pulses of electrons emerging from the anode may be amplified, shaped, and sent to a pulse counter. The last technique is known as *photon counting* or *pulse counting* (PC). A standard PC practice is to discriminate against low-amplitude thermal noise pulses; this improves the signal-to-noise ratio. The various techniques to handle the output of the PMT are described extensively in the literature (e.g., Kitchin, 1984). In this chapter, the pulse-counting technique is assumed.

The optical elements of the telescope and photometer, the sensitivity of the PMT, and the atmosphere combine to produce the overall spectral sensitivity of the system with wavelength. The integral equation describing this dependence is

$$\ell = \int_{\lambda 1}^{\lambda 2} F_\lambda \cdot t_a \cdot t_0 \cdot a_T \cdot \tau \cdot d\lambda \qquad (23.1)$$

where F_λ is the monochromatic flux from the star, t_a is the transmission of the atmosphere (this is supposed to be normalized out by techniques described later but such a correction can never be made exactly because of the varying character of the atmosphere from moment to moment), t_0 is the optical transmission factor of the telescope and photometer, including the filters and the Fabry lens, a_T is the effective collecting area of the telescope, and τ is the integration time. By adding the products of these functions for each wavelength interval, we integrate over the wavelength region of interest, λ_1 to λ_2.

23.3 Observing Techniques

The optics in the photometer head usually contains a low-power field eyepiece with good eye relief so that the field of the target can be identified and the object easily centered. A high-power eyepiece viewing the image through the diaphragm permits final centering of the image. Both eyepieces must be removed from the field of view if light is to pass to the PMT, and so mirrors which swing out of the way are often used to direct the light into the eyepieces (or viewing ports, since miniature television cameras may now be used instead of direct-vision eyepieces). The selection of the diaphragm is determined by the seeing and by the tracking of the telescope. As every observer knows or soon discovers, the tracking is affected by temperature, wind load, and other effects. The diaphragm is chosen so that the star can be assumed to be always within it, given the excursions that may occur during each observation. Normally, the diaphragm is selected so that the star image occupies one-third or less of the diameter of the diaphragm.

Pulse counting is the currently preferred mode of PMT operation, both for precision measurement of faint sources and for ease of data acquisition and subsequent reduction. This method does, however, create difficulties for observation of bright stars because many pulses may arrive too quickly to be counted separately and so the count underestimates the brightness of the star. Consequently, a *coincidence correction* must be applied to all observed counts. This correction is of the form

$$N_c \approx N_o \cdot (1 + N_o \cdot \tau) \qquad (23.2)$$

where N_c is the corrected number of counts, N_o is the observed number, and τ is the dead time of the system.[4]

Pulse-counting units are available on printed circuit boards in the form of cards which fit into the slots of personal computers, and software can be written in BASIC, FORTRAN, or other computer languages to access the data, interrupt the data streaming, obtain the time of each observation, and so on. Even this stage is not essential, because data acquisition packages, like *Measure* and *Asyst* (Chapter 13), are commercially available, and some of them can access even the serial port of the computer. This is window dressing, of course; one need only have a bank of numbers of counts and time (and a fast and accurate pencil hand) to record the basic data required for photometry: the object name, the signal, and the time. The idea of this kind of activity—working in a cold, dimly lit observatory—would make many modern "keyboard" astronomers pale, but this, not keyboard fingering, is what astronomy is really all about: obtaining the clues about nature from nature itself. It does, however, greatly improve efficiency if the data can be stored directly onto electronic media and hard copy, for systematic and timely data reduction and analysis.

Thus far we have not discussed the observing program. Variable stars are a passion of at least one

---

[4]This approximation is nearly exact for values of $N_o \cdot \tau$ < ~0.05.

of us, and so we strongly recommend them as worthy recipients of your observing time. Moreover, there are many more variable stars than there are variable star photometrists. Light curves obtained by careful and systematic observers will almost certainly prove valuable and worthy of subsequent analysis. Deciding on the specific target may take a bit more attention. Organizations like the American Association of Variable Star Observers (AAVSO) publish newsletters and circulars, among which are calls for observational data on specific targets. The AAVSO also can provide extensive sets of star charts, which are essential for fainter stars. Once a program star is selected, the observing program needs to be planned.

The observing procedure requires selection of a target star, a comparison star, and a check star (to check on the constancy of the comparison star). The comparison star, C, and check star, K, should not be known or even suspected variables; if either is, it will complicate the reduction and analysis of the light curve. One sequence of observations that you can follow is to observe the comparison star first, then the sky near it, s, the check star, its sky, the variable star, its sky, and so on, to provide a sequence like this:

C, s, K, s, C, V, s, V, s, C, s, V, s, V, s, V, C, . . .

with occasional additional observations of K. If the stars are bright and there is no twilight, moonlight, aurora, or the like, to worry about, the sky can be observed less frequently. Moreover, if the air mass is changing slowly, the variable star may be observed for a longer period. Prudence dictates that the observations be made at low air masses (i.e., when the stars are high in the sky).

Light curve analyses of data taken in more than one passband add weight to the determinations of the parameters that describe the physical state of the star or system and therefore are preferred to single passband observations. The case is especially compelling for eclipsing binary light curves, but it is true for most other variable stars too. The fewest that should be obtained is two, because corrections for atmospheric extinction and standardization require color indices. If V observations are made, B observations should be made also. Modern gallium arsenide detectors are capable of acquiring data from the ultraviolet to the near infrared (~ 1 μm). The passband observations are best made in a sequence like IRVBUUBVRI, for example. If you observe them in a symmetric set like this, data in all passbands refer to the same average instant. This is useful when plotting the data and in the reduction process, to be described later. In our example, observations at the shortest wavelengths are closest together. This is desirable because extinction variations are usually greatest for the shortest wavelengths. Thus if the variable star is varying sufficiently slowly, multiple passband sequences can be carried out for each of the star observations. How slow is "sufficiently slow"? The question can be answered empirically—the only way this can be done for irregular variable stars. The integration time of each observation (the number of seconds over which the counts are obtained) is usually decided by the desired signal-to-noise ratio. For $S/N \approx 200$ and a relative precision of 0.5%, the number of counts should be $\sim 40,000$ or more (see Appendix B, §B.3, for a full discussion). The integration time per filter to collect this many star counts, the time to change filters, and so on, must be added up. If the star varies perceptibly over this interval, some compromise, such as the reducing the number of filters, is in order. For periodic variables, you can also look at the phase variation over the interval required for a full set of multicolor observations. The *phase* of a variable star indicates which part of cycle of variation it is in at any instant, t:

$$C \cdot \phi = (t - E_o) / P \qquad (23.3)$$

where C is the integral number of cycles since the epoch, E_o, and P is the period of variation and s is a light-time correction to reduce the observation to the Sun (see Landolt and Blondeau 1972). For eclipsing binaries, E_o refers to an instant at which the midpoint of the deeper or primary eclipse occurred; for pulsating stars, it refers to the peak of the light curve, which is usually the sharpest part of the light curve. The three variables are in units of days, with t and E_o expressed in Julian day numbers and decimals thereof (see Appendix D). Therefore, the phase range, $\Delta\phi$, corresponding to a time interval Δt, is

$$\Delta\phi = \Delta t / P \qquad (23.4)$$

The light variation is particularly fast on the branches of minima for eclipsing variables and midway between the extrema for pulsating variables, but a reasonable rule is to keep the passband sequences less than $\sim 0.01^p$ (i.e., 0.01 of a cycle). If this cannot be done for the full passband set, gamesmanship is required: do you sacrifice some color information for good coverage, or good coverage for color information? Your program's priorities dictate the choice.

23.4 Corrections for Extinction

The cause and effect of atmospheric extinction were discussed in §19.4. The basic equation (19.8) can be rewritten in terms of the visual magnitude and can include the color term:

$$y = y_o + k'_y \cdot X + k''_y \cdot X \cdot (b - y) \quad (23.5)$$

where y is the observed visual (or yellow) magnitude, y_o is the magnitude outside the atmosphere (i.e., at zero air mass), k'_y is the extinction coefficient in magnitudes per air mass, X is the air mass in units of the vertical column of air above the observer, k''_y is the second-order or color extinction term, and $(b - y)$ is the observed color index. Essentially, $X = \sec z$, but Eq. 19.8 is a formula to correct for atmospheric curvature and is useful to $z \approx 80°$. Figure 19.3 illustrates the air mass concept. An expression similar to Eq. 23.1 can be found for the color index $B - V$:

$$(b - y) = (b - y)_o + k'_{by} \cdot X + k''_{by} \cdot X \cdot (b - y) \quad (23.6)$$

where $(b - y)_o$ is the system color index, or the color index outside the atmosphere. The color indices $(u-b)$, $(y-r)$, and $(y-i)$ can be expressed in terms of the extra-atmosphere colors and the air masses in a similar way. The first step is to find the coefficients k' and k''. The former is large and variable (but typically 0.1 to 0.2 at a good site), and k'' is typically about ±0.04 or less. Techniques for finding k' and k'' are discussed extensively by Hardie (1962) and Young (1974). In our treatment here, we assume for simplicity that the k'' coefficients are small and constant.

23.4.1 The Bouguer Method

The Bouguer method of obtaining k' is to plot m (or c) against X and measure the slope of the line. If two stars of contrasting color are used, the difference in slope can be attributed to and used to find the second-order coefficient k''. Equations 23.4 and 23.5 can be rearranged to bring the observed quantities to the left-hand sides of the equations:

$$[y - k''_y Xc] = y_o + k'X$$
$$[c - k''_c X] = c_o + k'X \quad (23.7)$$

where $c = b - y$, and $c_o = (b - y)_o$, respectively. The theory of least squares permits us to obtain the values of the unknowns k'_y, y_o, and $(b - y)_o$, which give the best fit to a straight-line relation between

the left- and right-hand sides of these equations. For n observations, the resulting *normal equations* become

$$n \cdot m_o + k'_y \cdot \Sigma X_i - \Sigma[y - k''Xc]_i = 0$$
$$m_o \cdot \Sigma X_i + k'_y \cdot \Sigma X_i^2 - \Sigma[y - k''Xc]_i \cdot X_i = 0 \quad (23.8)$$

and

$$n \cdot c_o + k'_c \cdot \Sigma X_i - \Sigma[c - k''Xc]_i = 0$$
$$c_o \cdot \Sigma X_i + k'_c \cdot \Sigma X_i^2 - \Sigma[c - k''Xc]_i \cdot X_i = 0 \quad (23.9)$$

The extinction coefficient k'_y and the outside atmosphere visual magnitude y_o may be obtained from these equations:

$$k'_y = \{n \cdot \Sigma[m]_i X_i - \Sigma X_i \cdot \Sigma[m]_i\} / \{D\} \quad (23.10)$$
$$y_o = \{\Sigma[m]_i \cdot \Sigma X_i^2 - \Sigma X_i \cdot \Sigma[m]_i X_i\} / \{D\}$$

where $[m]$ is the bracketed expression in Eqs. 23.8 and

$$\{D\} = \{n \cdot \Sigma X_i^2 - (\Sigma X_i)^2\}$$

The color extinction coefficient k'_c and the outside atmosphere color index $(b - y)_o$ can be obtained from these equations:

$$k'_c = \{n \cdot \Sigma[c]_i X_i - \Sigma X_i \cdot \Sigma[c]_i\} / \{D\}$$
$$(b - y)_o = \{\Sigma[c]_i \cdot \Sigma X_i^2 - \Sigma X_i \cdot \Sigma[c]_i X_i\} / \{D\} \quad (23.11)$$

where $[c]$ is the bracketed expression in Eqs. 23.9. An example of the Bouguer method was provided in §19.4.

23.4.2 The Hardie Method

Hardie's method is to find k' by using at least six pairs of standard stars. The stars are paired according to color index (i.e., the difference should be 0.2^m or less) but must be observed at different air masses. The observations can be done quickly, but there are drawbacks (see Young, 1974). The Bouguer method is convenient if you are already observing comparison and check stars at regular intervals, but a combination of data from comparison stars and standard stars provides the best indication of the behavior of the sky, and the more stars the better. The extinction determination programs at the Rothney Astrophysical Observatory compare the data from each pair of stars that agree within a certain range of color index. A substantial list of UBVRI faint standards suitable for pulse counting and distributed along the celestial equator was established by Landolt (1983). These are quite suitable for extinction work. Sometimes extraordinary events, like time variation in the extinction or system sensitivity, require special treatment (see, for example, Milone et al., 1980). Because the Hardie method is a differential method, outside atmosphere magnitudes and color indices are not directly determined. The equations of condition are

$$dy - dy_o - k''_y \cdot d(Xc) = k'_y \cdot dX \quad (23.12a)$$

$$dc - dc_o - k''_c \cdot d(Xc) = k'_c \cdot dX \quad (23.12b)$$

where $c = (b - y)$ and $c_o = (b - y)_o$, and where the prefix d denotes the differences between two stars of similar color index. Equation 23.12b can be particularized for as many color indices as apply. The outside atmosphere magnitudes and color indices may not be known in advance, so the values of dy_o and dc_o can be replaced by the standard values:

$$d(B - V) = \mu \cdot d(b - y)_o, \quad dy_o = dV - \varepsilon \cdot d(B - V)$$
$$(23.13)$$

The coefficients μ and ε may not be known in advance of the determination of the transformation coefficients and zero points (discussed in §23.5), so an iterative process is required. The extinction coefficients are determined with initial guesses for μ and ε (e.g., 1 and 0, respectively, will do); after the initial transformation coefficients have been found, the extinction coefficients are redetermined. Then the transformation coefficients can be redetermined, then the extinction coefficients, and so on. Convergence occurs quickly in most cases, but if the sky is nonphotometric, no number of iterations (or amount of data culling) will save the situation!

The normal equations, from which k' and k'_c are easily found, are

$$\Sigma[(dy)_i - (dy_o)_i - k''_y \cdot d(Xc)_i] \cdot dX_i = k'_y \cdot \Sigma(dX)_i^2$$

$$\Sigma[(dc)_i - (dc_o)_i - k''_c \cdot d(Xc)_i] \cdot dX_i = k'_c \cdot \Sigma(dX)_i^2$$
$$(23.14)$$

Once the extinction coefficients are found, Eqs. 23.4 and 23.5 are solved for the outside atmosphere values, y_o and $(b - y)_o$.

23.5 Standardization of the Data

Once the data have been reduced to outside atmosphere values, they should be placed onto a standard system that permits comparison with data obtained by other observers. No two photometric systems have identical sensitivity to brightness and color, and so data from the local system must be transformed into the standard system. For the UBVRI ... system, there is a set of primary and secondary standard stars by which a transformation can be effected. Careful work by Landolt (1983) on the stars in the *Harvard selected area*, regions of the sky has provided a large number of faint stars for standardization work as well as extinction. For this purpose, high-altitude (low–air mass) standards are ideal. The higher stars of the Hardie extinction pairs can serve this purpose too.

The transformation equations are of the form

$$V = y_o + \varepsilon \cdot (B - V) + \zeta_V \quad (23.15)$$

$$B - V = \mu \cdot (b - y)_o + \zeta_{BV} \quad (23.16)$$

where V and $B - V$ represent the Johnson system visual magnitude and color index, y_o and $(b - y)_o$ are the outside atmosphere magnitude and color index, ε and μ are the transformation coefficients, and ζ_V and ζ_{BV} are the zero points. Transformation equations for other color indices can be expressed in a way similar to Eq. 23.16. Plots of $V - y_o$ vs. $B - V$ and of $B - V$ vs. $(b - y)_o$ yield both coefficients (the slopes) and zero points, which can also be found analytically. The least squares solutions for the unknowns in Eq. 23.15 may be written

$$\varepsilon = \{n \cdot \Sigma[m']_i \cdot (c_s)_i - \Sigma[m']_i \cdot \Sigma(c_s)_i\} / \{D'\}$$

$$= \{\Sigma[m'] \cdot \Sigma(c_s)^2 - \Sigma(c_s)_i \cdot \Sigma[m']_i \cdot (c_s)_i\} / \{D'\}$$
$$(23.17)$$

where n is the number of observations, $c_s = (B - V)$, and

$$[m'] \pm V - y_o \quad \text{and} \quad D' \pm n \cdot \Sigma(c_s)_i^2 - [\Sigma(c_s)_i]^2$$

As recommended by Hardie, the color equation (23.16) may be rewritten to a form similar to the magnitude transformation equation:

$$(B - V) - (b - y)_o = \mu' \cdot (B - V) + \zeta'_{BV}$$
$$(23.18)$$

where

$$\mu' = 1 - 1/\mu \quad \text{and} \quad \zeta'_{BV} = \zeta_{BV} / \mu$$

The solutions of the normal equations are

$$\mu' = \{n \cdot \Sigma[c']_i \cdot (c_s)_i - \Sigma[c'] \cdot \Sigma(c_s)_i\} / \{D''\}$$

$$\zeta'_c = \{\Sigma[c'] \cdot \Sigma(c_s)_i^2 - \Sigma(c_s)_i \cdot \Sigma[c']_i \cdot (c_s)_i\} / \{D''\}$$
$$(23.19)$$

TABLE 23.1. Extinction data.

Star	X	$[y]$	$[b - y]$	$[u - b]$	V	$B - V$	$U - B$
HR8780	1.0010	0.774	0.022	1.464	4.66	1.06	0.88
HR8860	1.0111	0.736	−0.028	1.461	4.86	1.67	1.87
HR0194	2.0953	1.144	0.328	1.826	4.75	1.00	0.84
HR0315	2.0536	1.082	0.313	1.747	6.12	1.01	0.85
HR0417	1.0101	0.763	0.088	1.495	4.83	0.42	0.00
HR0235	2.3585	1.197	0.358	1.934	5.20	0.50	−0.01
HR1069	1.0062	0.747	0.069	1.453	5.32	0.41	−0.02
HR0708	3.0864	1.470	0.486	2.127	4.89	−0.02	−0.05
HR0740	4.3866	1.961	0.874	2.623	4.75	0.45	−0.01
HR0818	6.6745	2.711	1.536	3.449	4.46	0.48	0.00

where

$$[c'] = c_s - c_o \quad \text{and} \quad \{D''\} = n \cdot \Sigma(c_s)_i^2 - [\Sigma(c_s)_i]^2$$

From Eqs. 23.18, μ and ζ_{BV} can be recovered:

$$\mu = 1 / (1 - \mu') \quad \text{and} \quad \zeta_{BV} = \mu \cdot \zeta'_{BV} \quad (23.20)$$

Once the transformation parameters are found, the data can be placed on the standard system. With this step, the standardization process is complete.

23.6 The Challenge of Real Data

Data obtained by undergraduates in the senior astrophysics laboratory course at the University of Calgary can be used to illustrate the Hardie method. Observations made on December 12, 1982, with the 41-cm telescope are presented in the form of visual magnitudes and B–V, U–B color indices in Table 23.1. Column 1 lists the Bright Star Catalogue entry (Harvard Revised photometry designation) and column 2, the air mass corrected for the Earth's

atmospheric curvature by Eq. 19.7. The bracketed column headings indicate quantities that evaluate the right-hand sides of Eqs. 23.11, but with modifications found in Eqs. 23.2 and 23.12. The bracketed values represent input into the final iteration, with final transformation coefficients of

$$\varepsilon = -0.065, \quad \mu_{by} = 0.95, \quad \text{and} \quad \mu_{ub} = 0.95$$

The final three columns contain the Johnson magnitudes and colors.

From the data in Table 23.1, the pairings according to color index produce the intermediate results given in Table 23.2. Here the color index similarity criterion has restricted the pairings to only 13 in number. In addition to the differences in bracketed quantities, the final residuals between the beginning and computed values of the bracketed quantities have been entered in the last three columns. The sums of the terms needed for the least squares operation are

$$\Sigma dx_i^2 = 124.2246$$

$$\Sigma[y]_i \cdot dX_i = 42.5973$$

$$\Sigma[b - y]_i \cdot dX_i = 31.3625$$

$$\Sigma[u - b]_i \cdot dX_i = 42.4774$$

Therefore,

$$k'_y = 0.343, \quad k'_{by} = 0.252, \quad \text{and} \quad k'_{ub} = 0.342$$

The uncertainties in these results may be derived from the expression

$$e(k') = e_1 / \sqrt{\Sigma(dX)_i^2} \quad (23.21)$$

where $e_1 = \sqrt{[\Sigma r_i^2/(n - 1)]}$, the error in a single pairing.

TABLE 23.2. Intermediate results.

Pairing	dX	$d[y]$	$d[b - y]$	$d[u - b]$	$r(V)$	$r(B - V)$	$r(U - B)$
8780–0194	−1.094	−0.370	−0.306	−0.363	+0.005	−0.030	+0.012
8780–0315	−1.053	−0.307	−0.292	−0.283	+0.054	−0.026	+0.077
0194–0315	+0.042	+0.063	+0.014	+0.079	+0.048	+0.004	+0.065
0417–0235	−1.348	−0.434	−0.270	−0.439	+0.028	+0.070	+0.022
0417–1069	+0.004	+0.015	+0.019	+0.042	+0.014	+0.018	+0.041
0417–0740	−3.377	−1.199	−0.786	−1.128	−0.041	+0.066	+0.027
0417–0818	−5.754	−1.948	−1.448	−1.954	+0.025	+0.005	+0.014
0235–1069	+1.352	+0.450	+0.289	+0.481	−0.014	−0.053	+0.019
0235–0740	−2.028	−0.765	−0.516	−0.689	−0.069	−0.004	+0.004
0235–0818	−4.406	−1.514	−1.178	−1.515	−0.003	−0.066	−0.009
1069–0740	−3.380	−1.214	−0.805	−1.170	−0.055	+0.048	−0.014
1069–0818	−5.758	−1.964	−1.467	−1.996	+0.011	−0.013	−0.027
0740–0818	−2.378	−0.749	−0.662	−0.826	+0.066	−0.061	−0.013

TABLE 23.3. Uncertainties in the coefficients.

Quantity	y	$b - y$	$u - b$
Σr_i^2	0.0210	0.0246	0.0149
e_1	0.042	0.045	0.035
$e(k')$	0.004	0.004	0.003

The sums of squares of the residuals as obtained from Table 23.2, the error in a single pairing, the errors of the coefficients derived from these quantities, and the sum, $\Sigma(dX)_i^2$, are given in Table 23.3. The data and the results that they yield are characteristic of a night in which the extinction is higher than normal, and not as uniform as one might desire. This is reflected by the relatively high scatter, and the trends in the residuals suggest that the k'' values may not have been typical (~ 0 for all but k''_{by}). Nevertheless, they are more typical of sites that are not located on a mountaintop site surrounded by desert or ocean than would be data from pristine nights. It is instructive that extinction coefficients can be obtained with relatively high precision even for an imperfect night.

The challenge of this chapter is for you to put to the test your observing and data analysis skills by carrying out your own photoelectric program. The work involves all aspects of astronomy from the location of stars to data reduction, and a calculator or computer is a valuable aid. If you have access to your own photoelectric data, attempt to observe the stars listed here or those from a more extensive list of standard stars given in one of the references, and obtain extinction and transformation coefficients. By observing two of your standard stars (of contrasting color) many times, you can compare results with the Bouguer and the Hardie methods and gain an appreciation of the merits of each method. If those two standard stars are near each other, they may yield second-order coefficients as well. If you are unable to obtain such data soon, review the equations relating to the extinction and transformation equations of this chapter, and from the data provided in the example, verify the extinction coefficients and their errors. With justification, it can be argued that observations at very large air masses, like that of HR0818 (see Table 23.1), presuppose too homogeneous an atmosphere and their inclusion gives unrepresentative extinction values for the rest of the sky. Test this hypothesis for the current case by reworking the problem *without* HR0818 in the data set.

For differential variable star work, the extinction and transformation coefficients alone suffice to reduce the data. The transformation zero points are essential in order to do any absolute photometry—for example, to obtain UBVRI values for the comparison, check, and variable stars. They do not have to be obtained every time, however, and reliable zero point determination requires a very good sky, so only the best nights should be used for this purpose.

Ideas for photoelectric projects can be found in publications of the American Association of Variable Star Observers and of the International Association of Amateur and Professional Photoelectric Photometrists, and these organizations provide support for their members. Some local amateur organizations have communal photometers installed on their telescopes to which you may be able to gain access, since all local organizations are eager to increase their membership and to spread interest in astronomy.

23.7 Challenges of the Future

The process of computing linear extinction and transformation coefficients that are appropriate for broadband differential and absolute photometry has been justly criticized by, for example, Young (1974). The photometry of the future should be able to produce still more precise and accurate results than the nominal 1% photometry achieved today (see Schmidt-Kaler, 1984). Discussion of the techniques to achieve this are beyond our current scope, but some suggestions concerning absolute photometry are discussed by Young (1988, 1989). For differential photometry, the situation is somewhat easier.

This chapter has dealt thus far with obtaining coefficients from a single-channel system. Many two-star photometers in use in observatories around the world permit the simultaneous observation of two stars. Such systems can greatly lessen the difficulties of extinction variation between observations of the comparison and variable stars, and they may permit observations even under what a traditionalist would call nonphotometric conditions. The system at the Rothney Astrophysical Observatory is known as RADS (for Rapid Alternate Detection System). It uses an oscillating secondary mirror to move between two stars and to sample the sky between them at rates up to ~ 10 Hz. The pulse counting is gated so that the system waits until the mirror has settled before beginning to count. Up to four separate regions on the sky can be observed in this way, and the counts are stored in the four separate bins until the programmed total integration time is reached. A programmable filter sequence permits a series of observations to be made for essentially the whole night, but in practice, no more than three or four symmetric runs of up to five filters are carried out before the star images are checked for centering. Images may drift out of the diaphragm because of telescope drive drift, wind load, and so on, so unless these effects can be checked during the observations,[5] frequent checking between observations is desirable. Further details on this and other two-star photometers are found in Milone (1980) and discussions of how to reduce the data can be found in Milone and Robb (1981).

Two-star photometers provide some independence from extinction variation such as light cloud. Area detectors do this too, but only over a relatively small region of the sky (at present writing). Nevertheless, the charge-coupled devices (CCDs) have become the detectors of choice in the professional photometry community because photometry of extended regions can be done so conveniently. These devices are solid-state detectors consisting of a two-dimensional array of from tens of thousands to millions of light-sensitive elements, or pixels, in an area of order 1 cm². The devices are normally most sensitive in the spectral region ~ 600 to 700 nm, but they are sensitive to blue and yellow light, and by special coating, they can be made sensitive to the ultraviolet region also. They have excellent dynamic range and are linear over most of it. Read-out noise and sensitivity variation from pixel to pixel limit the precision with which photometry can be done, but this appears to be no worse than typical photoelectric photometry precision. Frequent dark, or "bias," frames must be taken to determine the zero points of the pixels, and dome or sky "flats" must be observed to determine the sensitivity of the pixels. Data reduction can be accomplished with a software package like DAOPHOT by P. Stetson, which has been added on to a more general data reduction package called IRAF (for Image Reduction and Analysis Facility), distributed by the Kitt Peak National Observatory. This package and a radio astronomy package called AIPS are widely used among professional astronomers for imaging work. With the increasing availability of area detectors, IRAF will probably become more widely available in the near future.

CCDs can also be used for variable star photometry (see Schiller and Milone 1990; Milone, Filhaber, and van Houten 1990 for discussons of the relative precision of CCD's vs. photoelectric differential photometry). The difficulty is that many frames are required, which means that vast amounts of data are accumulated, even if only a portion of the CCD chip (say 50 × 100 pixels) is read out after each exposure. The tedium of data reduction will almost certainly decrease as software procedures for handling massive amounts of data are developed; IRAF promises to do just this. The challenges of the future have to do with employing the highest efficiency in handling large amounts of data!

References and Bibliography

Hardie, R.H. (1962) Photoelectric Reductions. In W. A. Hiltner (ed.), *Astronomical Techniques*. University of Chicago Press, Chicago, pp. 178–208.

Johnson, H.L. (1962) Photoelectric Photometers and Amplifiers. In W. A. Hiltner (ed.), *Astronomical Techniques*. University of Chicago Press, Chicago, pp. 157–177.

Johnson, H.L. (1966) *Annual Review of Astronomy and Astrophysics* **4**, 193–206.

Johnson, H.L., and Morgan, W.W. (1953) *Astrophysical Journal* **117**, 313–352.

Kitchin, C.R. (1984) *Astrophysical Techniques*. Adam Hilger, Bristol.

Landolt, A. (1983) *Astrophysical Journal* **88**, 439–460.

Landolt, A., Blondeau, K.L. (1972) *Publications of the Astronomical Society of the Pacific* **84**, 784–809.

Milone, E.F., Chia, T.T., Castle, K.G., Robb, R.M., and Merrill, J.E. (1980) *Astrophysical Journal* (Supplement) **43**, 339–364.

Milone, E.F., and Robb, R.M. (1983) *Publications of the Astronomical Society of the Pacific* **95**, 666–673.

Milone, E.F., Filhaber, J., and van Hauten, C.J. (1990) Non-Simultaneous Photoelectric CCD Photometry of a Variable Star: VY Crucis. In A.G.: Davis Philip; D.S. Hayes, and S.J. Adelman (eds.), *CCDs in Astronomy II. New Methods and Application of CCD Technology*. L. Davis Press, Schnectady. pp. 149–158.

Robinson, L.B. (ed.) (1988) *Instrumentation for Ground-Based Optical Astronomy: Present and Future*. Springer-Verlag, New York.

Schiller, S.J., and Milone, E.F. (1990) Simultaneous Photoelectric and CCD Photometry of a Variable Star: D.Y. Herculis. In A.G. Davis Philip, D.S. Hayes, and S.J. Adelman (eds.), *CCDs in Astronomy II. New Methods and Applications of CCD Technology*. L. Davis Press, Schenectady. pp. 159–165.

---

[5]This could be accomplished by examining an image fed by a beam splitter or dichroic filter which passes light to a TV camera at the same time that light is received by the photometer.

Schmidt-Kaler, Th. (1984) In F. Praderie (ed.), *Space Research Prospects in Stellar Activity and Variability*. Observatoire de Paris, Meudon, pp. 169–174.

Wood, F. B. (ed.) (1963) *Photoelectric Astronomy for Amateurs*. Macmillan, New York.

Young, A.T. (1974) Observational Technique and Data Reduction In N. Carleton (ed.), *Methods of Experimental Physics*, Vol. 12. Academic Press, New York, pp. 123–192.

Young, A.T. (1988) Improvements to Photometry. I. Bettes Estimation of Perivatives in Extinction and Transformation Equations. In W. J. Borucki (ed.), *Second Workshop on Improvements to Photometry*. NASA Ames Research Center, Moffett Field, CA. pp. 215–245.

Young, A.T. (1989) Extinction and Transformation. In E. F. Milone (ed.), *Infrared Extinction and Standardisation*. Lecture Notes in Physics, Vol. 341. Springer-Verlag, Heidelberg, pp 6–14.

24

Spectral Analysis

24.1 Introduction

The refraction of light in glass objects and interference effects produced by gratings led to the early realization that white light consists of a continuum of color. The relationship between white light and color has been amply discussed in literature (e.g., by Goethe) as well as the physical sciences. It is therefore all the more remarkable that the spectral lines[1] of the solar spectrum were not discovered until the nineteenth century (by W.H. Wollaston, 1808). These absorption lines (so-called because they represent light taken from the continuum by atomic absorption in the solar atmosphere) were studied extensively by Joseph von Fraunhofer (ca. 1820), whose name they now bear. Fraunhofer counted around 600 lines; today's spectral atlases of sunlight list thousands.

Soon after the discovery of the spectral lines it was noticed that the spectra of other stars appear different from that of the Sun, some very greatly. A system of classification on the basis of stellar spectra was developed by Angelo Secchi (ca. 1865) in Italy and by Annie J. Cannon (the Henry Draper Catalogue in 1890) at Harvard. The catalogue and classification scheme established at Harvard is still in use. Letters were initially assigned according to the complexity of the spectrum, but today the letters are arranged in the order of decreasing effective surface temperature of the stars:

O, B, A, F, G, K, M

A slightly different set of letters is used to denote cooler stars in which the chemical composition differs sufficiently to cause the features to change character (the R, N, and S classes). Fine divisions are provided by the numbers 0 to 9, though not all are used in every class. Spectral classification was soon followed by an understanding of the origin of the spectral lines. The field of spectral analysis, founded by Kirchhoff and Bunsen (ca. 1860), enabled Sir William Huggins (ca. 1870) to identify the lines of the Balmer series of hydrogen in the spectrum of Vega and other stars. Today the main focus of investigation is not so much the identification of an individual spectral feature, although this is still of interest[2] because many features of stellar spectra have not yet been identified. Instead it is the intensity *profile*, or detailed distribution of brightness across the line, and the overall strength of the line in relation to others. Through such studies, many parameters of stellar atmospheres can be obtained, such as the pressure, temperature, and chemical composition.

24.2 The Origin of Spectral Lines

Kirchhoff's radiation laws, which are illustrated in Fig. 24.1, summarize the three empirical types of spectra:

1. A continuous spectrum is emitted by an incandescent solid, liquid, or very dense gas. If it is in addition a black body source (see §17.2 and §20.2), its radiation is governed by the Planck radiation law (Table 20.1).

---

[1]When light is focused onto a narrow slit before it is allowed to fall on the collimating, dispersive, and focusing elements of a spectrograph, each point in the spectrum at the output end is a miniature image of the slit. Differences in brightness between adjacent wavelengths show up as spectral lines or bands across narrow regions of the spectrum. The lines are caused by atoms, the bands by molecules.

[2]A prime example is the discovery of lines of the short-lived radioactive element technetium in certain stars.

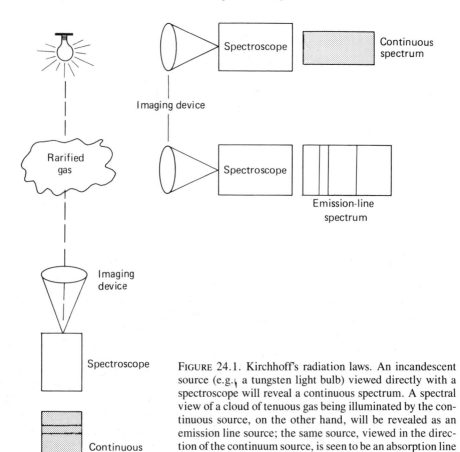

FIGURE 24.1. Kirchhoff's radiation laws. An incandescent source (e.g., a tungsten light bulb) viewed directly with a spectroscope will reveal a continuous spectrum. A spectral view of a cloud of tenuous gas being illuminated by the continuous source, on the other hand, will be revealed as an emission line source; the same source, viewed in the direction of the continuum source, is seen to be an absorption line source. The gas is seen to scatter the radiation from the continuous source by reemission in all directions, although it absorbs the radiation from only one direction.

2. An absorption spectrum can be seen in radiation from a rarefied gas. If the gas is illumined by a continuous source and an observer views the continuous source through the gas, absorption lines will be seen.

3. An emission spectrum can be observed from a rarefied gas also. If it is viewed obliquely, so that that continuous source is *not* visible through the gas, emission features will be seen, at the same wavelengths at which the absorption features appear.

In the case illustrated here, the gas absorbs the radiation and then reemits it, but in all directions; hence there will be fewer photons emerging in the original direction of the radiation. The preferential photon energies that are absorbed and then reemitted, $h \cdot \nu$, must correspond to an energy difference, ΔE, between two energy states of the absorbing atoms.[3] The *strength* of an absorption, by which we mean the total energy absorbed from the continuum, depends on the number of atoms in the excited state which can absorb the photons. The number of atoms in any one energy state at any given time depends on the energy of that state, the number of atoms in the ground or lowest energy state, and the temperature. It is given by the Boltzmann distribution:

$$N = N_1 \cdot e^{-\Delta E/kT} \qquad (24.1)$$

---

[3]"Energy state" refers not only to the size of the electron's orbit but also the subordinate states that are excited only under special conditions—such as in the presence of strong electric or magnetic fields. These states are characterized by quantum numbers, a more extensive discussion of which can be found in texts on atomic physics.

where N_1 and N are the number of atoms in the ground and the particular excited state, respectively ΔE is the energy difference between these two states, k is the Boltzmann constant, and T is the temperature in kelvins.

It is interesting to note that Eq. 24.1 has the same form as the equation describing the variation of atmospheric density, ρ with height, h in an isothermal (constant-temperature) atmosphere:

$$\rho(h) = \rho(0) \cdot e^{-(mgh)/kT} \qquad (24.2)$$

where (mgh) is the potential energy of a molecule of mass m under a gravitational acceleration g. The relations are analogous to a certain extent (although in the density relation the energies are not quantized as they are in the atom) and can be used to help understand the idea of populations of the various states.

The population of an excited state can be increased both through absorptions of photons, a process known as *photoexcitation* (illustrated in Fig. 24.1), and through collision with neighboring atoms, a process known as *collisional excitation*. The speeds of atoms and molecules increase as temperatures increase (indeed, temperature is a measure of the kinetic energy of a collection of such particles) and as atoms pass each other, they may exchange orbital energy for excitation energy.

Suppose we have an imaginary gas composed of 10 hydrogen atoms, in an environment in which the temperature is imagined to increase. Each atom consists of a proton and an electron and is thus electrically neutral. If the initial temperature of the gas is very low, so that $T \ll E/k$, then the number of atoms in the second level is 0. Although the atom is in a position to absorb photons of a certain energy (e.g., a Lyman α photon at wavelength 121.5 nm corresponds to the energy difference between the first and second energy levels of hydrogen), it cannot, at this instant, emit any until an absorption takes place. As the temperature rises, however, the second level begins to be populated, and emissions can take place. In the hydrogen atom, the larger the population of the excited states, the stronger the emission line when the atom is *deexcited*. The lifetimes of most energy levels are extremely short, so emission follows absorption quite quickly.[4] As the temperature con-

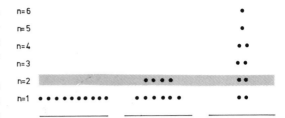

FIGURE 24.2. The populations of the energy levels (and the strengths of emission lines produced by downward transitions from these levels) depend on the temperature. If it is too low, the electrons remain in the ground level (*left*). At some optimal temperature, the second level will achieve its highest population (*middle*). At still higher temperatures, however, the other levels can compete for electrons, and the emission from the second level is again reduced (*right*). The population probability of any level thus varies with temperature, reaching a peak at some optimum value.

tinues to grow, Lyman α emission line from the second level again grows weaker. The explanation lies in the fact that the population of the second level is again decreasing because of the increasing populations of the higher levels. Figure 24.2 illustrates our thought experiment.

In §19.4 and §23.4, we described atmospheric extinction and its effect on starlight. In visible light, atmospheric extinction is brought about mainly by Rayleigh scattering (and in a later chapter we will discuss the scattering arising from the interstellar dust grains). In other regions of the spectrum, true absorption of photons does occur. In a stellar atmosphere, however, true absorption precedes reemission, which results in a redistribution of direction and wavelength of the photons. One can distinguish true scattering phenomena—electron, Rayleigh, and other forms—from absorption processes that also result in redirection of photons. The wavelength, too, can change in the absorption process: a high-energy photon may raise an atom to a very high energy state or ionize the atom entirely; the electron may then cascade down through the energy states of the atoms, emitting a sequence of photons each of less energy than the original exciting photon. Such effects are seen in planetary nebulae, for example, where ultraviolet radiation from the central hot, blue stars ionize atoms and give rise to sequences of emission lines. In stellar atmospheres, the absorption coefficient varies strongly with temperature, and the temperature varies with distance from the center

[4]The energy levels for which the lifetimes are very long are called metastable or even forbidden states. An example of the latter is the 21-cm line radiation of atomic hydrogen caused by a flip of the electron spin relative to that of the proton.

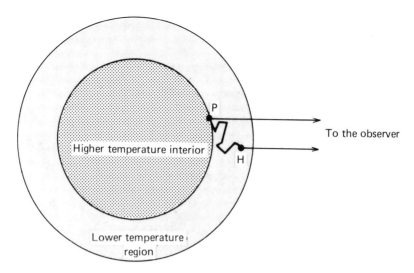

FIGURE 24.3. On the origin of absorption lines in stellar atmospheres. The temperature increases inward. Photons wend their way through repeated scattering, absorption, and emission from the deep interior to the surface. Greatly exaggerated path lengths for a greatly underesti-mated number of redirections illustrates the passage of a photon between point P and a higher point H. When the optical depth is low enough ($\tau \approx 1$ or less), such a photon can escape to illumine a grain on a spectrographic plate and thus contribute to the flux in an absorption line.

of the star. There are two types of absorption coefficient to consider: continuous and discrete.

The continuous absorption coefficient applies outside of the spectral lines, is relatively small, and varies only slowly with wavelength over most of the spectrum. In the continuum, the physical or *geometric depth* into the star to which the observer sees is greater than in the spectral lines. The geometric depth is controlled by the condition that the *optical depth*, $\tau \pm \int_0^R k \cdot dr \approx 1$, where k is the absorption coefficient and R is the geometric depth below the surface (see challenge 2 in §19.5). Since k is small for the continuum, R is large. At greater depths—that is, closer to the star's center—the temperature is greater, a fact that we will demonstrate in a later chapter when we discuss the limb-darkening in the

sun and stars. The greater temperature means stronger flux at each observed wavelength because the blackbody radiation curves are nested, with the higher temperature curves overlying the lower temperature ones (§20.2).

The discrete or line absorption coefficient increases greatly toward the center of the spectral line. This coefficient depends on the probability of absorption, which in turn depends on the population distribution of the states of the absorbing atom. At different geometric depths, the different temperatures create different population distributions. The appropriate population level for a particular atomic state to absorb a particular photon energy will be found over a range of temperature and thus over a range of geometrical depth. Hence

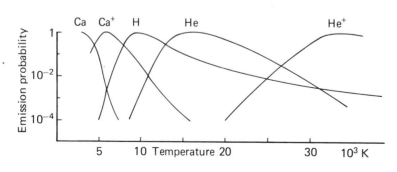

FIGURE 24.4. Emission probability for a few characteristic spectral features of stellar atmospheres. The variation with temperature explains the principal differences among stellar spectra.

absorption lines will originate at different physical depths depending on the atom and the wavelength. There is no single "reversing layer" in a star's atmosphere, contrary to statements in several early books on astrophysics. This does not necessarily mean that a photon of a certain wavelength will pass right through the atmosphere until it is absorbed at the optimum temperature level. Photons are repeatedly scattered and absorbed until they reach the upper atmosphere, undergoing many changes in direction and wavelength along the way. Figure 24.3 illustrates an imaginary path of a photon, derived from the stellar core in a nuclear reaction many, many interactions earlier, emerging from deeper layers at a certain moment at some point P. (The sharp boundary between higher and lower temperature regions is for illustrative purposes only. There is no sharp-edged "reversing layer".) After many more absorptions and reemissions, it may be high enough in the stellar atmosphere to actually escape the star and find its way into a spectrograph mounted on a telescope, appearing in a spectral line. If the line is very strong, the absorption coefficient is very large, and so the geometrical depth will be small. Thus high in the photosphere,[5] the temperature will be relatively low, and the blackbody flux will be lower: the flux removed from the core of the line by atomic absorption cannot be replaced by continuum emission from that atmospheric depth and so we observe an absorption line.

We close this section by noting that an atom can absorb a photon from any level. Thus, any of our four hydrogen atoms in the middle configuration of Fig. 24.2 may, before it deexcites, absorb another photon of sufficient energy to carry it to the third level (contributing to a Balmer α line absorption, at 625 nm), the fourth level (Balmer β, at 486 nm), and so on. In practice a stellar atmosphere contains so many atoms that all of these effects are seen, but it is interesting that the probabilities vary greatly with temperature. Thanks to this fact, we are able to classify stars according to their spectra.

24.3 The Basis of Spectral Classification

In Fig. 24.4, the variation of emission probabilities with temperature for selected atoms and ions is seen. These species play important roles in the visible spectra of stars and provide the main criteria for spectral classification. It can be seen, for example, that the Balmer lines of hydrogen (the

TABLE 24.1. Basic criteria for spectral classification of stars.

Spectral type[a]	Temperature (K)	Classification criteria[b]
O	40,000	Lines of highly ionized atoms: He II, Si IV, N III, . . . ; H relatively weak; occasional emission lines
B0	25,000	He II absent; He I strong; Si III, O II, H stronger
A0	10,000	He I absent; H at maximum; Mg II, Si II strong; Fe II, Ti II, Ca II weak
F0	7,600	H weaker; Ca II strong; ionized metals, e.g., Fe II, Ti II, peak ∼A5
G0	6,000	Ca II very strong; neutral metals, e.g., Fe I, Ca I, strong
K0	5,100	H relatively weak; neutral metals and others strong; molecular bands appear
M0	3,600	Neutral atoms, e.g., Ca I, very strong; TiO bands
M5	3,000	Ca I very strong; TiO bands stronger

[a]Subdivisions are provided through trailing integers (0–9) after the letters. Not all the integers are used for all classes.
[b]The Roman numerals represent stages of ionization: Ca I represents neutral calcium, Ca II singly ionized calcium, and so on.

only sequence of hydrogen lines to be seen between 300 and 700 nm) reach greatest strength at $\sim 10^4$ K surface temperature. They are seen in greatest strength in stars classified A0. In stars of both higher and lower temperatures, these features are weaker, or even absent. For the other features, the situation is the same, but in other temperature regions. Thus the rearrangement of the classes to provide a temperature sequence was a logical step.

The basic scheme for spectral classification is summarized in Table 24.1. Chemical composition differences among stars tend to show up more prominently at the low-temperature end of the sequence, and there are parallel classes to the K and M sequence: the R and N classes, characterized by strong carbon features; and the S class, in which zirconium oxide bands predominate and features due to molecules like LaO, YO, and SiH are also seen. There are many types of stars with

---

[5]This is the atmospheric layer of the sun and stars that gives us almost all of the light that we see in the visible part of the spectrum. Only during total eclipses, or in other regions of the solar spectrum where the optical depth is higher, can the tenuous outer, higher temperature, regions of the chromosphere and corona be seen.

TABLE 24.2. Luminosity classification.

Class	Type of star
Ia-0	Most luminous supergiant
Ia	Luminous supergiant
Ib	Less luminous supergiant
II	Luminous giant
III	Giant
IV	Subgiant
V	Dwarf
VI	Subdwarf

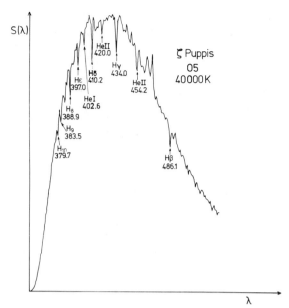

FIGURE 24.5. The Balmer series are among the most prominent lines in the spectrum of this hot star, even though they are weak relative to those seen in A stars. The lines of He I are visible and those of He II (this ion resembles the neutral hydrogen atom in having only one electron; its lines are members of the Pickering series, arising from the fourth energy level) are prominent.

anomalies; these include the very hot Wolf-Rayet (W-type) stars, which have a predominance of nitrogen (WN) or carbon and oxygen (WC) features, and the metallic line stars among the A–F classes where the metal lines are stronger than those for normal stars.

This is only one dimension of stellar classification, however. The appearance of many spectral features is also affected by the gas pressure in the line-forming region. This causes energy-level distortion, which contributes to spectral line broadening. Spectral lines of stars arising from less dense stellar regions appear much narrower than those from higher density regions. Between two stars of approximately the same temperature, such differences point to a difference in stellar size and indicate a second dimension of classification: the luminosity. Roman numerals following the spectral classification indicate the luminosity. The luminosity classes are listed in Table 24.2.

24.4 Stellar Spectra

A spectrum is a distribution of light with wavelength. It can be detected by spectrographic plate, by scanning spectrophotometer, or by electronic imager. A spectrographic plate is a photographic plate exposed at the image plane of the spectrograph. A scanning spectrophotometer (or spectral scanner) moves the dispersing element (prism or

TABLE 24.3. Scanner data stars.

Figure	Star	Spectral type	T_{eff} (K)
24.5	ζ Puppis	O5	40,000
24.6	α Eridani	B5	16,000
24.7	α Canis Majoris	A1	10,000
24.8	α Canis Minoris	F5	7,000
24.9	α Centauri	G2	6,000
24.10	α Arietis	K2	4,000
24.11	β Gruis	M3	3,000

grating) to produce the spectrum at an output slit of the spectrograph. A photomultiplier tube receives the light at this slit and the record of its output over time (the time to move the grating or prism) is the spectrum. Finally, electronic imaging devices such as reticons or CCD detectors can be used to replace the spectrographic plate.

Compared to photographic detection, photoelectric techniques and electronic imaging generally permit higher accuracy and so have largely replaced the older photographic methods at most observatories. However, data stored on spectrographic plates are still of value and are easily archived and examined, so the results of earlier work will certainly be with us for some time. Good photographic sequences of spectra may be found in several spectral atlases.[6]

---

[6]For example, those of Abt, Meinel, and Morgan (1968) or the more recent Morgan, Abt, and Tapscott (1978) for the hotter stars and Keenan and McNeil (1976) for the cooler stars. See §24.5 for references to another type of spectral atlas.

The spectra presented here (Figs. 24.5 to 24.11) were obtained with a spectrum scanner. Table 24.3 lists the stars whose spectra are displayed. The wavelength region runs from about 350 nm in the ultraviolet to 500 nm (green), which includes the region used for spectral classification. The scanner output for stars of the "earlier" spectral types (O–F; i.e., those of hotter stars) reveal glimpses of the stellar continuum, so that the envelope of the radiation curve can be recognized. The spectral lines are seen superimposed on this curve, which represents effectively the Planck distribution of the deeper, hotter regions of the stellar atmosphere. The distribution in each case has been modified substantially because of the wavelength dependence both of atmospheric and interstellar extinction and of detector sensitivity. Thus it is not possible to use Wien's law, for instance, to determine the temperature from the peak of the observed radiation curve. For the "later" spectral types, the atomic and molecular absorption features are so numerous that the continuum seems to disappear.

The scanner output spectra of stars seen in Figs. 24.5 to 24.11 represent a sequence of spectral types. The abscissas are approximately linear with wavelength but the relative fluxes on the ordinate scale are uncorrected for instrumental sensitivity or for atmospheric extinction. The wavelengths of the identified features are given in nanometers.

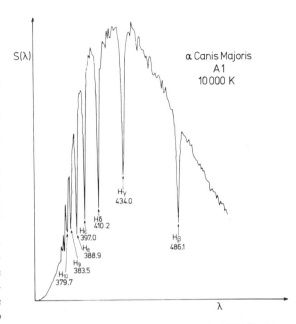

FIGURE 24.7. At a surface temperature of 10,000 K, the Balmer series is about at its strongest. The lines are noticeably broad in this luminosity class V star, due to temperature (Doppler motion) and pressure (Stark broadening) effects.

The figure captions provide further description of each spectrum.

Now we discuss instrumentation so that you can obtain your own spectra.

24.5 Constructing an Objective Prism Spectrograph

An objective prism camera is nothing more than a camera in which the objective lens is preceded by a narrow prism, as seen in Fig. 24.12. The prism, as well as the hand spectroscope mentioned in the challenges, can be obtained from mailorder science supply houses, like Edmund Scientific Corp. (Barrington, New Jersey). To give large enough images, the objective should be a telephoto lens of at least 100 mm focal length.

The camera and prism may be attached to a common piece of wood or plastic, with the prism tilted so that the spectrum strikes the image plane. Some trial and error is needed to do this and for focusing, but it does help considerably if the camera is a single-lens reflex (SLR) type. The aperture should be set at $f/2.8$ to collect light from a large field. With a prism made of crown glass, having a wedge

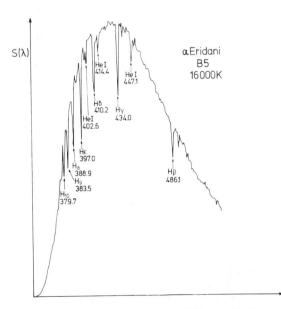

FIGURE 24.6. He II is no longer visible in this B-type star, but the Balmer lines are more prominent.

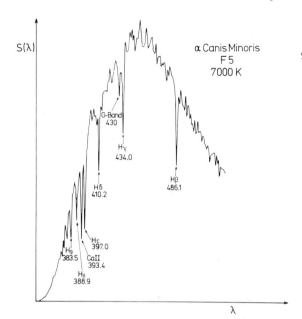

FIGURE 24.8. In F stars, the Ca II H and K lines (393.3 and 396.8 nm) are slightly stronger than those of hydrogen, which are weaker than in A stars. The G band is a blended feature of the radical CH and Fe lines.

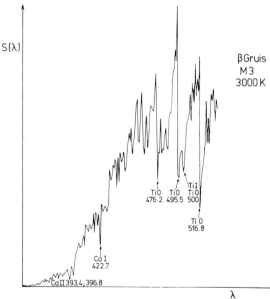

FIGURE 24.10. The spectra of K stars are dominated by neutral metal lines, but note the continued strength of the ionized calcium lines.

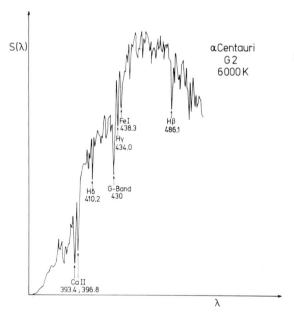

FIGURE 24.9. The solarlike spectrum of α Centauri shows a further weakening of the Balmer lines and increased strength of the metals. The continuum (at this resolution) is now difficult to see.

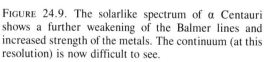

FIGURE 24.11. The spectra of very cool stars are dominated by molecular band absorption. In M-type spectra, this is due mainly to titanium oxide (TiO).

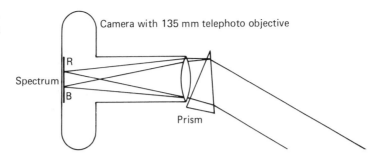

FIGURE 24.12. A simple but effective objective prism spectrograph: a camera and a prism.

Camera with 135 mm telephoto objective

R
Spectrum
B

Prism

angle of about 30°, and a focal length of 135 mm for the objective, a linear reciprocal dispersion of ~ 150 nm/mm can be achieved.

The length of the spectrum will be determined mainly by the film type but also by the camera lens, prism material, and of course the atmosphere, all of which will play a role in limiting the color region that is detected. With a panchromatic film and speed 400 ASA, the length will be ~ 2 mm. The camera–prism combination should be oriented so that the stars are allowed to trail perpendicularly to the dispersion direction, with a time exposure of a few minutes or so. This will give broad spectra in which the spectral features will be seen more easily.

Unless the camera is on a clock-driven equatorial mounting, its exposures must be brief and therefore its targets limited to relatively bright stars. Alternatively, fix the camera on a telescope mounted this way, piggyback style. The dispersion direction should be aligned north–south and the telescope gingerly rocked back and forth in hour angle (i.e., east-west) during the exposure. Such an exposure can go deep with a long exposure (an hour or so), but the rocking motion must be carefully controlled or the spectrum's width will be nonuniform. Guiding through the main telescope with a high-power eyepiece and reticle will help considerably.

The objective prism camera can reveal many details, and with it you will be able to carry out spectral classification.[7] Figure 24.13 shows the spectrum of the zero magnitude star α Lyrae (Vega) with its principal absorption lines and their wavelengths.

[7]Spectral atlases of objective prism spectra have been produced by Waltraut Seitter in Germany and by Nancy Houk in the United States. See, for example, Houk, Irvine, and Rosenbusch (1974) and Houk and Newberry (1984).

24.6 A Grating Spectrograph

In this section we consider a more elaborate instrument, the grating spectrograph installed in the teaching laboratory of the Astronomical Institute of the University of the Ruhr. The heart of the spectrograph is a reflection grating with 1200 lines/per millimeter and a length of 30 mm. Thus N, the number of lines, is 36,000, a quantity to which the resolving power is (to first order) related: $N = \lambda / \Delta\lambda$. Therefore, at 500 nm, two spectral lines of separation $\Delta\lambda = 0.014$ nm can be resolved, at least in theory.

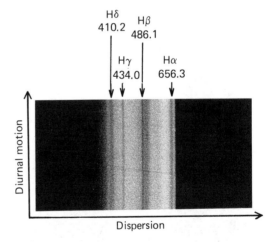

FIGURE 24.13. An objective prism spectrum of Vega (α Lyr). The camera was mounted so that the spectrum ran north–south, and the exposure was trailed in the east–west direction. The wavelengths of some of the Balmer lines of this A-type star are given in nanometers. The parameters were prism wedge angle of about 42°; objective focal length $f = 220$ mm; aperture (effective focal ratio) = $f/3.5$. The film speed was 400 ASA and exposure time was 8 min.

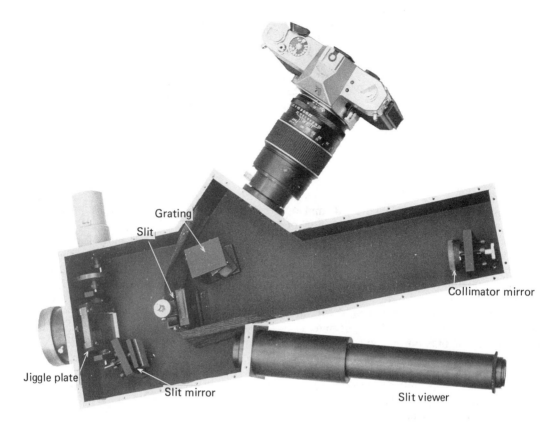

FIGURE 24.14. The grating spectrograph used for teaching at the Astronomical Institute of the University of the Ruhr, Bochum, Germany.

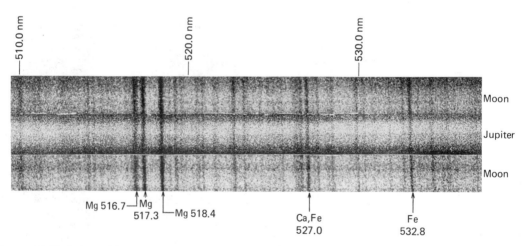

FIGURE 24.15. Prints of the spectra of Jupiter and the Moon from film obtained on June 2, 1979, with a grating spectrograph mounted on the 40-cm telescope of the Astronomical Institute of the University of the Ruhr. The film was Agfapan (400 ASA) and the exposures 2^m for the Moon, 3^m for Jupiter. The Jupiter spectrum overlies the Moon's spectrum, which therefore straddles it, so that the reflected solar lines are nearly coincident.

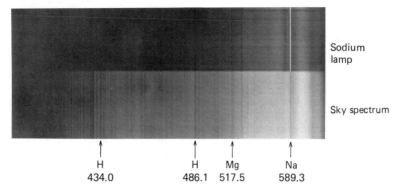

FIGURE 24.16. Proof of the presence of sodium in the Sun as revealed by a hand-held or pocket spectroscope and a sodium spectral lamp.

A spectrograph that comes close to achieving this resolution is seen in Fig. 24.14. The spectrograph is mounted on a telescope so that the light comes to focus at the slit plane. Behind the slit the light diverges again and falls onto the collimator mirror, which renders it parallel. This occurs because the slit is at the focal point of the collimator. From here the light reaches the grating, which disperses the light, still parallel, to the camera, which, set at infinity, can record the spectra. The broadening of the spectrum is accomplished with a "jiggle plate," (from the German "Wackelplatte") simply a piece of thick glass placed in front of the slit and perpendicular to the optic axis. It is jiggled up and down (perpendicular to the dispersion direction) and changes a point image into a line, thus filling the length of the slit. The slit viewer is used to center the star image on the slit. During exposure, the object can be guided by viewing the slit and moving the telescope when the image begins to reflect off of the slit jaws. Figure 24.15 shows the spectrum of Jupiter flanked by that of the Moon, both obtained with the grating spectrograph described here. Jupiter's spectrum was obtained with the slit placed along the planet's equator so that the points along the disk are Doppler shifted by different amounts, depending on the component of rotational velocity in the line of sight (see §9.2).

Appendix F contains a list of recommended photographic targets. Many of these are suitable for objective prism studies.

24.7 Further Challenges

1. A hand spectroscope or even a piece of transmission grating becomes a useful tool for examining the nature of light sources. Examine a tungsten light bulb with such a source and notice the continuum of color; mercury vapor and low-pressure sodium street lamps, on the other hand, are seen as emission line sources. Although tungsten lamps are rare outside of homes, they have been so numerous in the streets of large cities that astronomical spectra taken many kilometers away still show them. Astronomers are becoming particularly upset, though, at the increasing use of high-pressure sodium lamps. Examine the light from such a source and compare what you see to the other sources (artificial and natural) to find out why the pursuit of astronomical spectroscopy may be doomed if communities do not control their lighting better.[8]

2. One interesting terrestrial source is the extended source presented by a window that is illuminated by a fluorescent lamp. Instead of spectral lines (or dots from an unbroadened point source), you will see images of the window centered on the emitted wavelengths. This is analogous to what a slitless spectrograph produces when an emission nebula is examined, that is, multiple images of the nebula. By placing the hand spectrograph at a telescope eyepiece, one may get a glimpse of the Balmer series in a very bright A star, but spectral features are not readily resolved this way. Try to observe several stars this way and see if you notice any differences in the spectra of

[8]In fairness, we should add that some communities (such as Tucson, Arizona) have acted responsibly to reduce the effects of light pollution either by replacing such lamps or by properly shielding the sky, and thereby directing the light back down to the ground where it is useful. Such strategies can save money, too.

stars of different spectral types. You can try the technique of examining reflecting fluorescent sources, too, but be careful—there are laws against peeping toms in most jurisdictions and your motives may be misunderstood!

3. A hand spectroscope directed at the Sun will reveal the dark Fraunhofer lines, but only if a very narrow slit is used. One of the prominent features in the yellow-orange part of the solar spectrum is the sodium doublet at ~589.3 nm. Remember that it is dangerous to look at the Sun directly; a cloud or a white wall illuminated by the Sun will do just as well and be much safer. The identification of the lines can be verified by direct observation. For this, you will need to obtain a sodium spectral lamp, but if that is not available, the sodium lines can be seen by throwing ordinary table salt (sodium chloride) into the open flame of a Bunsen burner or gas range burner. First tape a piece of transparent paper across the lower half of the spectroscope slit. Then orient yourself so that both sources can be seen through the spectroscope—the bright sodium light through the lower half and the diffused sunlight through the top half. You should see the sodium emission line at the same wavelength as the sodium absorption line in the Sun, as Fig. 24.16 illustrates.

References and Bibliography

Abt, H. A., Meinel, A. B., and Morgan, W. W. (1968) *An Atlas of Stellar Spectra.* Kitt Peak National Observatory, Tucson, Ariz.

Houk, N., Irvine, N. J., and Rosenbusch, D. (1974) *An Atlas of Objective-Prism Spectra.* University of Michigan, Ann Arbor.

Houk, N., and Newberry, M. V. (1984) *A Second Atlas of Objective-Prism Spectra.* University of Michigan, Ann Arbor.

Keenan, P. C. (1963) Classification of Stellar Spectra. In K. Aa. Strand (ed.), *Basic Astronomical Data.* University of Chicago Press, Chicago, pp. 78–122.

Keenan, P. C., and McNeil, R. C. (1976) *An Atlas of Spectra of the Cooler Stars: Types G, K, M, S, and C.* Ohio State University Press, Columbus.

Morgan, W. W., Abt, H. A., and Tapscott, J. W. (1978) *Revised MK Spectral Atlas for Stars Earlier than the Sun.* Yerkes Observatory, Williams Bay Wisconsin, and Kitt Peak National Observatory, Tucson, Ariz.

Morgan, W. W., and Keenan, P. C. (1953) *Astrophysical Journal* **117**, 313.

Morgan, W. W., and Keenan, P. C. (1973) *Annual Reviews in Astronomy and Astrophysics* **11**, 29–50.

Seitter, W. C. (1970) *Atlas for Objective Prism Spectra,* Part 1. Dümmler Verlag, Bonn.

Seitter, W. C. (1975) *Atlas for Objective Prism Spectra,* Part 2. Dümmler Verlag, Bonn.

25

The Doppler Effect

25.1 Introduction

The Doppler effect, named for Christian Doppler (ca. 1842), has tremendous importance for science and technology. Thanks to it, air traffic controllers can distinguish among objects moving in the line of sight; on the ground, speeding cars can be identified before serious accidents occur; and in space, the velocities of the stars and the rotations and surface relief of the planets can be measured.

The change in frequency (relative to the rest frequency, ν) and in wavelength (relative to the rest wavelength, λ) caused by the relative velocity, ν, between the source and the observer is given by

$$\Delta\lambda / \lambda = - \Delta\nu / \nu = v / c \qquad (25.1)$$

The differences are in the sense: observed − rest, for both λ and ν. The velocity v is positive if the distance between the source and the observer is increasing, corresponding to an increase in the observed wavelength (a red shift); the velocity is negative if the distance is decreasing, corresponding to a decrease in the wavelength (a blue shift).

25.2 The Doppler Effect in the Microwave Region

Microwave apparatus is required for this section. If you have access to such equipment, you will be able to confirm the applicability of the Doppler effect to this part of the electromagnetic spectrum. The components of the apparatus used here are shown assembled in Fig. 25.1. They include Gunn diode, G (to operate at $\lambda = 3$ cm); a voltage supply, S; an emitting horn, H_1; a narrow, fixed metallic reflecting strip, R_1; a movable reflection strip, R_2; a detector diode, D; an oscilloscope, O; and a loud speaker and amplifier L.

When the diode is fired up, R_1 reflects back to the receiver horn H_2 a constant signal at wavelength λ (or frequency ν). If reflector R_2 is set in motion at a certain speed $v/2$, it too reflects back a signal, but at wavelength $\lambda - \Delta\lambda$ (or frequency $\nu + \Delta\nu$). The two signals are superposed at the receiver. The idea is to investigate the superposition of the two signals. Since these two signals are waves, we can represent them with sine and cosine functions and make use of the trigonometric identity:

$$\sin x + \sin y = 2 \cdot \sin[(x + y)/2] \cdot \cos[(x - y)/2] \qquad (25.2)$$

The arguments of the sine function, x and y, then express the phase of each of the two reflected waves: for the original signal, $y = 2\pi \cdot \nu \cdot t$, and for the (approaching) Doppler-shifted wave, $x = 2\pi \cdot (\nu + \Delta\nu) \cdot t$. By substituting these expressions in Eq. 25.2, the interference of the waves can be expressed as the sum of the two sine waves:

$$f(t) \pm \sin[2\pi \cdot (\nu + \Delta\nu) \cdot t] + \sin(2\pi\nu \cdot t)$$
$$= 2 \cdot \sin[2\pi \cdot (\nu + \Delta\nu/2) \cdot t] \cdot \cos[2\pi \cdot (\Delta\nu/2) \cdot t] \qquad (25.3)$$

Consequently, the diode detects a signal of frequency $\nu + \Delta\nu/2$ (virtually the same as ν) with amplitude modulated with frequency $\Delta\nu/2$. Figure 25.2 shows the superposition of the two waves.

The electronic system components that follow the detector and amplifier of the gigahertz receiver cannot respond fast enough to the signal at frequency ν, so a "null" mean value would be picked up by the oscilloscope and loudspeaker. Therefore, a diode is used to suppress negative excursions of

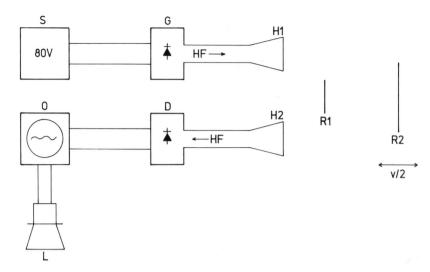

FIGURE 25.1. Experimental setup for a microwave Doppler effect experiment. A Doppler shift corresponding to a velocity v can be observed if the moving reflector R_2 has a velocity $v/2$ relative to both emitter and receiver.

the signal (Fig. 25.3). Thus you can measure the half-period of the *beat*. If the beat frequency, $\Delta v/2$, is small enough that it lies within the bandwidth of the oscilloscope or the loudspeaker, you can measure the Doppler shift (see Fig. 25.4). Other combinations of v and Δv that will work for either the oscilloscope (the first three) or the loudspeaker (last two) can be found from Table 25.1. From this table, a motion of reflector R_2 with respect to the emitter and the receiver horns H_1 and H_2 at 50 cm/s or more results in an *audible* Doppler frequency shift.

The experiment can be illustrated by the following example. In a laboratory in which just such an apparatus was set up, a procedure that relies on standing waves was used to determine the wavelength of a Gunn diode; it gave $\lambda = 3.268 \cdot 10^{-2}$ m. This wavelength corresponds to a frequency of

$$\nu = c / \lambda = (2.998 \cdot 10^8 \text{ m/s})/(3.268 \cdot 10^{-2} \text{ m})$$
$$= 9.174 \cdot 10^9 \text{ Hz}$$

Reflector R_2 was then moved uniformly with a speed

$$v/2 = 0.7889 \text{ m/s}$$

This velocity was maintained with a uniformly rotating disk, on the rim of which the reflector was placed. The resulting beat frequency was measured with an oscilloscope, the time sweep of which was calibrated with the electric main's frequency. The result was: $\Delta v = 49.02$ Hz, from which

$$\Delta v / v = 49.02 \text{ s}^{-1} / 9.174 \cdot 10^9 \text{ s}^{-1} = 5.343$$
$$\cdot 10^{-9}$$

Now

$$v / c = 1.578 \text{ m/s} / 2.998 \cdot 10^8 \text{ m/s}$$
$$= 5.263 \cdot 10^{-9}$$

so that the values agree to $\sim 8/530 = 1.5\%$.

The wave frequency and corresponding wavelength can be determined to high precision by counting a very large number of cycles and averaging (see §11.3 for an example of how this works). Typically, the precision of measurement of periodic phenomena like this is very high. This holds also for the velocity determination. The same cannot be said for the beat frequency, the precision of which depends on how well the oscilloscope screen can be read. In practice this is about 2 to 3%; in our case it is the major source of error. Since v varies linearly with Δv (Eq. 25.1), the determined quantities v/c and $\Delta v/v$ agree within the expected error.

25.3 The Doppler Effect in a Nova

In the summer of 1975, a "new" star, or *nova*, appeared in the constellation Scutum (Nova Scuti 1975). This in itself was not unusual, since novae occur in the galaxy at a rate of at least once per year, so they are neither novel nor new.[1] They do,

---

[1]They are certainly not new in any sense other than they may not have been seen before. It is now believed that novae are evolved stars in binary star systems that violently expel material hurled at them by their companion stars. The expulsion may recur, but the time scale of the recurrence varies widely.

FIGURE 25.2. Superposition of two wave trains in the case $\Delta \nu / \nu = 0.1$. The beat frequency is $\Delta \nu / 2$.

FIGURE 25.3. After rectification only the positive half of the wave remains.

FIGURE 25.4. The receiver can follow only the rms value of the frequency, or $\Delta \nu$, the period of which is measured on the oscilloscope trace. With a receiver with quadratic characteristics, the averaged tracing is again a sine curve.

however, increase greatly in brightness as the explosion expels the star's surface material into space. The Southern Hemisphere observatory station of the Astronomical Institute of the University of the Ruhr recorded observations of the nova near maximum light showing the great speed at which material was being ejected from the star. The material is assumed to have been ejected in the form of a shell. Although past novae outbursts provide much evidence that the ejected material is in the form of a nonuniform and irregular shell,[2] calculations are often carried out under the assumption of spherical symmetry because the degree of distortion from a sphere is not really known (in almost all cases), and the calculations are much easier to perform. This procedure is in some way an application of the principle of parsimony, which is that the simplest hypothesis that is consistent with experimental data is the preferred one.

If the ejected shell is moving outward from the star at some velocity v, a spectrograph on Earth will reveal very broad emission features of width $\sim 2v$. We see the shell as having a *spread* of velocities in the direction of the observer, because different regions on the shell have different components of v in our line of sight. The part

of the shell that is moving along the line of sight (and toward the observer) will have the largest negative radial velocity. The regions at the edges of the shell, as seen by the observer back on Earth, emit at the rest wavelength (actually, they retain the star's radial velocity). The shell continues to absorb the stellar radiation and to reemit it, but the part of the shell that we see *in front of* the star will contribute an absorption line.[3] Because of the shell's motion, the absorption line will appear to be extremely blue-shifted. Figure 25.5 illustrates the basic situation. The resulting combination—a narrow blue-shifted absorption line on the blue edge of a broad emission line—defines the classic P Cygni profile.[4]

The spectrum of the nova was scanned with a photoelectric spectrophotometer, as was that a bright star, α Aquilae (Altair), for comparison. In Figure 25.6 the output in the vicinity of three Balmer lines ($H\beta$, $H\gamma$, and $H\delta$) of these two objects are superimposed. The differences between the simple absorption profile of Altair (light line) and the P Cygni profiles of the nova (heavy line) are

TABLE 25.1. Velocities and frequency shifts for $\lambda = 3$-cm equipment.

v (cm/s)	$\Delta \nu$ (Hz)
3	1
10	3
30	10
100	30
300	100

[2]A variable-thickness triaxial ellipsoid is sometimes used to model the ejected material, but even this may be a crude approximation to the actual shape of the ejecta.
[3]See the discussion of Kirchhoff's laws in Chapter 24.
[4]After a variable star known for its shell ejections.

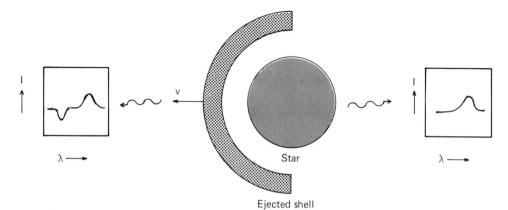

FIGURE 25.5. The spectrum of a nova is characterized by P Cygni profiles. The emission comes from the stars's surface and from the optically thick portions of the shell. The shell along the line of sight absorbs the underlying starlight, and because the line of sight component of the shell's speed is greatest in this direction, the absorption cuts into the short-wavelength side of the emission feature.

striking. Apparently the central star of the nova, like Altair itself, has only a small radial velocity relative to the observer compared to the ejected material. The Balmer lines of Altair thus serve as wavelength standards (the star itself has a radial velocity of 26 km/s, which corresponds to only about one-fifth of the width of a grating step and is thus ignorable). The separation between steps corresponds to a wavelength difference of 0.2 nm. With this information, and by counting the number of steps needed to superimpose the spectra, the velocity of the shell can be determined. Table 25.2

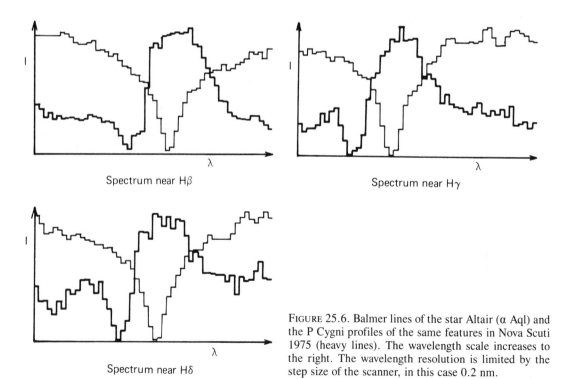

Spectrum near Hβ

Spectrum near Hγ

Spectrum near Hδ

FIGURE 25.6. Balmer lines of the star Altair (α Aql) and the P Cygni profiles of the same features in Nova Scuti 1975 (heavy lines). The wavelength scale increases to the right. The wavelength resolution is limited by the step size of the scanner, in this case 0.2 nm.

TABLE 25.2. Nova shell velocity data.

Line, λ	Steps	$\Delta\lambda$ (nm)	Shell velocity (km/s)
Hβ, 484.1	-9	-1.8	-1115
Hγ, 434.0	-9	-1.8	-1244
Hδ, 410.2	-8	-1.6	-1170
			$<-1176> \pm 37$

summarizes the Doppler shift in numbers of steps, corresponding wavelength shift, $\Delta\lambda$, and corresponding velocity. The average shell velocity is given at the base of the last column. The uncertainty in the result is controlled by the stepper resolution. The relative error increases with decreasing wavelength. If we take the worst case, we will have a conservative (i.e., worst case) error:

$$e_{\Delta\lambda} / \lambda = 0.2 \text{ nm} / 410.2 \text{ nm} = 4.88 \cdot 10^{-4}$$

This error is related to that in v, from a consideration of Eq. 25.1, by

$$e_v / c = e_{\Delta\lambda} / \lambda$$

Therefore, $e_v = 4.88 \cdot 10^{-4} \cdot 3 \cdot 10^5$ km/s, or $e_v = 146$ km/s.

25.4 Further Challenges

1. We have saved some space in this chapter by stating Eq. 25.2 without proof or demonstration; try demonstrating the truth of this trigonometric identity yourself.
2. With the help of tracing paper, copy the P Cygni profiles of the nova spectra in Fig. 25.6 and determine for yourself the spectral shifts required to fit the data. Then compute the shifts in wavelength and obtain an average velocity.
3. Here is a challenge for those who have studied the earlier chapters, especially those on optics and resolving power. Suppose the nova were 2000 pc away.[5] How long, at the computed rate of expansion, would it take until the shell was resolved by a ground-based telescope so superb that it could achieve, in the best possible site, a resolution of 0.2 arc sec? (Hint: The angular size of an object is given by

$$206{,}265 \cdot s / r \text{ arc sec}$$

where s is the linear size and r is the distance in the same units.)
4. How would you answer challenge 3 if a Space Telescope (with diffraction-limited optics and a diameter of 2.4 m) was substituted for the ground-based one? Assume a wavelength of 500 nm and ideal performance for the telescope.

[5]A parsec (pc) is a distance equivalent to $3.09 \cdot 10^{13}$ km. See Eq. 15.6 and the attendant discussion.

26

Binary Star Systems

26.1 Introduction

Star catalogue data indicate that only 30 of every 100 stars are alone (Allen, 1973). The rest are in double or multiple star systems. By itself, this finding would be sufficient justification for including binary star systems in a book designed to help its readers explore the nature of our cosmos. But there is much more. Binary stars are a means to obtain masses, sizes, temperatures, and surface details of stars other than the Sun. In other words, they are a major source of fundamental data. If scientists can be thought of as prospectors, binary stars are a rich lode of pure gold.

Binary stars come in a variety of types. Not all of them are equally well observable. The means for detection and study of a particular binary star depend on the separation of the stars, the orientation of the plane of the orbit relative to the plane of our sky, and the system's distance from us. In some cases the fact of binarity has strongly influenced the evolution of one and sometimes both components. The fundamental data that lead to such interesting conditions are obtained through the study of the positions, light, and motions of these star systems. In this chapter, we describe the visual, spectroscopic, and eclipsing binaries and the types of data that each is capable of providing. Finally, we describe a device with which the light variation of eclipsing systems can be simulated to study the shape and duration of eclipses.

26.2 Visual Binary Stars

Many stars appear to be double when they are viewed through a telescope. (Table 2.1 contains a short list of examples.) There is little in astronomy to compare to the beauty of a bright double star, especially if the components are of different colors. Some of these stars may be only *optical* doubles; that is, they may lie nearly along the same line of sight as seen from Earth, separated by a great gulf. Such stars move in straight lines, independently of each other, and so reveal that they are not joined by the bond of gravitational attraction. The motions of true binary stars, on the other hand, are curvilinear, and their paths are concave with respect to each other. The great visual star observers of the past, like W.H. van den Bos and G. van Biesbroeck, recorded the position of one star, usually the fainter, compared to the brighter, using a device known as a *filar micrometer*. Figure 26.1 shows how the coordinates were measured. The filar micrometer could be rotated to give a position angle relative to the north direction on the sky, and a movable crosshair could be used to record the separation of the two components. For both coordinates, precision screws with vernier scales would be used to read off the coordinates; repeated measurements could be used to improve the precision through averaging. Data gathered in this way provided *relative* orbits, where the motion of one component is shown relative to the other.

Figure 26.2 illustrates the apparent orbit of a visual binary; the true shape and the elements describing the orbit can be recovered by studying it.[1] The angular separation of the stars depends on the size of the orbit and the distance of the system. What is depicted in Fig. 26.2 is not the true shape

---

[1]For discussions about the recovery of the true orbit, see Aitken (1964), Binnendijk (1960), Heintz (1978), or Couteau (1978, 1981).

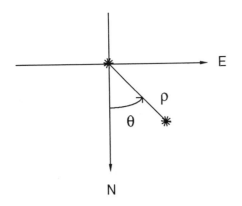

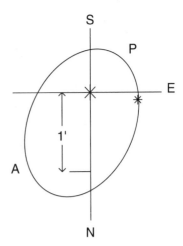

FIGURE 26.1. The measured parameters of separation and position angle, represented here by ρ and Θ, respectively.

FIGURE 26.2. The apparent relative orbit of the visual binary star 99 Herculis also known as ADS 11077. P and A mark periastron and apastron, respectively, on the true major axis; the smaller asterisk marks the position of the fainter component in 1994; the arrow indicating that star's direction of motion marks its relative position in A.D. 2000. (After Yan, Chu, and Pan, 1985, based on the orbital analysis of Makemson, 1948).

of the orbit, but a projection on the plane of the sky of the true orbit. By clever detective work, however, the shape of the true orbit can be found. First, the major axis can be identified by the extremes of the angular rate of motion in the relative orbit; it is also the line through the center of the ellipse and the primary star. In addition to these elements, the instant of time that star B is at periastron (or, when $e = 0$, at a node), T_o, must be given for subsequent prediction of position. Next the eccentricity, ε, which is determined by the ratio of the focus-to-center distance to the semimajor axis, is independent of the projection angle. The remaining elements are the inclination of the orbital plane to the plane of the sky (the projection angle), ι; the longitude of a node of the orbit (Ω), which is the position angle of one of the two points where the orbit intersects the plane of the sky ($\Omega < 180°$); the angle between that node and the periastron (ω), measured in the direction of the companion star's motion (if $\varepsilon = 0$, ω is undefined), and the period of the orbit, P_o. The methods used to recover the true shape and the elements, summarized in Table 26.1, may be either primarily geometric or primarily analytical. The end result is the determination of elements. These elements are analogous to those defined in Chapter 14, but the reference plane and directions are not the same. Given these elements, the true semimajor axis, a, is obtained from α and the distance r (see Chapter 17 for a discussion of techniques to obtain this important parameter):

$$a = \alpha \cdot r \qquad (26.1)$$

Figure 26.3 demonstrates the motions and orbits of both components of a hypothetical visual binary.

An interesting problem arises when only one component is seen but reveals orbital motion. Under some circumstances, the mass of the unseen companion can be found. The two relevent relations are

$$(m_A + m_B) = (a_A + a_B)^3 / P^2 \qquad (26.2)$$

$$m_A / m_B = a_B / a_A \qquad (26.3)$$

In these equations, A refers to the visible component and B to the invisible. Notice that Eq. 26.2 is just Kepler's third law. If the period (in years), P, mass of star A (in solar masses), m_A, and the mean distance of star A from the barycenter (in astronomical units), a_A, are known, then m_B and a_B can be found from Eqs. 26.2 and 26.3 by iteration.

TABLE 26.1. Elements of the orbit of a visual binary.

α	Angular semimajor axis
ε	Orbital eccentricity
ι	Inclination of the orbit to the sky
Ω	Longitude of the ascending node
ω	Argument of periastron
P	Period (time for complete circuit of the orbit)
T_0	Epoch when star B is at periastron, or where $\varepsilon = 0$, at a node

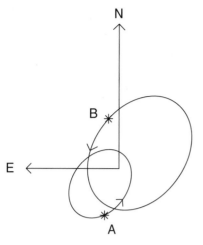

FIGURE 26.3. The absolute orbit and motions of the visual binary 99 Herculis. (After Binnendijk, 1960.)

Take, for example, the dog star—the bright shiny nose of Canis Major—α CMa or Sirius. Prior to the discovery of the "pup," companion to the dog star, by telescope maker Alvin Clark in 1862, Sirius was just such a system. Without that discovery, could we find the mass of the pup? Let's see. The distance to the system, from parallax measurements, is known to be about 2.65 pc. The orbital period is 50.1^y. The absolute semimajor axis of Sirius about the system center of mass is 2.5 ", and from a study of similar stars in binaries, $m_A \approx 2.3\ m_\odot$. From Eq. 26.1, we have

$$a_A = 2.5'' \cdot 2.65\ \text{pc} = 6.6\ \text{au}$$

We must find an initial value for m_B, which we know will be smaller than m_A. Let $m_B = 0$; then, from Eq. 26.2,

$$(a_A + a_B) \approx \sqrt[3]{[P^2 \cdot m_A]} = \sqrt[3]{[(50.1)^2 \cdot 2.3]}$$
$$= 17.9\ \text{au}$$

so that

$$a_B \approx 17.9 - 6.6 = 11.3\ \text{au}$$

Now, applying, Eq. 26.3, we obtain a first estimate for m_B:

$$m_B = m_A \cdot (a_A / a_B) \approx 2.3 \cdot (6.6 / 11.3)$$
$$= 1.3\ m_\odot$$

Tackling Eq. 26.2 once again, this time with a better value for m_B, we obtain,

$$(a_A + a_B) \approx [(50.1)^2 \cdot 3.6]^{1/3} = 20.8\ \text{au}$$

so that

$$a_B \approx 20.8 - 6.6 = 14.2\ \text{au}$$

Whence, from Eq. 26.3, a second estimate for m_B emerges:

$$m_B \approx 2.3 \cdot (6.6 / 14.2) = 1.1\ m_\odot$$

Again,

$$(a_A + a_B) \approx [(50.1)^2 \cdot 3.4]^{1/3} = 20.4\ \text{au}$$

so that

$$a_B \approx 20.4 - 6.6 = 13.8\ \text{au}$$

from which a third estimate for m_B emerges:

$$m_B \approx 2.3 \cdot (6.6 / 13.8) = 1.1\ m_\odot$$

This is the same as the second estimate (to two significant figures), and therefore the results have converged to a satisfactory solution.

The results can now be compared to the full astrometric solution:

$$m_A = 2.2 \pm 0.2\ m_\odot, \quad m_B = 0.9 \pm 0.1\ m_\odot$$

and can be seen to be in reasonable agreement.

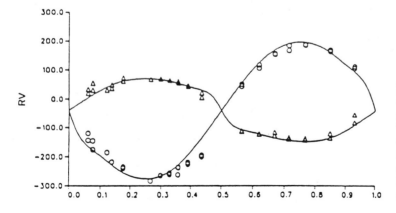

FIGURE 26.4. The radial velocity curves of the double-lined spectroscopic and eclipsing binary TY Bootis. The shape of the theoretical curve takes into account the effects of eclipses and light distribution across the neck of the common envelope of this contact binary star system.

Challenge 3 of §26.6 gives you the opportunity to attempt another such solution, this one involving another interesting and historical case.

26.3 Spectroscopic Binary Stars

The Doppler effect in stellar spectra enables us to obtain the radial velocity of the stars, as noted in Chapters 9 and 25. Evidence of radial velocity variation in a star (aside from that due to the Earth moving around the Sun) is a sign that the region of stellar atmosphere from which the lines arise is moving relative to the observer. In pulsating variable stars this motion reflects the rise and fall of the stellar atmosphere; in binaries, it indicates the motion of the component about the center of mass. If only one spectrum is seen, because of a great disparity in the luminosities of the components, a *mass function* only is obtained:

$$f(m_2) = m_2^3 \cdot \sin^3 \iota \, / \, (m_1 + m_2)^2 \quad (26.4)$$

where m_2 is the mass of the component whose spectrum is not seen. If, however, both components are visible, the system is known as a *double-lined spectroscopic binary*. The radial velocity curve of such a system is illustrated in Fig. 26.4. The ratio of the velocities with respect to the center of mass yields the mass ratio:

$$(V_1 - \gamma) \, / \, (\gamma - V_2) = M_2 \, / \, M_1 \quad (26.5)$$

where V_1, M_1 and V_2, M_2 are the radial velocities and masses of the more and less massive components, respectively, and γ is the radial velocity of the center of mass of the system.

Spectroscopic binaries (alone) may yield the following parameters:

Orbital period, P
Orbital eccentricity, ε
Periastron longitude, ω
$a \cdot \sin \iota$
$M_1 \cdot \sin^3 \iota$ and $M_2 \cdot \sin^3 \iota$
$M_1 \, / \, M_2$

Thus only lower limits to the true masses are obtainable from the radial velocities alone. Determination of the true mass of each component requires the inclination of the system, ι. If the system is a visual binary or an eclipsing binary as well, however, this parameter can be found, and with it, the mass of each component.

26.4 Eclipsing Binary Stars

Figure 23.1 displays the light curve – a plot of the brightness vs. the phase[2] – of an eclipsing binary star. The physical separation of the components of many binaries is so small compared to the distance from us that the two stars usually cannot be observed as a visual binary. Those binary stars which are oriented so that the orbital plane includes the line of sight to the Earth may be seen as eclipsing. The eclipse condition is extremely useful because it permits determination of basic properties of the system and of its components. If only the light curve is available, the following properties are obtainable:

Stellar radii, relative to the semimajor axis, r_1, r_2
Inclination of the orbit, ι
Eccentricity of the orbit, ε
Relative surface brightness of the components, F_1, F_2
The relative luminosities, ℓ_1, ℓ_2

The subscript one usually refers to the star eclipsed at primary minimum (i.e., the star with the greater surface brightness and the higher temperature). The relative luminosities are in units of power but are not bolometric quantities; they apply only over the bandpasses of the observations, which are assumed to have been properly reduced and standardized (see Chapter 23). If, in addition to the light curve, radial velocities are available and the system is seen to be a double-lined eclipsing binary, the semimajor axis, the mass ratio, and the individual masses of the components can be found. If the system is a known visual binary, the mass ratio can be found from the distances of the components from the center of mass of the system. Figure 26.5 illustrates a light curve similar to that of Algol, a bright eclipsing binary with approximately constant light (in the visual region of the spectrum anyway) outside of eclipses and a large difference between the depths of the eclipses.

The star visible at primary (deeper) minimum is star 2, and for illustrative purposes it is shown as the larger of the two components. The eclipse of

---

[2]The fractional portion of the cycle of variation. As noted in Chapter 23, $\phi = D\{[T - E_0]/P\}$, where $D\{ \, \}$ means the decimal part of, T is the instant of the observation, usually specified in Julian day numbers and decimals thereof, E_0 is the epoch, and P is the period length in days. Computer programmers can use the expression $D(x) = x - \text{INT}(x)$, where $INT(x)$ is the integer part of x.

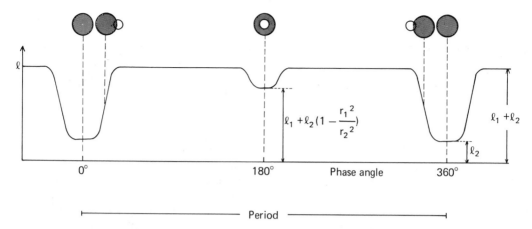

FIGURE 26.5. A representation of an Algol-like light curve, with the physical appearance of the binary demonstrated at selected phases. The greater depth at primary minimum indicates that the smaller component is the hotter star. Note that the depth, or light lost at mid-secondary minimum, depends on the sizes of the components: $\ell_2 \cdot (r_1/r_2)^2$

the smaller by the larger star is called an *occultation*; the eclipse by the smaller of the larger is called a *transit*. In Fig. 26.5, the primary eclipse is a total occultation, the secondary eclipse, an annular transit. The brightness of the system at mid-primary minimum eclipse, ℓ_{pm}, is therefore just that of star 2, while the brightness at the secondary minimum, ℓ_{sm}, involves the size ratio:

$$(\ell_{pm} = \ell_2)$$

$$\ell_{sm} = \ell_1 + \ell_2 \cdot [1 - (r_1 / r_2)^2] \qquad (26.6)$$

As we will show in the next section, flat light curve maxima and total eclipses make the determination of the relative sizes of the components much easier. In most eclipsing binaries, however, the stars are so near each other that they suffer tidal distortion in their shapes, leading to light changes as they rotate. This "oblateness effect" causes continuous variation in light. Moreover, the proximity of two stars causes mutual heating of the facing hemispheres, leading to a so-called *reflection effect*, which increases the brightness of the system before and after eclipse. This effect is particularly important at the secondary eclipse of a larger star. Finally, such close binaries suffer gravity darkening—at the surfaces of their longest diameters—and all stars show a certain amount of limb-darkening, caused by decreasing temperature with increasing radius through the stars' atmospheres. For all these reasons, light curve modeling is something of an art.

Analyses of eclipsing binary light curves are regularly carried out with a light curve modeling computer program. Beginning ca. 1912, H. N. Russell and his collaborators developed a technique for obtaining basic parameters of an eclipsing binary star system through analysis of the shapes

TABLE 26.2. Bright eclipsing binaries.

| | | Position, 2000 | | E_o | P | V magnitude | |
	Name	α	δ	2,400.000+	(days)	max,	min
ζ	Phe	01^{h}08.4^m	−55°15′	41,643.689	1.669	3.92,	4.42
β	Per	03 08.2	+40 57	40,953.465	2.867	2.12,	3.40
λ	Tau	04 00.7	+12 29	35,089.204	3.952	3.30,	3.80
ε	Aur	05 02.0	+43 49	35,629	9,892	2.92,	3.83
V	Pup	07 58.2	−49 15	28,648.304	1.454	4.7,	5.2
δ	Lib	15 01.0	−08 31	42,937.423	2.327	4.92,	5.90
u	Her	17 17.3	+33 06	44,069.386	2.051	4.6,	5.3
β	Lyr	18 50.1	+33 22	45,342.39	12.935	3.34,	4.34
VV	Cep	21 56.7	+63 38	43,360	7,430	4.80,	5.36

FIGURE 26.6. The schematic for a simple photodiode photometer to register the simulated light curve of a model eclipsing binary and (see Chapter 28) to measure solar limb darkening. The detector is Siemens photodiode model BPW 21; for slightly higher sensitivity, BPW 33 may be somewhat better).

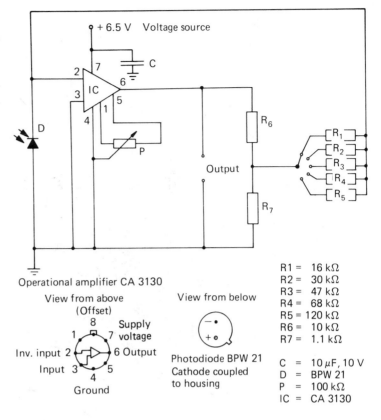

Operational amplifier CA 3130

View from above
(Offset)

8	Supply
1	7 voltage
Inv. input 2	6 Output
Input 3	5
4	
Ground	

View from below

Photodiode BPW 21
Cathode coupled
to housing

R1 = 16 kΩ
R2 = 30 kΩ
R3 = 47 kΩ
R4 = 68 kΩ
R5 = 120 kΩ
R6 = 10 kΩ
R7 = 1.1 kΩ

C = 10 μF, 10 V
D = BPW 21
P = 100 kΩ
IC = CA 3130

and depths of the eclipses. The most complete discussion of the method, together with a full set of tables and description of the nomographs, is found in Russell and Merrill (1952). The technique requires first the "rectification" of eclipsing binary light curves to rid them of such noneclipse effects as "oblateness" and "reflection." The rectification process transforms the light curve from one caused by stars assumed to be shaped like triaxial ellipsoids to a light curve with constant light maxima produced by spherical stars. The geometry of eclipses of spherical stars is captured by nomographs and tables for a range of limb-darkening coefficients of the components. The elements which the method yields are: $k = r_s/r_g$, r_g/a, ι, and L_1/L_2. Convenient for hand-calculation, the preliminary results which the technique yields can be refined by applying least squares to the observed and calculated light curves. However, the triaxial approximation for close binary star components is inferior to the Roche geometry assumed by modern analysis techniques. Prior to the 1970s the Russell–Merrill technique was the standard analytical method, and it is still used occasionally.

Subsequently, other light curve programs, capable of synthesizing the light curves of systems as complicated as those of W UMa (a contact system) or β Lyrae (a semi-detached system with a thick disk around an unseen component), have been created. One of the best known is that of Wilson and Devinney (1971).

26.5 A Simulated Eclipsing Binary

Here we demonstrate how to simulate an eclipsing binary. First we need a detector. In Chapter 23 we discussed the use of photomultiplier and charge-coupled diode detectors for astronomical observations. If you have access to such equipment, the challenge of real data is always greater and more exciting than simulated data, which, however, do have heuristic value. In Table 26.2 we list some bright systems that can be tackled with a small telescope equipped with a photometer. If you are unable to borrow or buy a photometer, there is always the challenge of building one. Figure 26.6 is a schematic of photodiode circuitry to support a

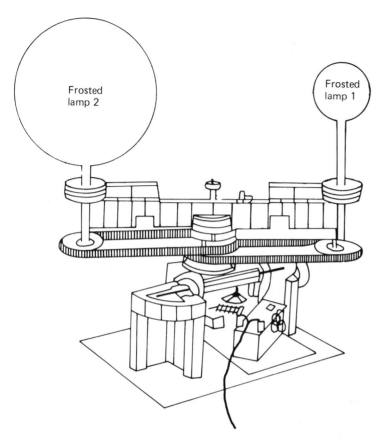

FIGURE 26.7. Artificial eclipsing binary star system made from components of commercially available model kits.

low-voltage, low–dark current photodiode manufactured by Siemens. More recent devices are described from time to time in the *International Amateur–Professional Photoelectric Photometry Association Communications* (see especially No. 14, December 1983), in the publications of the Astronomical Society of the Pacific, and in *Sky & Telescope*. The photocurrent of the device lies in the range of nanoamps to microamps. Digital ammeters to measure such currents are commercially available. The detection limit for such a device permits only relatively bright objects—the Sun, Moon, bright stars and planets, and light bulbs, for instance—to be observed; but for the present case (and for solar limb darkening in Chapter 28), this is perfectly adequate. Some modest shielding can be placed around the detector, and filters can be taped in place to cut down the light and to change the relative depths of the observed eclipses.

If you are successful in obtaining a detector, you might like to try to construct the simulated eclipsing binary seen in Fig. 26.7. The stars are represented by frosted lightbulbs of different voltages and thus temperatures and surface brightnesses. A similar device has been on display at the mountaintop museum of the Kitt Peak National Observatory near Tucson, Arizona. Only one lamp is required to revolve, the other may rotate or remain motionless.

The two frosted spherical light bulbs of the device illustrated in Fig. 26.7 are driven by a motor to revolve in some period *P*. The motor is not essential, of course; one can measure the brightness of a particular configuration of the system as seen at the detector and move the system by hand from one position to another. The configuration must be reproducible, however. If a rotation of the bulbs can be effected also, by some kind of gear-chain linkage, variations on the bulb surfaces will provide a touch of realism to the simulation since real stars do have variations in brightness over their projected disks, as we will see in Chapter 28. The distance between the detector and the light bulbs should be at least 15 times the separation of the bulbs, so that geometrically induced departures from the inverse square distance law will be tolerable.

Figure 26.8 is an observed light curve of the simulated eclipsing binary. Note the light variation

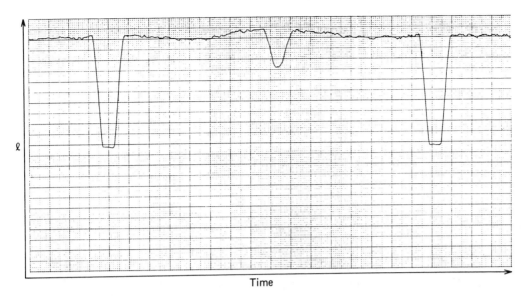

Time

FIGURE 26.8. A strip-chart output of the photodiode detecting the light of the simulated eclipsing binary of Fig. 26.7. Note the departure from flatness around the minima. Compare this light curve with the ideal representation of Fig. 26.5.

around the secondary minimum compared to the primary minimum.

The radii of the stars can be determined by the durations of the partial and total eclipse phases. Figure 26.9 illustrates the region around primary minimum. The phases of contact are shown. Note that the ratios involving the durations (in either time or phase measure) yield the radii:

$$(r_2 - r_1) / (r_2 + r_1) = \Theta_i / \Theta_e \quad \text{or} \quad r_1 / r_2 = (\Theta_e - \Theta_i) / (\Theta_e + \Theta_i) \quad (26.7)$$

where Θ_i and Θ_e are the phases of third and fourth contacts, respectively. The radii themselves may be obtained if the relative velocity, v, is known:

$$2r_1 = v \cdot (t_2 - t_1) = v \cdot (t_4 - t_3)$$
$$2r_2 = v \cdot (t_3 - t_1) = v \cdot (t_4 - t_2) \quad (26.8)$$

These formulas assume that the orbital plane includes the line of sight. If this is not the case, the factor sin ι must be included on the left-hand sides of Eqs. 26.8 (the factor is already incorporated in the velocities).

26.6 Further Challenges

1. Figure 26.8 demonstrated continuous variation in the secondary minimum and departures from flat maxima; what can you conclude about the bulbs used in the experiment?

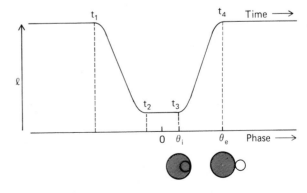

FIGURE 26.9. Enlargement of the primary eclipse region of the light curve representation of Fig. 26.5. Inner (third) and outer (fourth) contacts are identified.

2. Equations 26.4 to 26.6 are presented without proof. Try to derive these expressions.
3. In §26.2, we described the astrometric binaries and mentioned Sirius as an early example. Another interesting challenge is that of Barnard's Star, the star with the largest known proper motion (motion across the sky in arc sec per year), at ~ 10 arc·s/y. Although Gatewood has cast some doubt on the reality of the periodic variation detected by van de Kamp in the motion of Barnard's Star, the calculation of the planetlike mass in that system is an interesting challenge. Assume that the orbital period is 25^y and that the semimajor axis of the visible component, star A, is $a_A = 0.028''$. The distance of the system is known from parallax studies to be 1.83 pc. The spectral type of star A is M5; the mass of such a red dwarf star, from a study of other binaries, is $\sim 0.15\,M_\odot$. By iteration, find the mass, m_B, and orbital radius, a_B, of the possible companion.
4. What happens when the orbits are eccentric? Try to illustrate the visual binary orbits, the radial velocity curves of a double-lined spectroscopic binary, and the eclipsing binary light curves of systems with eccentric orbits. Consider two cases: periastron along the line of sight and periastron in the plane of the sky. For simplicity, let the inclination for the visual binary be 0° and for the other two types, 90°.
5. Table 26.2 lists bright eclipsing binary systems that may be within reach of a photometer on a small telescope (two are suitable only for the Southern Hemisphere). Try your hand at differential photometry using a bright nearby standard. The *General Catalogue of Variable Stars* (Kholopov, 1985–1987) and the *Catalogue of Suspected Variable Stars* (Kholopov, et al.

1982) are the most complete references to be consulted to avoid selecting another variable star as the comparison star. The guidelines of photometry discussed in Chapter 23 should be reviewed before venturing on this challenge. You can keep track of the phase of observations by converting the date and time of each observation to Julian day numbers and decimals thereof, and using Eq. 23.3 to compute the phases.

References and Bibliography

Aitken, R. G. (1964) Binary Stars. Dover, New York. Original ed., 1935: McGraw-Hill, New York.
Allen, C. W. (1973) Astrophysical Quantities, 3rd ed. Athlone Press, London.
Binnendijk, L. (1960) Properties of Double Stars. University of Pennsylvania Press, Philadelphia.
Couteau, P. (1978) *L'Observation des étoiles doubles visuelles*. Flammarion, Paris, (1981); English, revised version (translated by A. H. Batten) *Observing Visual Double Stars*, MIT Press, Cambridge, MA.
Heintz, W. D. (1978) Double Stars. Reidel, Dordrecht.
Kholopov, P. N. et al. (ed.) (1982) Catalogue of Suspected Variable Stars. NAUKA, Moscow.
Kholopov, P. N. (ed.) (1985–1987) General Catalogue of Variable Stars, 4th ed., 3 vols. NAUKA, Moscow.
Makemson, M. K. (1948) *Astronomical Journal* **63**, 41.
Wilson, R. E. (1979) Astrophysical Journal **234**, 1054–1066.
Wilson, R. E., and Devinney, E. J. (1971) Astrophysical Journal **182**, 539–547.
Yan, L., Chu, Z., and Pan, D. (1985) *General Catalogue of Ephemerides and Apparent Orbits of 736 Visual Binary Stars*, Shanghai Science and Technology Press, Shanghai, p. 225.

27

Fundamental Properties of Stars

27.1 Introduction

How do stars differ from one another? Can we make use of their similarities to learn something about the nature of the universe in which we live? If you have attempted the challenges presented in some of the earlier chapters, you may already know the answers to these questions. In Chapter 15, for example, you could find the distance to a star by assuming that it had the same intrinsic brightness as the sun. In Chapters 20 and 24, you could see the spectral criteria by which stars are classified according to temperature and luminosity. In Chapters 19 and 23, you could explore the effects of the atmosphere on the photometry of astronomical sources and the methods for reducing the data to extra-atmosphere magnitudes and colors and for transforming the data to a standard system. Finally, in Chapter 26, you could get a hint of the techniques for the determination of masses and radii in binary stars. In this chapter, techniques to obtain the fundamental parameters of stars are described and we examine the relations among the stellar properties that give us parameters which cannot be measured directly.

It is standard to assume apparent magnitudes and color indices have already been treated for effects of the earth's atmospheric extinction and have been transformed to a standard system like that of Johnson. The choice of the UBVRI photometric system is not universal, but it is widespread. The idea is to examine the connection between the intrinsic and the observed properties of a star and to establish relations among the fundamental parameters and the readily observed ones like apparent brightness and color. The final stage, relating the intrinsic properties to the evolutionary state and chemical composition of the stars, is discussed in Chapter 29.

27.2 Stellar Luminosities

The apparent brightness of a star is due both to its intrinsic brightness and to its distance from the earth, r. If we designate the total power radiated by the star as its *luminosity*, $\mathcal{L}$, and the observed flux outside the terrestrial atmosphere as F, then, as discussed previously (Eq. 15.1),

$$F = L / (4\pi \cdot r^2)$$

The magnitude corresponding to the observed flux is the *apparent magnitude*,

$$m = -2.5 \cdot \log (F / F_0) \qquad (27.1)$$

where F_0 is the flux of a zero-magnitude star. As defined here, F is the *total flux*, across all wavelengths, and the magnitude corresponding to it is known as the *bolometric* magnitude, m_{bol}.

Although there are infrared detectors known as "bolometers," we do not actually measure the flux at *all* wavelengths (at least not yet). What we can measure with precision is the flux in well-defined passbands of the spectrum. Magnitudes corresponding to fluxes observed through a limited passband can be expressed in a way similar to Eq. 27.1:

$$m_\lambda = -2.5 \cdot \log (F_\lambda / F_{\lambda,0}) \qquad (27.2)$$

Table 27.1 lists the zero-magnitude fluxes for the Johnson system; note that these zero points permit the outside atmosphere flux to be computed for any magnitude that has been properly transformed to the standard system (see Chapter 23). The zero-magnitude fluxes are given per unit frequency interval, F_ν, as well as per wavelength interval, F_λ. The relation between these quantities is

$$\nu \cdot F_\nu = \lambda \cdot F_\lambda \qquad (27.3)$$

TABLE 27.1. Zero-magnitude fluxes for the Johnson system.

Filter	λ_{eff} (µm)	$F_{V,0}$ [W/(m²·Hz)]
U	0.36	$1.88 \cdot 10^{-23}$
B	0.44	$4.65 \cdot 10^{-23}$
V	0.55	$3.95 \cdot 10^{-23}$
R	0.70	$2.87 \cdot 10^{-23}$
I	0.90	$2.24 \cdot 10^{-23}$
J	1.25	$1.77 \cdot 10^{-23}$
H	1.62	$1.09 \cdot 10^{-23}$
K	2.2	$6.3 \ \cdot 10^{-24}$
L	3.4	$3.1 \ \cdot 10^{-24}$

Now the bandwidths of the Johnson passbands are not 1 Hz or 1 µm wide, but the flux in a particular passband is nevertheless computed per unit interval centered on the passband. It is in this sense that we are speaking of monochromatic fluxes and the corresponding magnitudes. The actual passbands are of order 100 nm (1000 Å, or 0.1 µm) wide, so that compared to bolometric quantities they do approximate the "monochromatic" state.

The connection between the total flux and what is observed is the *bolometric correction*, (BC), which is defined as the difference between the visual and bolometric magnitude:

$$\text{BC} \pm V - m_{bol} \qquad (27.4)$$

The bolometric correction defined in this way should be positive (sometimes it is defined as $m_{bol} - V$, in which case it is always negative). It is close to zero for stars like the Sun but increases to both earlier and later spectral types.[1] There is never any ambiguity because m_{bol} must be *smaller* than V. Even when we find the total flux, however, we still do not know the luminosity. To get this, we must find the distance or use some kind of luminosity indicator. First we examine the effect of distance on magnitude.

The difference in the magnitude of two stars of the same luminosity but at different distances is given by

$$
\begin{aligned}
m_1 - m_2 &= -2.5 \cdot \log [(\mathcal{L}_1 / r_1^2) / (\mathcal{L}_1 / r_2^2)] \\
&= -2.5 \cdot 2 \cdot \log (r_2/r_1) \\
&= + 5 \cdot \log (r_1/r_2) \qquad (27.5)
\end{aligned}
$$

where r_1 and r_2 are distances of the two stars. If r_2 is set equal to 10 pc, m_2 is called the *absolute magnitude*, designated M. Notice that the luminosity

[1]BC $\approx$ 2.3 for B0V-type stars and $\approx$ 3.9 for M8V-type stars.

has dropped out of Eq. 27.5 and so we can consider the differences between magnitudes in any passband instead of having to use the bolometric magnitudes. If $m = V$ of the UBV system, $M = M_V$, the V absolute magnitude. We can drop the subscript 1 so that Eq. 27.5 becomes

$$
\begin{aligned}
V - M_V &= 5 \cdot \log (r / 10) \\
&= 5 \cdot \log r - 5 \qquad (27.6)
\end{aligned}
$$

This is known as the *distance modulus*, because it relates the distance to the brightness properties of the star. In principle,

$$m_{bol} - M_{bol} = V - M_V = B - M_B = m_{pg} - M_{pg}, \text{ etc.,}$$

in the bolometric, visual, blue, photographic passbands, respectively. In practice, there is a wavelength dependence for the distance modulus because of the effects of interstellar dust between the star and the observer. Equation 27.6 is strictly correct only if the interstellar extinction[2] is negligible. The evidence is that in the visible to near infrared region of the spectrum interstellar extinction obeys the Whitford law, $A_\lambda \propto 1 / \lambda$. A_V is usually obtained from the color excess, defined as

$$E_{B-V} \pm (B - V) - (B - V)_0 \qquad (27.7)$$

where $(B - V)_0$ is the intrinsic color index, and the selectivity is

$$R = A_V / E_{B-V} \qquad (27.8)$$

There is some evidence that R varies from place to place, but an average value around 3.3 is usually used; it can be derived from studies of the reddening of stars of early spectral types in galactic clusters or by comparing the magnitudes of reddened and unreddened stars whose spectra are otherwise identical. In the general interstellar medium, $A_V \approx k \cdot r$, where k is the average extinction per parsec of distance, r; and $k \approx 1^m / 1000$ pc. The extinction is not uniform with distance from the Sun, however, and may vary with galactic longitude as well. For stars further than about 100 pc, Eq. 27.6 should be corrected by subtracting A_V from the left-hand side. When this is done, the expression is called the *corrected distance modulus*:

[2]The interstellar extinction (sometimes called absorption, and written as A_λ, even though it is principally due to scattering by interstellar dust) is related to the change in color because of the reddening effect of the interstellar medium.

$$V - M_V - A_V = 5 \cdot \log (r / 10) \text{ or } V_0 - M_V$$
$$= 5 \cdot \log (r / 10) \qquad (27.9)$$

where $V_0 \pm V - A_V$.

The corrected distance modulus then yields the absolute visual magnitude. For many purposes, this quantity is sufficient. If we need to obtain the true luminosity, however, the absolute bolometric magnitude, M_{bol}, may be obtained if we know the bolometric correction. We can write, analogous to Eq. 27.4, that

$$M_{bol} = M_V - BC \qquad (27.10)$$

independent of distance. The relation between M_{bol} and $\mathcal{L}$ may be expressed in terms of solar values:

$$M_{bol} = M_\odot - 2.5 \cdot \log (\mathcal{L} / \mathcal{L}_\odot) \qquad (27.11)$$

If one can obtain the distance, say, by trigonometric parallax measurements, then one can find the absolute magnitude of a star, and thus its intrinsic brightness.

The determination of luminosities thus requires four pieces of information: observational magnitudes and color indices, distances, the appropriate corrections for interstellar absorption and reddening, and the bolometric correction. Classes of objects whose luminosities have been carefully determined are known as *standard candles*. Examples of such standard candles are RR Lyrae stars or Cepheid variables[3] of a particular period. The value of such standard candles is seen when they are found in an ensemble of stars like a star cluster or galaxy. The luminosities of other objects in these ensembles can then be obtained and secondary standards can be established. The difficult question of the effect of chemical composition on the criteria makes the process more complicated than astronomers would like. We discuss distances to star clusters again in Chapter 29.

27.3 Size and Temperatures

The sizes of individual stars can be measured through a variety of techniques. The direct method involves phase or intensity interferometry (including speckle interferometry, discussed in Chapter

---

[3]Studies have shown that there are at least two types of Cepheids—members of populations I and II, respectively—and that the period–luminosity relations are different for these two types. The difference led to a revision of the distance scale in the early 1950s.

22). Other techniques involve lunar occultations—where the Moon's edge slices across the stellar disk giving a diffraction pattern modified by the finite disk of the star. The determination of the radii of the component stars from the analysis of eclipsing binary light curves was mentioned in Chapter 26. For certain pulsating stars, the average radius may be found from a consideration of the phases at which the color index is the same but the overall brightness is not. The technique, known as the *Baade-Wesselink method*, requires both radial velocity curves, in order to find the distance traveled by the photosphere, and the light curves. The method can be summarized in the expression

$$\ell_V \propto \sigma_V \cdot R^2 \qquad (27.12)$$

where ℓ_V is the monochromatic luminosity in the visual region of the spectrum, σ_V is the surface brightness or radiant flux in the same region, and R is the radius. In terms of magnitudes,

$$m_V = -2.5 \cdot \log_{10} \ell_V + \text{const}$$

so that

$$m_V = -2.5 \cdot \log_{10}(\sigma_V \cdot R^2) + \text{const}$$
$$= -2.5 \cdot \log_{10} e \cdot \log_e(\sigma_V \cdot R^2) + \text{const}$$

where $\log_{10}$ (elsewhere written just log) and $\log_e$ are the base 10 logarithm and the natural, or base e, logarithm function, respectively. Since $\log_{10} e = \log (2.7183) = 0.4343$,

$$m_V = -1.086 \cdot \log_e(\sigma_V) - 1.086 \cdot \log_e(R^2)$$
$$+ \text{const}$$

This leads to the expression for the difference in V between any two phases:

$$\Delta m_V = -1.086 \cdot \Delta \ell / \ell = -1.086 \cdot \Delta \sigma_V / \sigma_V$$
$$+ 2.172 \cdot \Delta R / R \qquad (27.13)$$

At the phases where the color index is the same, the assumption is that $\Delta \sigma_V = 0$ and so, where $\Delta R / R$ is small,

$$\Delta R / R = 0.461 \cdot \Delta m_V \qquad (27.14)$$

From many such phase pairs, Wesselink obtained $<R> = 48 R_\odot$ for the population I-type Cepheid prototype star, δ Cephei, with period about 5.366^d.

The connection between the radiant flux, the size, and the temperature suggests that if both photometry and spectroscopy were carried out, and the distance was otherwise known, the radius could be derived. This is indeed so, because of the temperature scales that have been established

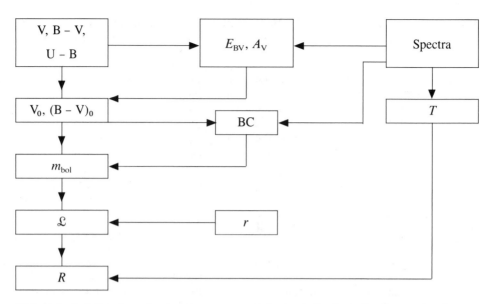

FIGURE 27.1. A logical flowchart showing the steps to obtain the radius from the distance by photometry and spectroscopy.

between temperature and spectral class for a particular luminosity class (e.g., giants or main sequence stars). The logic flowchart is illustrated in Fig. 27.1 and is considered later. The method is not a highly precise one because the distance and the corrections are rarely known with high precision.

The B − V color index and the spectral type are both related to the surface temperature of the star; thus if you plot the absolute magnitudes of stars against their color indices, you can expect to see some kind of ordered progression, with stars of higher temperature having greater luminosities, since by the Stefan-Boltzmann relation (Table 20.1) the total flux increases strongly with the temperature. It is also true that by the Planck relation the monochromatic flux increases with temperature (remember that the Planck radiation curves are nested).

Figure 27.2 illustrates the relation between brightness and temperature for stars of known intrinsic brightness. A plot of M_V vs. spectral type is known as the *Hertzsprung-Russell diagram*, after Eijnar Hertzsprung (1873–1972) and Henry Norris Russell (1879–1957), who pioneered the study of fundamental properties of stars through observation. A plot of magnitude vs. color index is referred to as a *color-magnitude array*. Figure 27.2 incorporates the features of both types of plots, with logarithmic scales for the monochromatic luminosity and the surface temperature as well. The correspondences among the astronomical (color indices, spectral types, M_V) and physical units ($\mathcal{L}$, T) vary with luminosity class. The relations among the variables are illustrated by the *main sequence* of stars of luminosity class V. The dashed curve represents the theoretical relation between the luminosity and surface temperature as predicted by the Planck function for $T > 6000$ K. The fact that the main sequence rises above it indicates that the stellar radius must increase for main sequence stars of early spectral types.

If all main sequence stars had the same surface areas and therefore the same radius, the Planck function (Table 20.1) would describe the monochromatic surface flux (or radiated power per unit surface are), at any particular wavelength, for a spate of temperatures:

$$\ell_\lambda = [\text{const}/\lambda^5]/[e^{hc/k\lambda T} - 1] \quad (27.15)$$

This is not the case. The dashed curve in Fig. 27.2 represents the hypothetical case for stars hotter than the Sun, and it clearly lies below the main sequence. Although one could argue that the cause of the departure could be a systematic departure from black body radiation at the higher temperatures, an alternative, simpler, and it turns out more correct hypothesis[4] is that the radius increases with

---

[4]See §20.6 for a discussion of the degree to which stars depart from black body radiation. The question is not *if* they do, but *by how much*.

FIGURE 27.2. The Hertzsprung–Russell diagram combined with a color-magnitude array, showing the relationship between stellar surface temperature and radiated power, as well as the surface temperature and luminosity. The shaded variables are presented for completeness only and are not required for this exercise.

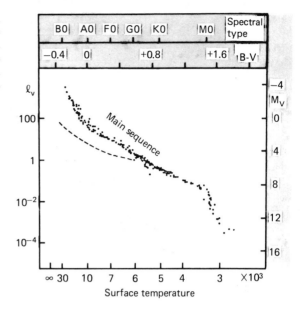

temperature. As a first approximation, Eq. 27.15 must be modified to the form of a monochromatic luminosity

$$\mathcal{L}_\lambda = [\text{const} \cdot R^2 \,/\, \lambda^5] \,/\, [e^{hc/k\lambda T} - 1] \quad (27.16)$$

Three stars selected from Fig. 27.2 will reveal the extent of size differences among the stars of the main sequence:

1. A late B star with $L_V = 100\,L_\odot$, $T = 12{,}000$ K.
2. A G-type star with $L_V = 1\,L_\odot$, $T = 6000$ K (the solar case)
3. An M star star with $L_V = 0.01\,L_\odot$, $T = 3200$ K.

At $\lambda = 550$ nm, the monochromatic radiated power per unit surface area, by Eq. 27.15, is for stars 1, 2, and 3, in proportion,

$$9.85 : 1 : 0.022$$

Then, since the monochromatic luminosities are in proportion,

$$100 : 1 : 0.01$$

the surface areas, by Eq. 27.16, must be in proportion:

$$100/9.85 : 1 : 0.01/0.022$$

from which the B-star radius is found to be 3.19 $R_\odot$ and the M star radius is 0.67 $R_\odot$. So we learn

the interesting fact that across a factor of 10^4 in monochromatic luminosity, the radius changes by about a factor of 5.

27.4 Masses and Luminosities

The monochromatic luminosity M_V of a star can be found when we have (1) the observational V magnitude and the B - V color index; (2) the distance; and (3) the interstellar extinction correction, A_V. The luminosity is obtained from the monochromatic luminosity and the bolometric correction. In general neither the bolometric correction nor A_V is known to high precision. Therefore, the most accurate and precise luminosities we possess are for nearby solarlike stars.

The masses of stars are determined fundamentally from binary star analyses (see Chapter 26). A carefully compiled selection of the most reliable of these determinations has been provided by Popper (1980). The masses of visual binaries with well-determined orbits and spectroscopic binaries which are also eclipsing binaries were included in his study. Using main sequence data from this selection, Griffith, Hicks, and Milone (1988) recently reexamined the relationship between mass and luminosity and provide the following convenient relation between these two basic parameters:

$$\log \mathcal{L} = 4.20 \cdot \sin (\log M - 0.281) + 1.174 \quad (27.17)$$

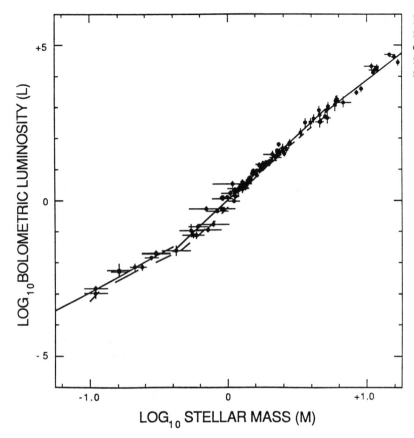

FIGURE 27.3. The mass–luminosity relation for linear fittings to the data. (Reproduced with permission from the Journal of the Royal Astronomical Society of Canada.)

This relation between the luminosity and mass (both in units of the Sun) is not as precise a predictor as linear fittings, but it covers the entire range of the useful data: $0.1 \leq M \leq 17$. The best linear fittings are given by the following relations in three regimes of $\log M$:

$$\log \mathcal{L} = -0.52 + 2.44 \cdot \log M, \quad \log M < -0.4$$

$$\log \mathcal{L} = 0.006 + 4.16 \cdot \log M, \quad -0.4 < \log M < 0.7$$

$$\log \mathcal{L} = 0.37 + 3.51 \cdot \log M, \quad \log M > 0.7 \tag{27.18}$$

Figures 27.3 and 27.4 illustrate the relation for the linear and the sinusoidal fittings, respectively. Figure 27.3 also shows the results of theoretical modeling by VandenBerg et al. (1983) and VandenBerg and Bridges (1984). We note without explanation that the theoretical lines fit the general shape but run consistently below the data and the empirical curves that represent them.

27.5 Further Challenges

1. *Why* must a star's bolometric magnitude always be smaller than its visual magnitude? Now for a harder question: *How* can the bolometric magnitude be obtained for a star? In the past this has been done from theoretical models and with ground-based infrared measurements. For the Sun, an additional source of information has been used. What is it, and why are ground-based observations limited in providing accurate and precise measurements of the bolometric corrections, especially for very hot and very cool stars?

2. With a solar projection screen, like that described in Chapter 28 and shown in Fig. 28.5, millimeter scale graph paper, and a stopwatch, you can time the passage of the Sun across the surface of the graph paper and arrive at a measure of the solar diameter. Orient the graph paper east–west, then start the timing when the western limb of the Sun reaches a marked line

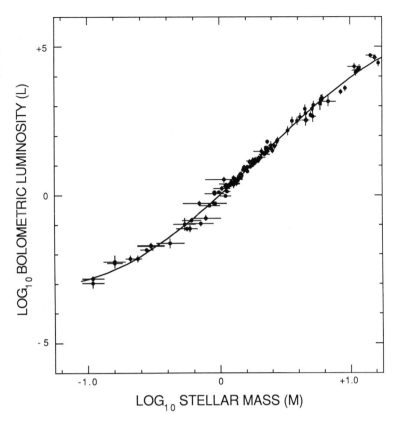

FIGURE 27.4. The mass–luminosity relation for a sinusoidal fitting to the data. (Reproduced with permission from the Journal of the Royal Astronomical Society of Canada.)

and stop when the eastern limb reaches the same line. The time interval, Δt, can be measured many times in this way, and a statistical sample, which follows the distribution seen in Fig. 27.5, will be obtained. The distribution is Gaussian[5] and represents the results of small, unavoidable, random errors intruding on the measurements.

The data were obtained on August 26, 1980, at about 12:30 P.M. LCT at a site in western Europe. The relationship between the measurement and the angular size, α, is

$$\alpha = 15'' \cdot \cos \delta \cdot \Delta t^{s} \qquad (27.19)$$

On this date the declination of the Sun was $\delta = 10°17'9''$. The values of the mean, the standard deviation, and the standard deviation of the mean, are given in terms of both Δt and α in the following table:

---

[5]See Appendix B for the background on, and mathematical representation of, this type of distribution.

Quantity	Mean $\pm$ σ_M	σ
Δt	$128.897^{s} \pm 0.014^{s}$	0.150^{s}
α	$31'42.39'' \pm 0.213''$	$2.215''$

The almanac value for α at this time, $31'42.66''$, indicates that this method is capable of yielding both accurate and precise results. Note the importance of the large number of trials. The uncertainty in a single observation here was $\sim 2.2''$. Try a run of measurements yourself. But first a few words of warning. Not every set of observations will go as well as this one. If you have a projection eyepiece with color aberration, or the telescope is subject to wind, or you have bad seeing, the resulting fuzziness in the image will degrade the precision of your measurements. Be careful of systematic errors too: try to observe each time in the same way, and try to make all measurements at low air mass (i.e., when the Sun is high in the sky so the effects of seeing and differential refraction are minimized). If you

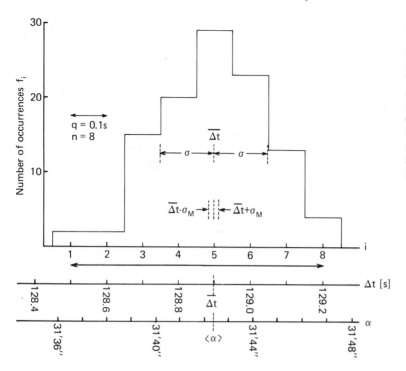

FIGURE 27.5. The distribution of 108 timing measurements of the passage of the solar diameter across a solar projection screen. The corresponding angular diameter of the Sun is also shown. The timing resolution is 0.1 s, so this is the bin size and there are eight of them. The mean, Δt, standard deviation, σ, and the standard deviation of the mean, σ_M, are all shown.

have done all this, you are now in a position to determine the radius of the Sun. By Fig. 12.1,

$$R = r \cdot \sin(\alpha/2)$$

In this example, $r = 1.01033$ au, so

$$<R> = 1.01033 \cdot 0.0046122 = 0.0046598 \text{ au}$$

Taking the value for the astronomical unit as $1.4959787 \cdot 10^{11}$ m, we have

$$<R> = 6.971 \cdot 10^8 \text{ m}$$

The uncertainty in $<R>$ is

$$e_R = r \cdot \cos(\alpha/2) \cdot \Delta(\alpha/2) = 1.01033 \cdot 0.107 / 206265$$

$$\approx 0.522 \cdot 10^{-6} \text{ au} = 7.80 \cdot 10^4 \text{ m}$$

Therefore, our determination of the solar radius is

$$<R> = (6.971 \pm 0.001) \cdot 10^8 \text{ m}$$

3. Compare the predictions of the mass–luminosity relation formulas of §27.4 for the three sample types of stars of §27.3.

References and Bibliography

Griffiths, S. C., Hicks, R. B., and Milone, E. F. (1988) *Journal of the Royal Astronomical Society of Canada* **82**, 1–12.

Johnson, H. J. (1966) *Annual Review of Astronomy and Astrophysics* **4**, 193–206.

Popper, D. M. (1980) *Annual Review of Astronomy and Astrophysics* **18**, 115–164.

VandenBerg, D. A., Hartwick, F. D. A., Dawson, P., and Alexander, D. R. (1983) *Astrophysical Journal* **266**, 747–754.

VandenBerg, D. A., and Bridges, T. J. (1984) *Astrophysical Journal* **278**, 679–688.

28

The Face of a Star: The Sun

28.1 Introduction

A striking aspect of photographs of the Sun is the darkening of the solar image toward the limb (see Fig. 28.1). The decrease in brightness, which is accompanied by a reddening of the solar image in color photographs, is not uniform. Near the center of the disk the change in brightness is small but within a few arc seconds of the limb the brightness falls off so steeply that the edge of the solar disk appears sharp. This is the *limb darkening* of the Sun, sometimes called the *center-to-limb variation*. An incandescent solid, on the other hand, will have a constant surface brightness. A glowing tungsten or platinum wire, for example, shows no limb darkening. Limb darkening is the most visible sign of the gaseous nature of the outer layers of the Sun.

The solar atmosphere consists of three regions: the photosphere, the chromosphere, and the corona. The *photosphere*—literally the sphere of light—contributes most of the light that we see in the visible disk of the sun. The temperature of the photosphere amounts to ~6100 K at disk center but only 5770 K averaged over the disk (due to limb darkening). In sunspots, like those seen in Fig. 28.1, the temperature lies about 1000 K lower. Like the solar limb, the sunspots would appear bright if seen projected against the sky; they are dark only in contrast to the brighter, unspotted, disk center.

The layer above the photosphere is the *chromosphere*, the sphere of color, a name derived from its appearance during total solar eclipses. Its narrow, irregular red rim around the darkened disk of the Sun has been compared to a prairie fire. The red color is due to the emission of hydrogen α (the Balmer α line, marking the energy transition between the third and second energy levels of the hydrogen atom), although the other Balmer lines are visible spectroscopically. Except in these lines, in the visible region of the spectrum, the optical depth in this region is low. The thickness of the chromosphere is very thin—only about 7000 km compared to the solar radius of 700,000 km—but throughout this relatively thin region, the temperature rises rapidly from a temperature minimum (~4200 K) until it reaches the temperature of the solar corona. For this reason, the upper chromosphere is sometimes referred to as a transition region.

The *corona* is the overlying region of high temperatures (~$2 \cdot 10^6$ K) and very low densities (particle densities are only ~10^{12} m^{-3} compared to photospheric values ~10^{23} m^{-3}). The high temperature explains why the corona extends to several solar radii, and it explains as well the appearance in the spectrum of the extremely high ionization stages of atoms [e.g., Fe XIV — an iron atom ionized 13 times]. The photospheric Fraunhofer lines reflected from the dust component of the corona (the F corona) also show the effects of those high temperatures in the great broadening of the lines by the rapid Doppler shifts of the atoms and electrons. Yet despite the very high temperatures and huge physical extent, the optical depth (outside of the emission lines, in the visible part of the spectrum) is very low. The pearly light of the corona is visible only during an eclipse (although this may be simulated by an occulting disk at a high-altitude site or from space), and its total light amounts to only as much as that of the full Moon. The light of the corona has a peculiarly structured appearance—coronal plumes—controlled by magnetic fields. Since the magnetic field structures undergo changes through the solar cycle, the shape of the corona varies from maximum to minimum along with the level of solar activity. Condensations of higher density in the corona,

FIGURE 28.1. A partial solar eclipse. The sharp bright edge of the lunar disk contrasts with the limb-darkened solar disk. The exposure was made using a telescope with focal length of 1.5 m and an aperture ratio 1:15; with an exposure of 1/60 s through a neutral density filter (attenuation ~ 10,000 ×) on film with speed ASA 50.

called *prominences*, are also bound by magnetic fields because of the ionized state of the matter and, like iron filings near a magnet, reveal the magnetic field structure.

The two selective aspects of the sun which we examine in this chapter, limb darkening and prominences, are characteristic features of the *quiet Sun* and the *active Sun*, respectively.

28.2 The Photosphere Explored Through Limb Darkening

The passage of radiation through layers of gaseous atmosphere is connected to the absorption and emission coefficients of the gas. The absorption coefficient, $k_\lambda(r)$, is a function of the wavelength (indicated by the subscript) and of the depth to which we see into the atmosphere, r. Generally, it increases inward from the solar radius toward the center and essentially describes the power of a cross section of matter to absorb energy. The total amount of energy removed from a beam depends on the physical extent of the gas, hence the use of the notion of *optical depth*, τ_λ (introduced in challenge 2 of §19.5, on the astronomy of fogs and mists). In the case of the penetration of light through a terrestrial fog, τ_λ can be written

$$\tau_\lambda = \int_{r=0}^{r=R} k_\lambda(r) \cdot dr \qquad (28.1)$$

The use of integration should be familiar at this point. Equation 28.1 can be thought of as merely the addition of the areas of many narrow rectangles of width dr and height $k_\lambda(r)$ over the range of r from 0 to R. When k_λ is constant, Eq. 28.1 becomes

$$\tau_\lambda = k_\lambda \cdot R$$

The depth to which one can peer into the solar atmosphere is obtained by similar reasoning. The process of absorption must be balanced by the process of reemission.[1] The reemission process is somewhat analogous, but in the case of the fog, the absorption is actually produced by the scattering of light out of the beam, and the radiation scattered by a fog volume element of size (1 m²) · dr is proportional to the "absorption coefficient." Similarly, the corresponding emission volume element in the solar atmosphere produces, according to Kirchhoff's law, an amount of radiation proportional to $k_\lambda(r)$. The *emissivity* is given by

$$\varepsilon_\lambda = 1 \text{ m}^2 \cdot B_\lambda(T) \cdot k_\lambda(r) \cdot dr \qquad (28.2)$$

where $B_\lambda(T)$ is the Planck function (or sometimes, the Kirchhoff–Planck function). Because T, r, and τ_λ all increase inward, B_λ can be considered a function of τ_λ. Therefore, Eq. 28.2 can be rewritten:

$$\varepsilon_\lambda = (1 \text{ m}^2) \cdot B_\lambda(\tau_\lambda) \cdot d\tau_\lambda \qquad (28.3)$$

To avoid rewriting the volume element each time, we hereafter assume that the emissivity is given in SI units *per square meter*.

The effects of absorption and emission processes on a beam of radiation emerging from the outer layers of the photosphere are sketched in Fig. 28.2. At the upper boundary, the contribution to the intensity of radiation per square meter may be summed up in the following integral:

$$I_\lambda = \int_{r=0}^{r=R} B_\lambda(T) \cdot k_\lambda(r) \cdot \exp[-\int_{s=0}^{s=r} k_\lambda(s) \cdot ds] \cdot dr$$

or

$$I_\lambda = \int_{\tau=0}^{\tau=\infty} B_\lambda(\tau_\lambda) \cdot e^{-\tau} \cdot d\tau_\lambda \qquad (28.4)$$

Mathematical techniques permit a run of values of $k_\lambda(r)$ and $T(r)$ to be traced back from measured

---

[1] We must recognize from the existence of spectral absorption lines that it may not *look* that way when there is a flux of radiation through a gas. When the radiation emitted in *all* directions is taken into account, there is a balance between total absorbed and total emitted photon energy. If there was not, the atmosphere could not be considered stable. Such instabilities are found in pulsating and eruptive variable stars.

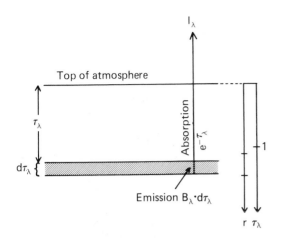

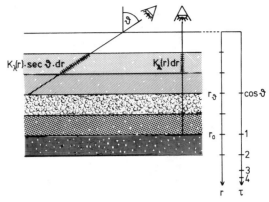

FIGURE 28.2. Radiation from the solar atmosphere. The atmosphere may be considered a series of parallel layers, each of optical depth $d\tau_\lambda$, and contributing an amount $B_\lambda \cdot d\tau_\lambda$ to the beam of light in the line of sight. On the way to some arbitrary "top" of the atmosphere (really a reference level defined by some suitably small particle density), the contribution is weakened by a factor $e^{-\tau}$.

FIGURE 28.3. A demonstration of absorption (gray areas) and emission (white points) in a series of parallel glass plates.

values of I_λ, but the process is not trivial and requires more discussion than we can provide here.

As in the discussion of fog in Chapter 10, we note that, to a sufficient degree of approximation, the intensity that we see from an emerging beam of light originates from optical depth $\tau_\lambda = 1$. To clarify the process, we can explore the effects of absorption and emission in a collection of stacked glass plates. We are going to do this as a kind of "thought experiment", but it makes an interesting challenge to actually carry it out. Figure 28.3 shows the basic layout. Each plate represents an infinitesimal layer of the solar atmosphere, with its own absorption coefficient, $k_\lambda(r)$, which increases with depth r. Each plate has tiny lamplets[2] embedded in it, representing the emissivity of the solar atmosphere. Moreover, the brightness of the lamps increases with r, simulating the increasing flux due to rising temperature and, at the same time, shifting the peak wavelength of the radiation to shorter wavelengths. As we look at the stack of plates along the normal to their planes, the amount of light we see is determined by the lamplets at $\tau_\lambda \approx 1$ (in our

approximation). The physical depth corresponding to this optical depth of unity is found from the equation

$$\tau_\lambda = \int_{r=0}^{r=R} k_\lambda(r) \cdot dr = 1 \qquad (28.5)$$

If we now look along a slant line, that is, a line which makes an angle Θ with respect to the normal, the light that is seen is again determined by the lamplets at optical depth $\tau_\lambda = 1$. But, *along this direction*, the path length through each plate of thickness dr becomes

$$dr / \cos \Theta \quad \text{or} \quad \sec \Theta \cdot dr$$

and setting the corresponding physical depth along the normal to r_Θ, the expression for the optical depth becomes

$$\int_{r=0}^{r=r_\Theta} k_\lambda(r) \cdot \sec \Theta \cdot dr = 1 \quad (28.6)$$

for $r_\Theta \leq r_0$, where r_0 is the physical depth at which the light ray originates. When we look along the normal, what is the optical depth to which we see, at the physical depth r_Θ? This value is given by

$$\tau_\lambda^* = \int_{r=0}^{r=r_\Theta} k_\lambda(r) \cdot dr$$

or

$$\tau_\lambda^* = (1/\sec \Theta) \cdot \int_{r=0}^{r=r_\Theta} k_\lambda(r) \cdot \sec \Theta \cdot dr$$

so that, comparing to Eq. 28.6,

$$\tau_\lambda^* = (1/\sec \Theta) \cdot 1$$

Therefore,

$$\tau_\lambda^* = \cos \Theta \qquad (28.7)$$

_____

[2]Simulation of the lamplets is the tough part in any attempt to carry out the experiment. Of course, scatterers in the material can provide both absorbers and reemitters, as in the case of the fog.

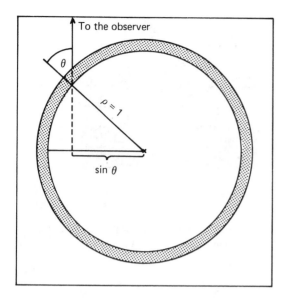

FIGURE 28.4. The geometry of limb darkening is seen in the relation between the line-of-sight direction defined by Θ and the relative distance from the center of the solar disk, $\rho \cdot \sin \Theta$, where ρ, the solar radius, is taken as unity.

Our thought experiment has thus produced the following insight. The direction of light path and the depth to which our line of sight penetrates into the solar atmosphere are related. The light that emerges at an angle Θ with respect to the surface normal permits us, as always, to see down to an optical depth of unity along that line of sight. However, the corresponding physical depth along that line (r_Θ) is shallower than the physical depth from which a normally directed light ray originates (r_0). At the shallower depth, the temperature is lower; thus, by Planck's law the monochromatic flux is less, by the Stefan–Boltzmann law the total (bolometric) flux is less, and by Wien's law the radiation is predominantly redder than the conditions that prevail at greater depth. Finally, the radiation originates from that layer where the optical depth in the normal direction is

$$\tau_\lambda^* = \cos \Theta$$

Figures 28.3 and 28.4 show that it is indeed the case that increasing the slant angle of view of our plate glass stack is directly analogous to viewing the solar atmosphere closer to the limb of the Sun. Thus we have a probe to study the temperature structure of the solar atmosphere. The actual physical depth in the photosphere corresponding to the condition $\tau_\lambda^* = \cos \Theta$, for each measured

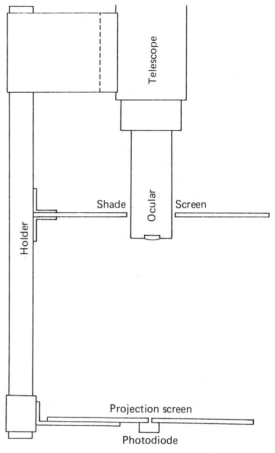

FIGURE 28.5. Basic components of a solar photometer for the measurement of solar limb darkening. With a projection screen, the apparatus can also be used to demonstrate sunspots, faculae, and other solar phenomena. (After J. Isserstedt, Astronomical Institute, Würzburg University, Würzburg, Germany.)

value of Θ, will be taken up in a succeeding section. First we need some means of measuring the limb darkening.

28.3 Measuring the Limb Darkening

In Chapter 26, a description of a photodiode photometer was provided for the purpose of measuring intensities of an artificial eclipsing binary star system. If you can attach this photodiode firmly to a solar projection screen behind a small telescope, you will have the means to measure the intensity of the solar disk. The direct light from the sky which

FIGURE 28.6. The distribution of surface brightness over the Sun's disk in the blue and near infrared regions of the spectrum.

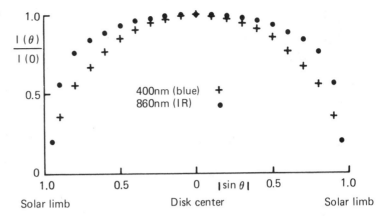

falls on the detector is not catastrophic for the purposes of this experiment because the contribution to the total signal will be constant. But you can improve the contrast, and avoid varying shadowing effects from the Sun, by adding the shade screen seen in Fig. 28.5. You may wish to fix it to the telescope tube itself to permit more flexible movement.

The diameter of the light-sensitive portion of the photodiode should be only 1% of the size of the projected solar image. Since the latter usually amounts to 10 cm or more, and a photodiode is typically ~1 mm across, the condition is readily fulfilled. The brightness of the Sun can be decreased if necessary with neutral density filters, without badly altering the effective wavelength of the passbands of the filter work that is planned. For the present experiment, two filters were used: one with effective wavelength 400 nm in the blue, and the other ~860 nm in the near infrared. You may wish to carry out the experiment with your own filters; but use of at least two filters, with a wide wavelength difference between them, is desirable to illustrate the difference of limb darkening with wavelength.

The idea is to permit the diameter of the Sun's image to pass squarely over the photodiode. The full diameter must be used so that the maximum intensity at the disk center is measured. The scans can be repeated for improved precision, but they should be done for each of the passbands. The focus will be different for the infrared and the blue filters, and care must be taken to have the sharpest image on the photodiode for each scan. The infrared image can be best focused by performing several scans at different focus values and determining which one registers the sharpest edge for the Sun's image. The actual scan is accomplished by locking the telescope in declination and setting

the telescope west of the Sun's image, permitting the Sun's diurnal motion to carry it across the detector. Each scan will therefore take only about 2 min.

The continuous trace produced by a strip chart (similar to the one that produced the light curve in Chapter 26) will yield the light distribution across the disk. By subtracting off the background (registered prior to and following the solar trace), you can measure the light intensity at several points along the trace. The points can be converted into fractions (decimals) of the radius on either side of center (marked by maximum light). The intensities can then be divided by the intensity at maximum to give a series of points that can be plotted as in Fig. 28.6. Note that the abscissa is given as the absolute value of sin Θ. As Fig. 28.4 demonstrates, this is the same as the fractional radius of the disk.

28.4 The Temperature Variation Through the Photosphere

While Fig. 28.6 shows the limb darkening that we already see so clearly on black and white photographs, it also demonstrates the pronounced reddening toward the limbs, and thus the fact that limb darkening varies with wavelength. From each value of sin Θ, the value of Θ and of cos Θ can be computed. The data can then be plotted as in Fig. 28.7. This figure clearly shows that the surface brightness is directly proportional to cos Θ and that the slope of the relation is a function of wavelength.

The temperature as a function of depth can be derived from Fig. 28.7. The fact that the blue intensities fall further and further below the infrared intensities shows that the temperature is decreasing with cos Θ. To obtain the actual temperatures at the

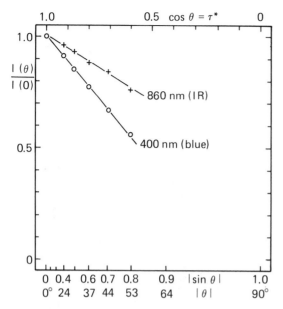

FIGURE 28.7. The center-to-limb variation plotted against the absolute values of Θ and sin Θ and also against cos Θ (upper scale). The relative surface brightness in both spectral regions is linearly proportional to cos Θ.

depths corresponding to the values of cos Θ, we must assume a value for the temperature at disk center ($\Theta = 0$). We adopt a value $T_{\Theta=0} = 6100$ K.

The Planck law,

$$I(T,\lambda) = [2\pi hc^2 / \lambda^5]/[e^{hc/k\lambda T} - 1] \quad (28.8)$$

can now be used to compute the expected intensity ratio for the blue and near IR regions of the spectrum. Because the calculation of a spread of values of $I(T,\lambda)$ is tedious, we have tabulated values of the Kirchhoff–Planck function in units of 10^{12} J/(m$^2 \cdot$ s)

per unit bandwidth (Table 28.1). We begin by calculating the ratio

$$Q_0 \pm I(6100 \text{ K}, 860 \text{ nm}) / I(6100 \text{ K}, 400 \text{ nm})$$
$$\approx 55 / 101 = 0.54$$

The quantity Q_0 serves to normalize the intensity ratio at other temperatures and allows us to evaluate quantitatively the reddening of a black body source with decreasing temperature. The ratio

$$Q \pm [I(T, 860 \text{ nm}) / I(T, 400 \text{ nm})] / Q_0 \quad (28.9)$$

can now be calculated for the temperature range 5300 to 6100 in steps of 100 K (but note the slight amount of interpolation required in Table 28.1 to get the appropriate intensity for 860 nm). But Q is just the ratio

$$[I(\Theta) / I(0)]_{860\text{nm}} / [I(\Theta) / I(0)]_{400\text{nm}} \quad (28.10)$$

which can be obtained from the data in Fig. 28.7. The ratio can be plotted on Fig. 28.7 using a new scale on the right-hand side of the diagram.[3] The values of Q computed from Eq. 28.9 and Table 28.1 can now be located on the line representing Eq. 28.10, and the values of cos Θ can be read off. Figure 28.8 is the plot of T vs. cos Θ. The absolute value of the difference between $r_\odot$ and r_0 has been marked on the upper scale corresponding to values of cos Θ below. Notice that even though cos Θ has been traced only to 0.4, the corresponding depth change is ~ 40 km and that over this distance the temperature has declined by ~ 600 K. The strong variation of r with τ accounts for the relatively sharp edge to the solar limb, a result which is not necessarily expected of a gaseous sphere.

---

[3]A new scale is required because Eq. 28.10 is 1 at $\Theta = 0$ and increases with increasing Θ.

TABLE 28.1. Values of the Planck function for a range of T and λ.

T (K)	λ (nm)											
	350	400	450	500	550	600	650	700	750	800	850	900
6100	84	101	108	108	103	96	88	79	71	63	56	50
6000	75	91	99	100	96	90	83	75	67	60	53	47
5900	67	82	90	92	89	84	78	70	64	57	51	45
5800	59	74	82	84	83	78	73	66	60	54	48	43
5700	53	66	74	77	76	73	68	62	56	51	46	41
5600	46	59	67	71	70	67	63	58	53	48	43	39
5500	40	53	61	64	64	62	59	54	50	45	41	37
5400	35	47	54	58	59	57	54	51	47	42	38	35
5300	30	41	49	53	54	53	50	47	43	40	36	33
5200	26	36	43	47	49	48	46	44	40	37	34	31

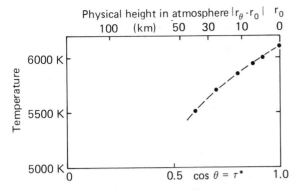

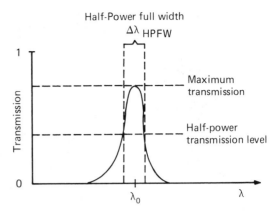

FIGURE 28.8. The temperature variation with height in the solar atmosphere. A range of temperatures was used to look up the intensities at two wavelengths in Table 28.1; the ratio of these intensities for 860 and 400 nm was then used to find the corresponding values of cos Θ from Fig. 28.7. The upper abscissa scale gives the difference $|r_\Theta - r_0|$ or the corresponding physical distance above r_0, the physical depth corresponding to $\tau_\lambda^* = 1$. The level where $\tau_\lambda^* = 1$ is the reference level to which we see at the center of the solar disk.

FIGURE 28.9. Interference filter characteristics for an incident on-axis beam.

28.5 Solar Prominences as Probes of the Corona

The limb darkening of the photosphere characterizes the "quiet Sun" and is not much affected by the 11- or 22-y[4] solar activity (sunspot) cycle. The "active Sun" involves increase in the number and latitude distribution of sunspots, increase in frequency of occurrence of faculae or plages (bright areas), prominences, and flares. As an interesting target of active-Sun study, we recommend solar prominences.

Prominences have densities, temperatures, and spectra that more closely resemble those of the chromosphere than those of the corona; indeed, in eruptive prominences the lower atmosphere does seem to be their source. There are some prominences in which material seems to "rain" out of the corona, however. Clearly the situation is more complicated than it might first appear.

Prominences are seen as dark, stringy regions called *filaments* when projected onto the disk of the Sun, but they are seen in their full, glowing splendor off the solar limb. They change faster than do

sunspots, and they are far more likely to be seen than are the short-lived solar flares. Moreover, like many other solar phenomena, they are accessible with a relatively small telescope. A narrowband interference filter is required to study them, however.

The clouds of ionized hydrogen that form the solar prominences appear red because of the very strong emission of Hα photons (at 656.3 nm), but other emission lines can be seen as well, especially those of the other members of the Balmer series. Images of the Sun in very narrow passband filters, called *spectroheliograms*, reveal the distribution of the hydrogen ions, which are strongly coupled to the local solar magnetic field, and thus trace its shape. An *interference filter* passes only a very narrow spectral region and suppresses the rest of the spectrum; it improves the contrast of objects such as prominences, which emit mainly in emission

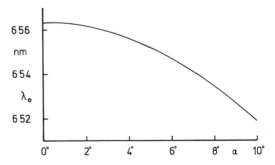

FIGURE 28.10. The shift of transmitted wavelength as a function of incident angle of the beam onto the interference filter.

---

[4]The magnetic polarities of the north and south solar hemispheres reverse every 11 years, so that a complete cycle is 22 y long, on average.

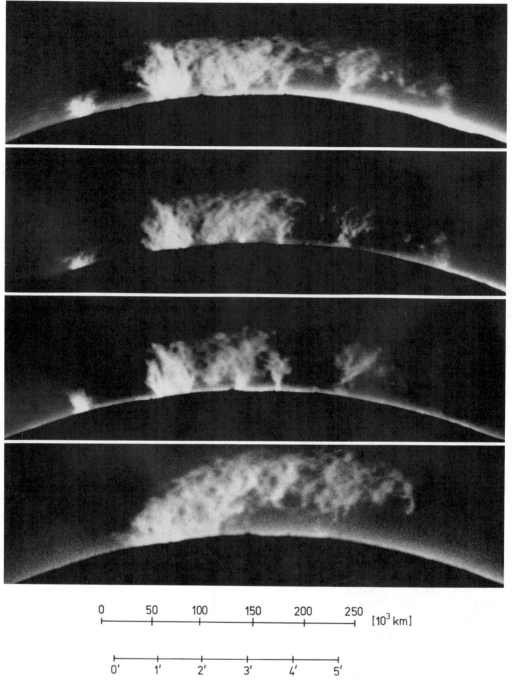

FIGURE 28.11. Four exposures of a solar prominence, originally recorded on color film (Kodak Ektachrome) with speed ASA 400 and exposure times of 1/30 s. The exposures were obtained, from the top, on September 7, 1980, at 10:30, 13:00, 16:00, and on September 8 at 13:00 LCT, respectively. Commercially available red-sensitive films, such as Kodak TP 2415, can also be used.

lines, by removing the scattered light of the continuum. The bandpass characteristics of such a filter are described by the sketch of Fig. 28.9 and by Fig. 28.10. Figure 28.9 shows how the transmission varies with wavelength for a beam along the optic axis of the filter, while Fig. 28.10 shows that the central wavelength of the transmitted light changes as a function of angle away from the optic axis.

Figure 28.10 contains a series of spectroheliograms produced with an interference filter centered on the Hα line. In general, an Hα filter has a half-power full width (HPFW) of ~0.6 nm, but spectroheliograms can be obtained at still narrower passbands, at greater expense.[5] The central wavelength can be "tuned" by a heater and thermostat, because the central wavelength of transmission varies monotonically with increasing temperature.

An examination of the exposures of Fig. 28.11 shows that changes are occurring on time scales of hours or less. As a challenge, you can examine the motion of a filament of the prominence, in terms of its change of position. The change in position can be gauged by a chord drawn across the solar photosphere, indicated by the dark occulting disk in Fig. 28.11. From the length of this chord, you can deduce the solar radius in these units. The final calibration to meters can be done by setting this number proportional to the solar radius in meters, $7 \cdot 10^8$ m. Can you convert this scale to arc-minutes? Are the scales provided in Fig. 28.11 correct?

---

[5]Available from several sources which are advertised in the popular astronomy magazines like "Sky and Telescope" and "Astronomy."

29

Star Clusters

29.1 Introduction

Star clusters are families of stars that originated sufficiently near each other to share similar space motions and, for the most part, a common age and chemical makeup. Stars within a cluster differ mainly because they have different masses. Since stars of differing masses have differing rates of evolution, the properties of the stars of a cluster provide a kind of snapshot of the evolutionary development of stars at a certain common age. If we can find objects that are standard candles (see Chapter 27) within a cluster, the distance to the cluster itself can be obtained. This means that the absolute magnitude of the entire main sequence of the cluster can be calibrated, in principle, because the observed magnitudes of all the observed stars can be placed on the absolute magnitude scale. Since the faster rate of evolution of more massive stars causes a bend in the color–magnitude array, a calibrated main sequence will reveal a "turnoff" point at a precise magnitude and color index. The turnoff point proceeds down the main sequence with time, like the burning down of a candle and, like the diminishing height of the candle, provides an index of age.

There are two main types of clusters. *Globular clusters* are spherical in shape and are very rich (contain many stars); *open* or *galactic clusters* are irregular in shape and usually much less rich. The latter are found predominantly in the plane of the Milky Way galaxy, whereas globular clusters are much more evenly distributed around the center of our galaxy. Plate 29.1 illustrates the two types.

Although a cluster may differ markedly from others of its type, the principal differences among these two types are:

1. Space distribution and motion, with the globular clusters having high velocities (including z velocities, normal to the galactic plane).
2. Populations[1] of stars (with the globular clusters having mainly older, metal-poor, extreme population II stars).
3. Gas and dust content, found, for the most part, only among the young open clusters of the Milky Way.

29.2 Moving Clusters

One of the identifying properties of clusters is the common motion of the stars within them. The motion of any star has two components: on the plane of the sky and in the line of sight. *Proper motion* is the apparent motion on the plane of the sky in units of arc sec per year. It is determined astrometrically by careful measurement of position relative to the positions of distant standard stars (or as the Dutch-American astronomer Jan Schilt proposed, galactic nuclei, because of their

---

[1]Modern studies suggest at least five populations of stars: extreme and intermediate population I, disk, intermediate and extreme population II. For our purposes, it is sufficient to identify two basic populations: I and II. Population I objects are confined to the plane of the galaxy, have higher metal (i.e., elements heavier than helium) abundance, and are more closely associated with the gas and dust clouds that are the spawning places of the stars. Population II objects have a more uniform and spherical distribution around the center of the galaxy, are much older, and reflect the chemical composition of earlier generations of stars. The globular cluster stars are examples of extreme population II; the open cluster stars are of population I, typically.

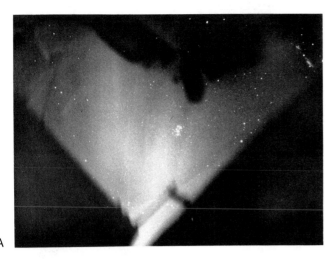

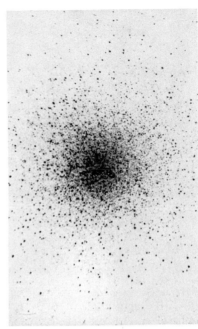

PLATE 29.1. Photographs of an open cluster (the Pleiades) obtained with a 35 mm camera at the Rothney Astrophysical Observatory, and a globular cluster (M13). The latter is a reproduction of the first astronomical photograph taken with the Canada France Hawaii Telescope (courtesy of the CFHT Corporation).

great distances and immeasurably small shifts on the plane of the sky over historic periods of time). *Radial velocity* (V_R) is the motion in the line of sight; it is measured from the Doppler shift of spectral features in units of kilometers per second. Figure 29.1 illustrates the significance of the proper motion in terms of the distance and the motion of the star *across* the line of sight, which is known as the *transverse velocity* (V_T). From the definition alone and a careful consideration of units, it may be shown that

$$V_T = 4.74 \cdot \mu \cdot r \qquad (29.1)$$

where V_T is the *transverse* (or *tangential*) velocity (km/s), μ is the proper motion (arc sec/y), and r is the distance in (pc).

In general, V_T is not known. When many stars of a particular type are observed across the whole sky, the average motions of the stars in any given direction may be assumed to be the same as that in any other direction. Under assumptions such as this, the mean radial velocities may be used instead, and the mean distance to the collection of stars found. Under certain special circumstances, the geometric relation between V_T and V_R can be found. This is the case for *moving clusters*, that is, nearby

clusters whose proper motions can be clearly perceived and measured with precision.

Figure 29.2 illustrates the connection between the radial velocity, the transverse velocity, and the *space velocity*, which is the combination of the two. Note that

$$V_R^2 + V_T^2 = V_S^2 \qquad (29.2)$$

In the case of a nearby cluster, it is possible to determine the distance by carefully observing the proper motions and radial velocities of all the known members of the cluster (this is a bit circular, because the kinematic evidence is one way of determining membership). By plotting the proper motions of the stars on a chart of their positions, you can see that the cluster will appear to be either converging toward or diverging from some point in the sky, much as a pair of railway tracks will appear to converge in the distance. Figure 29.3 demonstrates such a case.

The convergent or divergent point of the cluster can be determined from such a plot, and with this, the geometric relation between the radial velocity and the transverse velocity becomes known:

$$V_T = V_R \cdot \tan \alpha \qquad (29.3)$$

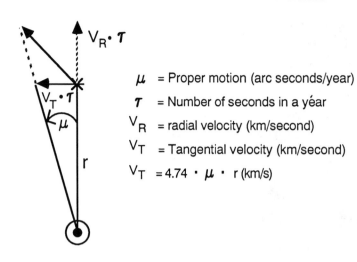

μ = Proper motion (arc seconds/year)

τ = Number of seconds in a yéar

V_R = radial velocity (km/second)

V_T = Tangential velocity (km/second)

V_T = 4.74 · μ · r (km/s)

FIGURE 29.1. The relationship among the proper motion, transverse velocity, and distance of a star. The quantity τ here is the number of seconds in a year ($3.16 \cdot 10^7$), so that $V_T \cdot \tau$ is the distance in km traveled across the line-of-sight by the star in one year.

where α is the angle of convergence (or divergence). By substitution in Eq. 29.1, the distance can be found. The distances of few clusters have been successfully determined this way, unfortunately, because the proper motions depend on distance, and most clusters are too far away to permit accurate determinations of the proper motions of their member stars. Fortunately, there are other methods.

29.3 The Magnitude–Color Array and Distances to Clusters

The relation between the luminosity and the radiant flux (Eq. 15.1),

$$F = \mathcal{L} / (4\pi \cdot r^2)$$

enables us to determine the distance to a cluster. This formula may be rewritten so that luminosities

are in units of the Sun's luminosity and distances are in units of parsecs:

$$F = 3.20 \cdot 10^{-30} \cdot (\mathcal{L} / \mathcal{L}_\odot) / r^2 \text{ W/m}^2 \qquad (29.4)$$

Here we illustrate the technique of using stars whose properties are believed sufficiently well known that the stars are considered *standard candles*. In some cases, this belief is not well founded because the chemical composition, which is known to vary from cluster to cluster and may vary even within a cluster, can affect both the observed spectral distribution and the evolution of a star. Nevertheless, as an illustration of the technique we present examples for three clusters. Figures 29.4 to 29.6 depict the main sequence of the star clusters M44 (Praesepe), NGC 188, and M92. Table 29.1 lists the properties of two or three stars for each cluster. The fluxes are in units of watts per square meter and the distances (and their averages) have been rounded off to three significant figures.

Once the luminosities and the total apparent flux of member stars are known, we can obtain the distance. However, you will recognize that the total flux is not directly measured and that tables of bolometric correction for runs of values of the temperature, spectral type, and color indices are required. The color–magnitude array more generally uses M_V vs. $(B - V)_0$. The observed values V vs. $(B - V)$ for a given cluster are compared to the zero-age main sequence plotted in the M_V vs. $(B - V)_0$ domain and the difference $V - M_V$ is read off the difference in scales. Thus we see both scales

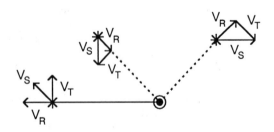

FIGURE 29.2. The relationship among the radial and transverse velocities and the space velocity.

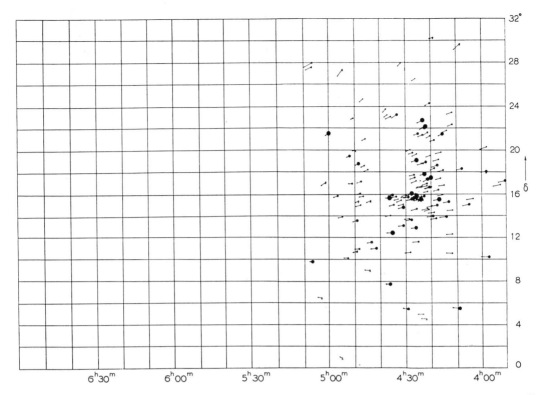

FIGURE 29.3. The convergence of a moving cluster in Taurus, the Hyades. The lengths of the arrows indicate the proper motions of the individual stars over an interval of about 15,000 years. (After H. van Bueren, *Bulletin of the Astronomical Institutes of the Netherlands* **11**, 392, 1952.)

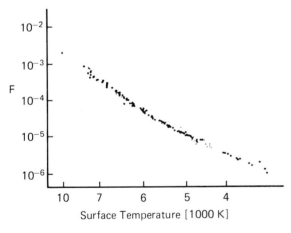

FIGURE 29.4. The main sequence of the open cluster Praesepe (M44) in Cancer.

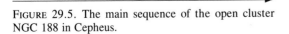

FIGURE 29.5. The main sequence of the open cluster NGC 188 in Cepheus.

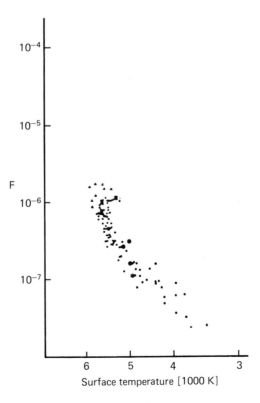

Surface temperature [1000 K]

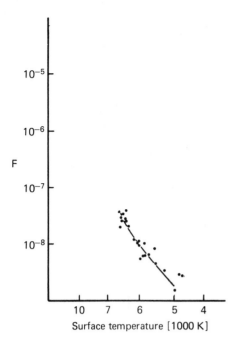

FIGURE 29.6. The main sequence of the globular cluster M92 in Hercules.
◄

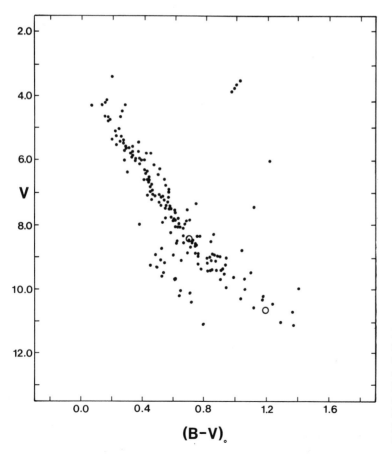

FIGURE 29.7. The color–magnitude array for the Hyades. The open circles represent the two components of the eclipsing binary HD27130 (From Schiller and Milone, 1987, adapted from Hagen, 1970, reproduced with permission from the Astronomical Journal.).

TABLE 29.1. Properties of stars on the main sequences of three star clusters.

T (K)	F	$\mathcal{L}/\mathcal{L}_{\odot}$	r (pc)
	M44		
4,000	$1.33 \cdot 10^{-35}$	0.1	155
6,000	$1.83 \cdot 10^{-34}$	1.26	148
7,000	$1.39 \cdot 10^{-33}$	10	152
			$<152> \pm 2$
	NGC 188		
4,000	$1.48 \cdot 10^{-27}$	0.1	1,470
5,000	$6.88 \cdot 10^{-27}$	0.38	1,330
			$<1,400>$
	M92		
5,000	$5.70 \cdot 10^{-29}$	0.38	14,600
6,000	$3.2 \cdot 10^{-28}$	1.26	11,200
			$<12,900>$

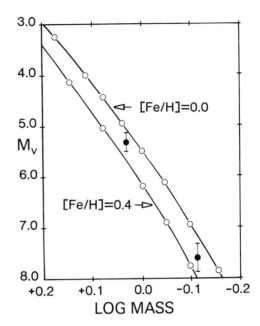

FIGURE 29.8. The absolute visual magnitude vs. mass relation for two theoretical values of the metallicity, [Fe/H], for zero-age main sequence stars. The crosses indicate the positions of the components of the Hyades eclipsing binary HD27130, implying a value [Fe/H] $\approx$ 0.2, corresponding to 1.6 times the solar value. (Reproduced with permission from the Astronomical Journal.)

in the color magnitude array for the Hyades in Fig. 29.7. This cluster is important because it is so close to us and is a major link in the establishment of the extragalactic distance scale. The difficulties in determining the distance to the Hyades are nicely discussed by Paul Hodge (1988).

Photometric data from the entire main sequence can be used in this way to establish the distance of the cluster. But this method too has its limitations. The exact location on the array of the zero-age main sequence is dependent on the chemical composition. Stars with higher metal abundance (metal-rich stars) will have stronger absorption lines from the contributing elements. The lines are particularly numerous at the blue end of the spectrum, so that the metal-rich stars tend to be fainter than the metal-poor stars in the ultraviolet and blue regions of the spectrum.

Yet another method of finding a cluster's distance depends on the existence of a double-lined spectroscopic and eclipsing binary system (see §26.3 for a discussion of these objects) in the cluster. With light curves and radial velocities, it is possible to determine fundamental parameters, such as the monochromatic luminosities, of the component stars. By this means, the distance to the system can be obtained. The double-lined spectroscopic binary HD27130, which was discovered to be an eclipsing system by Robert McClure in 1980, provides a useful example. It is located in the Hyades and was subsequently studied by Schiller and Milone (1987), who obtained a distance for the cluster of about 43.6 ± 4 pc. In Fig. 29.8, the absolute magnitude and mass of the component stars of the system are plotted along with theoretical zero-age main sequence curves for two values of metal

content of the cluster from the work of Vandenberg and Bridges (1984). A good indicator of metallicity, or metal content, is the ratio of numbers of iron to hydrogen atoms, defined by the index

$$[Fe/H] \equiv \log (Fe/H) - \log (Fe/H)_{\odot} \qquad (29.5)$$

In Fig. 29.8, the stars' positions are straddled by two values of [Fe/H]: 0.0 (solar value) and 0.4 (~2.5 times the solar value). The implied value of [Fe/H] is ~0.2 corresponding to a relative iron abundance ~1.6 times the solar value, in agreement with independent determinations. This suggests that the distance to the cluster has been determined correctly, assuming that other aspects of the modeling of both binary and cluster are correct. The drawback to this technique is that any individual star or binary may not be at the center of the cluster but on the near or the far side of it. In the case of the Hyades, its spread on the sky is ~12% of its distance, and therefore its depth can be expected to be about the same. Thus we have a technique of arriving at a distance, but it is not clear how this relates to the rest of the cluster. In

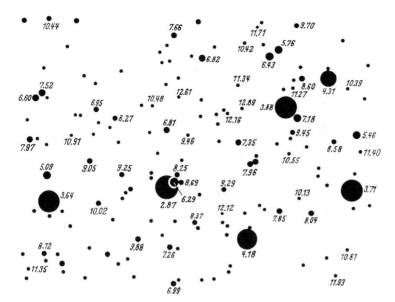

FIGURE 29.9. A chart of the central part of the Pleiades showing the visual magnitudes of selected stars.

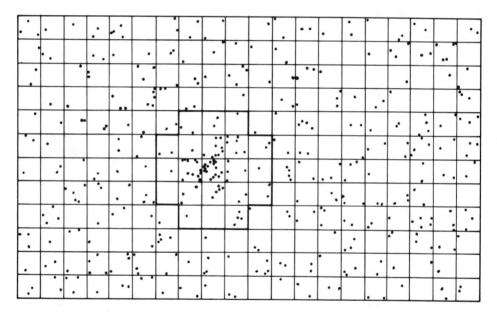

FIGURE 29.10. The region of sky around the open star cluster M41 as seen in the Bonner Durchmusterung star chart. The region of the star cluster is delineated from the rest of the field by darker boundaries on the $1/4°$ grid zone boundaries. The stars outside the cluster boundary are seen to be Poisson-distributed (see Appendix B, §B.3, for the definition and description of the Poisson distribution function). Near the core of the cluster, the frequency of stars is much higher than elsewhere.

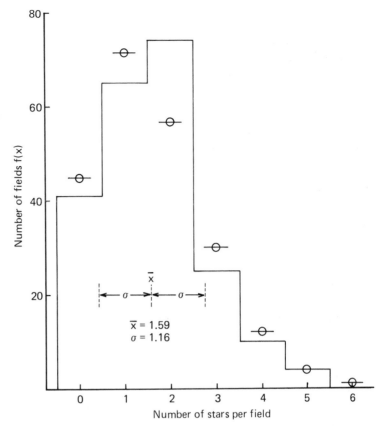

FIGURE 29.11. The distribution of the numbers of fields with x numbers of stars per field. The fields are $1/4°$-square regions on the sky outside of the M41 star cluster region of Fig. 29.10. Both expected and observed means are shown, as the standard deviation. The predicted frequency of occurrence of 0 to 6 stars/field is marked by the symbol. The distribution is necessarily skewed because there cannot be fewer than zero stars in any field.

the case of HD27130, this is not so much of a problem: the locations of the components in the color–magnitude array are squarely on the midline of the observed main sequence of the Hyades, shown in Fig. 29.7; along with proper motion and radial velocity data, one may conclude that the binary lies near the cluster center.

29.4 Further Challenges

1. Try to derive Eq. 29.1 using basic definitions and taking care with units. Next try to determine from Fig. 29.3 the convergence angle of the cluster and its uncertainty.
2. On a clear autumn or winter night the Pleiades are an attractive sight in the sky. The cluster appears as a small asterism located near the larger, V-shaped asterism, the Hyades (which contains the bright orange star Aldebaran). The Pleiades could be mistaken for a "little dipper" by those who do not know that this much larger asterism is located close to the North Celestial

Pole with Polaris as the last star in the handle. Although most people have difficulty seeing more than six or seven stars, a pair of binoculars reveals many more stars. An open star cluster is a collection of point sources, so from §21.2, the light-gathering power of the binoculars over the eye alone is

$$(D/\delta)^2 \qquad (21.1)$$

where D is the diameter of the objective and δ is that of the eye. For a 7×70 pair of binoculars, $D = 70$ mm and the magnification is $7 \times$. If for a dark-adapted eye $\delta \approx 7$ mm, then $(D/\delta)^2 \approx 100$. This means that the binoculars can see 5 magnitudes deeper into the sky than can the unaided eye and pull out objects that are as faint as tenth or eleventh magnitude. Check this conclusion by means of the chart of the Pleiades given in Fig. 29.9.

3. The existence of a cluster in a star field can be demonstrated both kinematically (using radial velocity and proper motion information) and

photometrically (by demonstrating that a main sequence can be seen when the stars are plotted in a color–magnitude array). An obvious way to check on its physical reality is to examine the likelihood of an apparent clustering of stars among the stars in the field. Figure 29.10 shows the star field in the vicinity of M41 upon which has been placed a grid of lines $1/4°$ apart, creating a total of 240 small field regions. A region encompassing the cluster contains 21 such regions, leaving 219 outside. In those 219 regions are 348 stars. The frequency distribution of the stars per region is summarized in the histogram of Fig. 29.11. From Eq. B.1, we know that the expected population of each region is

$$<x> = 348 \text{ stars} / 219 \text{ fields} = 1.59 \text{ stars/field}$$

The observed standard deviation, from Eq. B.2, is

$$\sigma = \sqrt{\{[\Sigma (<x> - x_i)^2]/(N - 1)\}} = 1.16 \text{ stars}$$

in good agreement with that predicted by Eq. B.6:

$$s = \sqrt{<x>} = 1.26 \text{ stars}$$

It is interesting to ask if the existence of a clustering of stars could be determined by statistical means or whether some additional information is needed to demonstrate its existence. In Fig. 29.10 there is one region with 16 stars in it. From Eq. B.5, you can calculate the expected frequency for such a number:

$$f(x) = m \cdot (<x> / x!) \cdot e^{- <x>} = f(16)$$
$$= 3.53 \cdot 10^{-9}$$

where m is the number of measurements. This means that $1/f(16)$ or 283 million regions would have to be examined, on average, until one with 16 stars could be found. This is very improbable, and so you can conclude that the object we designate M41 is not an accidental apparent clumping of stars on the sky but very likely a real physical association of stars. This conclusion is borne out by studies of the cluster, which show the stars to be at the same distance and at the same age. Examine some of the regions adjacent to the central 16-star region and investigate the likelihood that they too are more than a random clumping of stars. How do you assess the probability that two additional clumpings would be found next to a third?

References and Bibliography

van Bueren, H. (1952) *Bulletin of the Astronomical Institutes of the Netherlands* **11**, 392.

Hagen, G. L. (1970) *An Atlas of Open Cluster Colour-Magnitude Diagrams*. Publications of the David Dunlap Observatory, No. 4. David Dunlap Observatory, Toronto.

Hodge, P. (1988) *Sky & Telescope* **75**, 138–140.

Schiller, S. J., and Milone, E. F. (1987) *Astronomical Journal* **93**, 1471–1483.

VandenBerg, D. A., and Bridges, T. J. (1984) *Astrophysical Journal* **278**, 679–688.

30

Olbers's Paradox: Cosmology and the Night Sky

30.1 Introduction

Why is the night sky dark? Well, you say, because the Sun is set! But with the persistence of a pre-schooler, we ask again: *why* is it dark at night? Put this way, the question is profound — in fact, it is one of the most basic questions of cosmology.

The question poses a challenge to cosmological models, especially those that involve an eternal universe of infinite extent, in which galaxies of stars are distributed uniformly throughout. The properties of such a "steady-state" universe are constant in both space and time, so that the mean density, $<\rho>$, has been, is now, and forever will be the same. We will see that the answer to our question about the night sky will show, in support of other data, that this model of the universe is no longer viable.

Following similar ideas expressed earlier by Kepler (ca. 1610), Halley (ca. 1720), and Loys de Chéseaux (ca. 1744), the German physician Heinrich W. Olbers (ca. 1820) demonstrated that such an infinite universe, filled with stars, should be as bright as a sky full of suns.

30.2 The Paradox

Figure 30.1 shows a pyramidal section of space bounded by a one degree square angular region on the sky of an Earthbound observer. Using this geometry, we will proceed to sum up the contributions to the total brightness of all stars out to some distance, r, which we will extend without limit, $r \rightarrow \infty$. The cross-sectional area at any distance r from the observer is

$$f = a^2, \quad a = 2r \cdot \tan(\tfrac{1}{2}^\circ) = 1.75 \cdot 10^{-2} \cdot r$$

Therefore,

$$f = 3.05 \cdot 10^{-4} \cdot r^2 \qquad (30.1)$$

becomes our unit of area measure. The corresponding volume over a depth dr is

$$dV = f \cdot dr = 3.05 \cdot 10^{-4} \cdot r^2 \cdot dr \quad (30.2)$$

The number of stars in this volume element is on average

$$dN = <\rho_N> \cdot dV = 3.05 \cdot 10^{-4} <\rho_N> \cdot r^2 \cdot dr \qquad (30.3)$$

where $<\rho_N>$ is the mean number density of stars. Suppose now that each star had the same intrinsic brightness or luminosity, $\mathcal{L}$. This assumption is not in accord with the facts because stars do differ in intrinsic brightness. However, there are upper and lower bounds on stellar luminosity, both empirically and theoretically, so in practice our constant star brightness has real meaning as an average value, and our arguments will be unaffected by any numerical multiplier required to correct our constant to the true average brightness. Given this assumption, the apparent brightness will then differ from star to star only because of distance from us. By the inverse square law,

$$\ell = \mathcal{L} / r^2 \qquad (30.4)$$

The numbers of stars in our sample will then contribute a total brightness

$$d\ell = dN \cdot \ell$$

or

$$d\ell = 3.05 \cdot 10^{-4} \cdot <\rho_N> \cdot r^2 \cdot (\mathcal{L} / r^2) \cdot dr$$

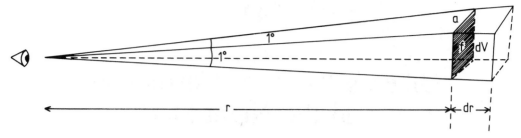

FIGURE 30.1. A $1° \times 1°$ region on the surface of the celestial sphere encompasses a pyramid whose cross-sectional area, at some distance r, is $f = a^2$, where $a = 2r \cdot \tan(1/2°)$. The numbers of stars and their total brightness, as well as the mean depth of penetration into space by the line of sight, can be determined by adding up or integrating over all volume elements, $dV = f \cdot dr$.

which simplifies to

$$dl = 3.05 \cdot 10^{-4} <\rho_N> \cdot \mathcal{L} \cdot dr \quad (30.5)$$

Now let us assume further that the contribution to brightness *from each volume element* is the same and independent of distance. With these assumptions, the total contributions from all such volume elements, out to a distance, r_0, becomes[1]

$$l_{tot} = \Sigma\, dl = 3.05 \cdot 10^{-4} \cdot \mathcal{L} \cdot <\rho_N> \cdot \Sigma\, dr$$

or

$$l_{tot} = 3.05 \cdot 10^{-4} \cdot \mathcal{L} \cdot <\rho_N> \cdot r_0 \quad (30.6)$$

As r_0 becomes larger and larger, l_{tot} increases without limit, that is,

$$\text{as } r_0 \to \infty,\ l_{tot} \to \infty$$

The assumption of a universe filled with an infinite number of stars thus leads to the interesting conclusion that the sky should be infinitely bright, day and night!

This conclusion must be modified slightly because stars have finite sizes, after all, and they are not transparent, so a particular line of sight cannot intercept more than one stellar disk. Now the size of the stellar disk is inversely proportional to its distance. Therefore, the celestial sphere ought to be filled with the images of overlapping stellar disks of differing but finite sizes. As a consequence, the day and night sky should have a surface brightness similar to that of a familiar average star—the Sun. The resulting increase in flux would be enormous, as would the corresponding equilibrium temperature. The Earth could never have formed under such conditions, and in its present state would instantly evaporate. In such a universe, filled with light, planets could not exist and therefore life as we know it would be absent.

Now we see that the paradox is worse than we thought. We do not seem to require a universe of infinite extent, only one uniformly populated by stars out to some distance r_0; and in such a universe, we ourselves could not exist to contemplate it.

The average distance of a stellar surface in our assumed universe is readily derived in a way similar to the total brightness. First we must determine what proportion of our volume element is occupied by the dN stars contained in it. If the linear radius of a star is R, the disk area seen by the observer is just πR^2; The dN stars have a total projected surface area of $df = dN \cdot \pi R^2$, or, substituting for dN from Eq. 30.3,

$$df = 3.05 \cdot 10^{-4} \cdot \pi \cdot <\rho_N> \cdot r^2 \cdot R^2 \cdot dr \quad (30.7)$$

We now introduce a proportionality factor, b, which gives the fraction of the area f occupied by the stellar disks. It has the value 0 if f is completely free of stellar disks and 1 if f is completely covered by the disk surfaces. With this definition of b, we have

$$df = db \cdot f$$

Combining this with Eqs. 30.1 and 30.7, we have

$$db = [3.05 \cdot 10^{-4} \cdot \pi \cdot <\rho_N> \cdot r^2 \cdot R^2 \cdot dr]\ /$$
$$[3.05 \cdot 10^{-4} \cdot r^2]$$

---

[1]When adding up a large number of contributions of very small increments, the mathematical process of integration is regularly used. In this case the form of the equation becomes

$$l_{tot} = \int dl = 3.05 \cdot 10^{-4} \cdot \mathcal{L} \cdot <\rho> \cdot \int dr$$

Taken over the limits from $r = 0$ to $r = r_0$, the result is the same as Eq. 30.6.

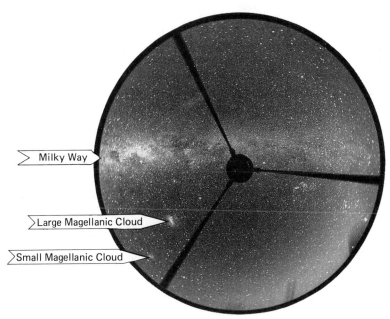

FIGURE 30.2. A wide-angle exposure of the Milky Way between Auriga and Centaurus. The diameter of this photograph spans 140° and includes the Large and Small Magellanic Clouds, the nearest extragalactic systems. (From *Atlas of the Milky Way*, by W. Schlosser, Th. Schmidt-Kaler, and W. Hünecke.)

Milky Way

Large Magellanic Cloud

Small Magellanic Cloud

or

$$db = \pi \cdot <\rho_N> \cdot R^2 \cdot dr \qquad (30.8)$$

Total coverage of each $1° \times 1°$ area of the sky will occur, if the universe has at least a radius r_0, when all the volume elements are added up:

$$\int_{r=0}^{r=r_0} db = 1$$

or, with Eq. 30.8,

$$\int_{r=0}^{r=r_0} \pi \cdot <\rho_N> \cdot R^2 \cdot dr = \pi \cdot <\rho_N> \cdot R^2 \cdot r_0$$

so that

$$r_0 = 1 / (\pi \cdot <\rho_N> \cdot R^2) \qquad (30.9)$$

In a subsequent section, we calculate the limiting value of r_0 and learn something about the universe in the process. First we should ask if we have forgotten anything that could resolve the paradox and thus preserve the notion of an infinite and eternal universe filled with stars. You might raise the question of interstellar matter.

Does the presence of interstellar nebulosity affect the paradox? At first glance, you might think so. Photographs of the Milky Way do show the presence of dark nebulae (see Figure 30.2),[2] and we know that clouds of such material restrict the depth to which optical astronomy can explore the galaxy around us to a few kiloparsecs. On further reflection, the

presence of interstellar matter in our infinite universe (or even one uniformly filled with stars out to a finite distance r_0 sufficient to completely fill each cross-sectional area f) fails to resolve the paradox. An example illustrates why this is so.

Imagine that we are peeking through a peephole into a thoroughly and uniformly heated coal furnace. Inside we see incandescent lumps of coal, ash at the bottom, and the walls of the furnace; all are at the same temperature and are aglow. Now suppose another piece of coal is thrown in. At first it looks black against the other surfaces, since, obeying Planck's radiation law, it radiates less heat energy per surface area than its hotter surroundings. After a while, however, it will begin to approach equilibrium with those surroundings and will glow brighter. After still more time, it will achieve the same temperature and glow equally brightly with the other local radiators. We can expect the same of any interstellar material in a universe filled with starlight. Soon the material would be heated to several thousand degrees and glow as brightly as the stellar background. Thus Olbers's paradox is unresolved by the presence of interstellar matter.

---

[2]In the dense Bok globules, investigated by Bart J. Bok (1906–1983), the absorption, A_V, may total about 27 magnitudes and the reddening, E_{BV}, about 9 magnitudes.

Other attempts to circumvent the paradox involving hierarchical universes were made by R. Proctor (ca. 1878) and C. von Charlier (ca. 1921). Just as the sun moves in an orbit around the Milky Way galaxy, and the galaxy in turn belongs to a local group, and this in turn is a member of the Virgo supercluster, the hierarchical universe was conceived as a series of higher ordered universes but arranged so that the mean densities decreased at each higher step. Indeed, if $<\rho_N>$ were to decrease at the right rate, it is possible that the divergence of Eq. 30.6 or even the more constraining finite radius for a uniform universe, Eq. 30.9, could be avoided. However, this contradicts what the observations tell us about the isotropic distribution of matter and suggests a favored place for the earthbound observer. The latter is a presumption that astronomers no longer want to make: the geocentric universe yielded to the heliocentric universe in the fifteenth century; the solar system was "displaced" from the galactic center in the twentieth century by Harlow Shapley's investigation of globular cluster distribution; and later in the twentieth century, the extragalactic nebulae "became" galaxies in their own right and no longer peculiar fringe members of the Milky Way. Every historical step has pointed away from a favored status for the earth, and the isotropy of matter around us does not lend support to the idea of density decreasing away from us.

Moreover, the empirical red shift of galaxies, discovered by Vesto M. Slipher at the Lowell Observatory early in the twentieth century, does not help much. The extreme reddening of the light of very distant objects cannot prevent the divergence of Eq. 30.6; it merely slows it down.

30.3 Cosmological Challenges

Our discussion of the paradox and its resolution has thus far followed historical developments based on the increase in starlight to be anticipated in a universe of stars spaced uniformly throughout. The universe is not quite organized in this way, and while not quite hierarchical either, the confinement of stars to galaxies and of galaxies to clusters of galaxies does raise questions about the true den-

sity of stars throughout the universe and the consequent effect on the mean free path of starlight. The actual density, as far as we know, is about three solar masses per cubic kiloparsec, a number equivalent to about

$$3 \cdot 2 \cdot 10^{30} \text{ kg} / (10^3 \cdot 2 \cdot 10^5 \cdot 1.5 \cdot 10^{11} \text{m})^3$$
$$\approx 2 \cdot 10^{-28} \text{ kg/m}^3$$

This number is lower than the mean density of stars in the solar neighborhood, reflecting the high concentration of stars in the Milky Way: ~1 solar mass/pc^3, or ~$7 \cdot 10^{-20}$ kg/m^3. Despite this tendency for stars to pile up in galaxies, it is more appropriate that we concentrate on the galaxies, and not on the individual stars of which they are composed, because this gives a more realistic picture of the universe as a whole. In this spirit, we formulate our question in a slightly different way and ask what mean distance, r_0, is required to fill a $1° \times 1°$ region with galactic disks[3] of surface brightness similar to the Milky Way. Figures 30.2 and 30.3 cover extensive regions of the sky. The wide-angle photograph of Fig. 30.2 shows also the two nearest galaxies, the Large and Small Magellanic Clouds, and you can see the similarity of the surface brightness of these objects to the Milky Way itself. The intensity or surface brightness, as noted elsewhere in this book (e.g., Chapter 19), is constant, regardless of the distance. Figure 30.3 also shows the remarkably wide separation of two slightly more distant galaxies, M31 in Andromeda and M33 in Triangulum.

A study of similar photographs reveals that there is a total of only seven galaxies of at least 0.1° in extent which have approximately the same surface brightness as the Milky Way galaxy. The seven are listed in Table 30.1, which gives their types (S for spiral; SB for barred spiral, and S0 for lenticular), positions, sizes (out to the level of the Milky Way's intensity), and distances (in megaparsecs).

---

[3]Since galaxies need not be spherical (there are spiral, irregular, as well as ellipsoidal galaxies, and only the E0 class ellipsoidals are completely spherical), the "disks" will on average appear slightly elliptical. The irregular galaxies are relatively few in number and will not greatly affect our argument.

FIGURE 30.3. The band of the Milky Way from Cepheus to Cassiopeia as well as the two brightest northern galaxies M31 (the Andromeda galaxy) and M33 (in Triangu-

lum). This photograph demonstrates the sparseness of the galaxies in space, and thus the plausibility of a mean free path length of gigaparsecs for extragalactic light.

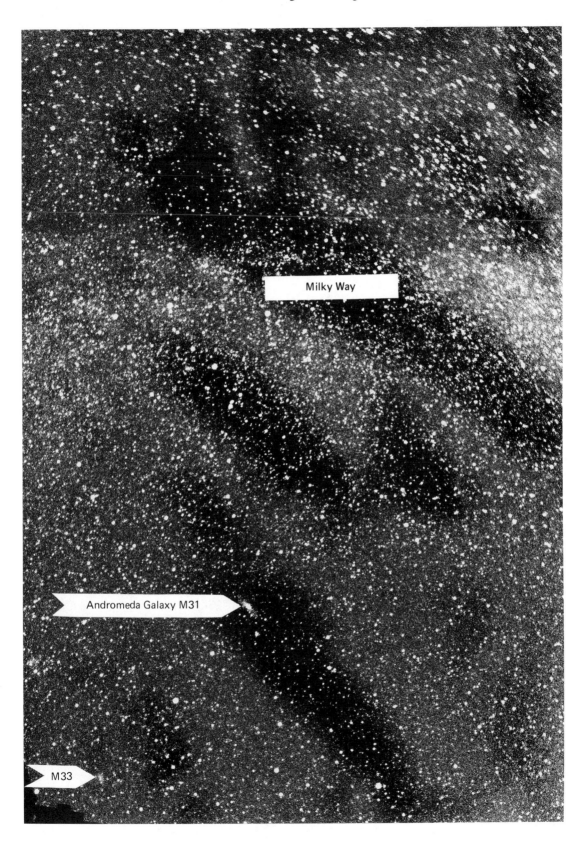

TABLE 30.1. The seven most prominent galaxies.

Galaxy	Type	Position, 2000 α	δ	Size (deg)	Distance (Mpc)
LMC	SB	05^{h}23.6^m	−70°26′	3 × 6	0.052
SMC	SB	00 53.7	−72 53	2 × 3	0.063
M31	S	00 42.8	+41 16	0.5 × 2	0.67
M33	S	01 34.8	+30 39	0.2 × 0.3	0.73
NGC 253	S	00 47.6	−25 17	0.1 × 0.3	2.4
M81	S	09 55.4	+69 04	0.1 × 0.2	3.2
NGC 5128	S0	13 25.7	−43 02	0.2 × 0.2	4.4
				Total 25.15 deg²	

30.3.1 Extragalactic Photography

You too can do a cosmological experiment. In this brief section we discuss photography of some of the bright targets of Table 30.1 and the use of the images to compare surface brightnesses and to determine the mean depth of penetration, r_0, of an extragalactic ray of light. The four galaxies M31, M33, NGC 253, and M81 are visible from midlatitudes to the equator in the Northern Hemisphere, while almost all are visible from midlatitudes to the equator in the Southern. There is no difficulty in photographing these objects with fast film (ASA 400, for example), but the long exposure times (~1 h typically) at focal ratio $f/2$ with a 50-mm focal length lens, require a clock drive mounting for the camera. Since the objects are faint, care must be taken to exclude bright objects; this means photography from a dark site, with no Moon. If the Milky Way can be included, however, the comparison of surface brightnesses is facilitated. Try this experiment, verifying the angular extents of the galaxy or galaxies that you are able to photograph, and work through the relations we discuss later.

30.4 The Resolution of the Paradox and the Limits of the Universe

Table 30.1 shows that our sample extends to about 4 Mpc, the distance to the last two galaxies in the list. The total area of the seven sources amounts to 25.15 deg². Since there are 41,253 deg² in the entire sky,[4] the factor db of Eq. 30.8 becomes

$$db = 25.15 / 41,253 = 6.1 \cdot 10^{-4}$$

As we know from Eq. 30.8, this fraction of the cross-sectional area f, which is filled with extended sources, is proportional to the distance. Complete coverage of f is achieved at what we may refer to as the *critical distance*,

$$r_0 = R / db = 6.6 \text{ Gpc}$$

where 1 Gpc ± 1 gigaparsec ± 10^9 pc = $3.1 \cdot 10^{25}$ m. This is a lower limit for r_0 because a glance at Table 30.1 shows that the Magellanic Clouds are dominant contributors to the total area. Omitting them gives a total area of only 1.15 deg². In this case, $db = 1.15 / 41,253 = 2.8 \cdot 10^{-5}$, and thus

$$r_0 = r / db = 145 \text{ Gpc}$$

We conclude that the brightness of the night sky can reach that of the Milky Way, in a static universe, only if the critical distance lies within ~145 Gpc. In cosmology, such a number can be interpreted in other ways. Because of the real possibility that we live in an open universe, perhaps only the corresponding time T for the universe to achieve this size, has meaning:

$$T = r_0 / c \qquad (30.10)$$

where c is the velocity of light. Using the values of db with and without the Magellanic Clouds, we obtain, respectively,

$$T = 22 \cdot 10^9 \text{ y} \quad \text{and} \quad T = 470 \cdot 10^9 \text{ y}$$

These values are upper bounds for the age of the universe in its present form, that is, *as we know it*. The universe could certainly be younger than $22 \cdot 10^9$ y but it is also true that we have no information about the universe prior to the Big Bang. In the course of the last 60 years cosmology has been placed on a firm observational foundation. The discovery by Slipher that the spectra of many galaxies were greatly red-shifted, and the subsequent determination of an empirical correlation between distance and the size of the red shift—the Hubble[5] relation—led to important advances. The relation is

$$V_r = H \cdot r \qquad (30.11)$$

where V_r is the radial velocity in kilometers per second, r is the distance in megaparsecs (Mpc),

---

[4]In the complete sky there are 4π steradians and the number of square degrees is $(180°/\pi)^2$ per steradian, giving in total $4\pi \cdot (180°/\pi)^2 = 41,253$ deg².

[5]Named for Edwin Hubble (1899–1953) for his pioneering work begun with Milton L. Humason (1891–1972) in the 1930s.

and H is the Hubble constant in kilometers per second per megaparsec. The value of H is empirically determined, with published values appearing from ~30 to ~100 km/(s·Mpc). The inverse of the Hubble constant is a measure of the age of the universe. Taking the middle of this range, $H = 65$ km/(s·Mpc), we obtain an age

$$1 / H = 1.5 \cdot 10^8 \text{ km/au} \cdot 2 \cdot 10^5 \text{ au/pc}$$
$$\cdot 10^6 \text{ pc/Mpc} / 65 \text{ km/s}$$

or

$$T \approx 3.0 \cdot 10^{20} / 65 = 4.6 \cdot 10^{17} \text{ s} \quad \text{or} \quad 1.5 \cdot 10^{10} \text{ y}$$

The range of ages corresponding to the range of determinations of H is 0.9 to $3.2 \cdot 10^{10}$ y. These ages are again overestimates since we have assumed a constant speed of recession for the galaxies of the expanding universe. Because we do not live in an empty universe, gravitational acceleration must decrease the speed of the expanding galaxies with time. Observations of the rates of recession of the most distant galaxies are not quite good enough to tell us if the acceleration is insufficient to halt the expansion, in which case the universe will expand forever, leading to the "Big Chill"—a heat death in which all the stars will burn out and all the black holes will evaporate—or, if the acceleration is sufficient to stop the expansion, eventually leading to a "Big Crunch" and perhaps even a repeating cycle of universes. One theory, however, is no longer supported by the data.

The steady-state theory—which maintained that the mean density of the universe remained constant because new stars and galaxies were spontaneously and continuously created in the voids left by the expanding galaxies—has been shown to be incompatible with modern data. The perfect cosmological principle[6] has been demonstrated to fail, at least in the realm of the universe accessible to optical

and radiotelescopes. There are three main observational lines of evidence against it:

1. The existence of the three-degree black body radiation, which is interpreted as the remnant of the Big Bang.
2. The existence of quasars, which were most numerous at a certain past epoch of the universe rather than either before or after.
3. A decrease over time of the number density of distant radio galaxies.

Each of the three offers evidence of the non-homogeneity of space and of the change in the universe with time. A contrary argument which could be raised—that such things mark local inhomogeneities in a vastly greater and more uniform universe—is without observational support.

Current theories about the universe evolving from particle theory physics suggest that the universe is just at the critical density needed to stop the expansion of the universe, $\langle \rho \rangle \approx 5 \cdot 10^{-27}$ kg/m^3. This is, however, about an order of magnitude greater than the sum of all the known visible matter in the universe. Therefore, either the theory is wrong, or is there a great deal of material of an unseen kind that populates the universe. No one can await the next developments in this field without a profound sense of the wonder of the cosmos and without a realization that we remain but children of the universe. This is the ultimate challenge of astronomy for humanity: to discover how we got here and where we are going.

---

[6]The *cosmological principle* (called by some astronomers in the 1950s a "presumption") is that the universe looks the same from anywhere. A corollary to it is the *perfect cosmological principle*, that the universe looks the same throughout all time as well. The steady-state universe was predicated on the truth of these principles.

Appendix A

Coordinate Systems and Spherical Astronomy

A.1 Definitions

This short summary is not sufficient to provide a comprehensive knowledge of spherical astronomy. It furnishes only a few basic rules of spherical trigonometry and of the principal coordinate systems of astronomy. The interested reader should examine some of the references to spherical astronomy provided at the end of this appendix for more complete details and many more examples.

A.1.1 The Celestial Sphere

Any cut through a sphere which passes through its center produces a *great circle* on the surface of the sphere. The arcs produced by the intersections of three great circles constitute a spherical triangle. The arc lengths making up the sides of the spherical triangle can be considered as angles measured from the center of the sphere. This is because of the relationship between the arc length and angle: $r \cdot \Theta = s$, where r is the radius of the sphere and Θ is the subtended angle of arc s. In the astronomical context, the radius of the sphere is usually taken as unity. Consequently, $s = \Theta$. Figure A.1 illustrates these basic ideas.

In Fig. A.2, the spherical triangle sides are labeled a, b, and c. A, B, and C mark the angles of intersection of the great circle arcs on the surface of the sphere. They are the angles of the spherical triangle and by convention are drawn opposite to sides a, b, and c. The relations among the sides and angles of a spherical triangle can be condensed, for most purposes, into two essential laws: the sine and cosine laws of spherical trigonometry.

A.1.2 The Sine and Cosine Laws

The Sine Law

$$\sin A \,/\, \sin a = \sin B \,/\, \sin b = \sin C \,/\, \sin c \tag{A.1}$$

The Cosine Law

$$\cos a = \cos b \cdot \cos c + \sin b \cdot \sin c \cdot \cos A \tag{A.2}$$

As with the sine law, the variables may be rotated to produce two other formulas:

$$\cos b = \cos c \cdot \cos a + \sin c \cdot \sin a \cdot \cos B$$

and

$$\cos c = \cos a \cdot \cos b + \sin a \cdot \sin b \cdot \cos C$$

A.1.3 Coordinate System Definitions

The basic components of any coordinate system located on the surface of a sphere are the poles, the reference circles (primary and secondary), the small circles, and the coordinates (and the sense or direction in which each is measured). These components are described in detail in this section and are illustrated in Fig. A.3.

The Poles

The poles are points that are equidistant from a great circle and that are on the surface of the sphere. While every great circle has two poles, the poles of a coordinate system, P and Q, are those of the great circle designated as the reference secondary circle. An example of a pole is the North Pole of the terrestrial coordinate system.

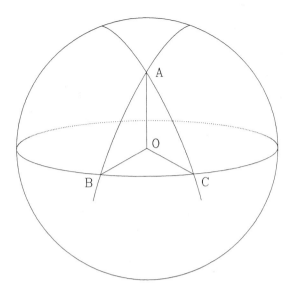

FIGURE A.1. A sphere with several great circles and the spherical triangle *ABC* created by their intersections. Note that the sides of the spherical triangle are equivalent to the angles that they subtend at the center of the sphere.

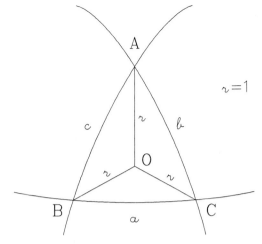

FIGURE A.2. A spherical triangle is on the surface of a sphere. Note that the sides (*a*, *b*, and *c*) are equal to the angles that subtend them at the sphere's center. The angles *A*, *B*, and *C* are located on the surface of the sphere, opposite sides *a*, *b*, and *c* respectively.

Reference Circles

These are great circles that establish the zero point of the system, the point from which the coordinates of any position are measured. The circle on which the points *P*, *G*, *C*, and *Q* are located, *PGCQ* designates what we call the *primary reference circle*, from which one of the coordinates is measured. It can be measured along *ABCD*, the *secondary reference circle*. The other coordinate is measured from this circle toward one of the two poles.

The use of the terms primary and secondary in this context should not be construed as standard; the usage varies in the literature. Sometimes, for example, what we call the secondary reference circle is called the fundamental circle. In the terrestrial coordinate system, the primary reference circle is the prime meridian through Greenwich. The coordinate measured from it is the *longitude*, measured either east or west. In this system the secondary reference circle is the equator, and the *latitude* is measured from it, either north or south.

Secondary Circle

A small circle in a plane parallel to that through the secondary reference circle is called a secondary circle. An arc on such a circle is related to the corresponding arc on the reference secondary

circle by the cosine of the latitudelike coordinate. For example,

$$EFGH = ABCD \cdot \cos \Theta$$

where Θ is the secondary coordinate measured from the secondary reference circle. In the terres-

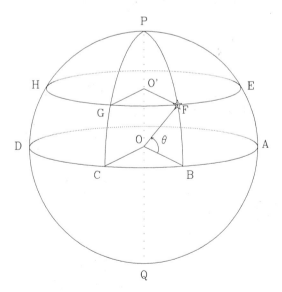

FIGURE A.3. The basic components of a spherical coordinate system.

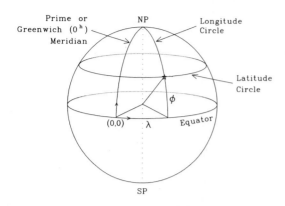

FIGURE A.4. The terrestrial coordinate system.

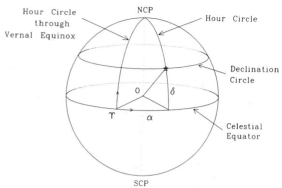

FIGURE A.5. Right ascension equatorial system.

trial system, such a secondary circle is called a *latitude circle* because all points along a particular latitude circle have the same latitude.

Note that

$$\text{arc } FG \pm \text{length } O'F \cdot L\ FO'G = \text{length } OF$$
$$\cdot \cos \Theta \cdot L\ FO'G$$

so that

$$\text{arc } FG = \text{length } OB \cdot \cos \Theta \cdot L\ BOC$$

and therefore

$$\text{arc } FG = \text{arc } BC \cdot \cos \Theta$$

Also, by Eqs. A.1 and A.2, we can show that the angle at the pole, $L\ FPG$ (or BPC), is equal to $L\ BOC$ and to arc BC:

$$\sin BPC\ /\ \sin BC = \sin BCP\ /\ \sin BP$$

but

$$\sin BCP\ /\ \sin BP = \sin 90° \ /\ \sin 90° = 1$$

Therefore

$$\sin\ L\ BPC = \sin BC$$

Moreover,

$$\cos BC = \cos BP \cdot \cos CP + \sin BP \cdot \sin CP$$
$$\cdot \cos L\ BPC$$

and

$$\cos BC = \cos 90° \cdot \cos 90° + \sin 90° \cdot \sin 90°$$
$$\cdot \cos L\ BPC$$

Therefore

$$\cos BC = \cos L\ BPC$$

Both conditions are fulfilled only if $BC = L\ BPC$.

Polar angles are very important in spherical trigonometry and are in fact essential in the spherical triangles of interest to astronomers. In the coordinate systems discussed here, all but one are on the celestial sphere and are centered on the observer. Observer-centered coordinates are said to be *topocentric*. In many contexts, coordinates are referred not to the location of the observer but to the center of the Earth or Sun. These are *geocentric* or *heliocentric* coordinates, respectively.

Because the details of the systems differ, they are defined one by one. The analogy with the more familiar terrestrial coordinate system will be clear by examining that system in the same way as the others, the equatorial (right ascension and hour angle) systems, the horizon or altazimuth system, the ecliptic system, and the galactic system.

A.2 Coordinate Systems

A.2.1 Terrestrial Coordinate System

Poles: North, South poles on the rotation axis.
Circles: Longitude, latitude circles.
Reference circles: *Prime meridian* (through Greenwich, U.K.); *equator.*
Coordinates: *Longitude* (λ) measured E (+) or W (−) from prime meridian, in HMS units of time (hours, minutes, and seconds) or in degrees; *latitude* (ϕ) measured N (+) or S (−) from the equator, in degrees. See Fig. A.4.

A.2.2 Equatorial Systems

A.2.2.1 Right Ascension Equatorial System

Poles: North, South Celestial poles (extension of the rotation axes into the celestial sphere).

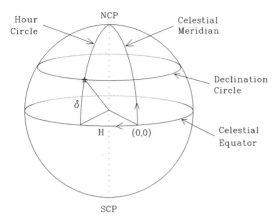

FIGURE A.6. The hour angle equatorial system.

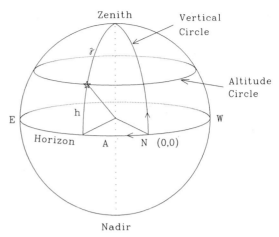

FIGURE A.7. Horizon or altazimuth system.

Circles: Hour, declination circles.

Reference circles: *Hour circle* through a point called the vernal equinox; *celestial equator.*

Coordinates: *Right ascension* (RA or α) measured E from vernal equinox, in HMS; *declination* (DEC or δ) measured N (+) or S (−) from celestial equator, in degrees. See Fig. A.5.

The vernal equinox is also called the "first point of Aries," after its location in that zodiacal constellation thousands of years ago, when our terminology came into usage. The vernal equinox is also the name given to the time of year when the Sun crosses the celestial equator, moving northward. Since this time of year is autumn in the Southern hemisphere, many astronomers prefer to call it the March equinox, leaving the term vernal equinox to mean the location on the sky only.

A.2.2.2 Hour Angle Equatorial System

Poles: North, South Celestial poles (extension of the rotation axes into the celestial sphere).

Reference circles: *Celestial meridian* (hour circle through the zenith); *celestial equator.*

Coordinates: *Hour angle* (HA or H) measured W from the celestial meridian, in HMS; *declination* (DEC or δ measured N(+) or S (−) from the celestial equator, in degrees. See Fig. A.6.

A.2.3 Horizon or Altazimuth System

Poles: Zenith and nadir (plumb bob directions above and below, respectively).

Circles: Vertical and altitude circles.

Reference circles: *Vertical circle* through the north celestial pole; *horizon.*

Coordinates: *Azimuth* (A), measured E from N point of the horizon, in degrees; (alternatively, especially for navigation: W from S. point), *altitude* (*h*), measured from horizon toward zenith, or *zenith distance* (*z*) measured from zenith toward the horizon), in degrees. See Fig. A.7.

A.2.4 Ecliptic System

Poles: North, south ecliptic poles (extension of axis of the Earth's mean orbital plane into the celestial sphere).

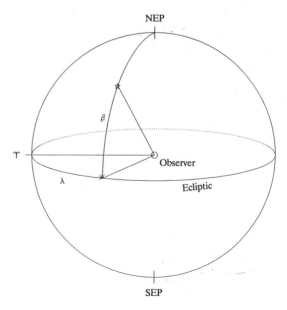

FIGURE A.8. The ecliptic system.

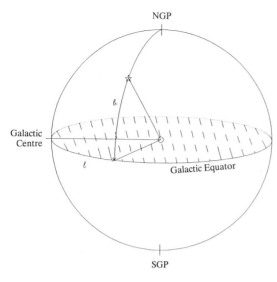

FIGURE A.9. Galactic coordinate system.

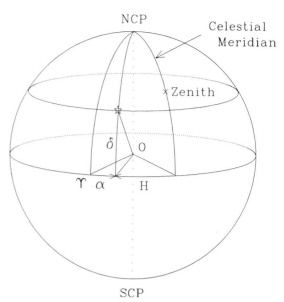

FIGURE A.10. The two equatorial systems linked by the local sidereal time.

Circles: Celestial longitude, latitude circles.

Reference circles: *Celestial longitude circle* through the vernal equinox; *ecliptic* (intersection of the mean orbital plane of the Earth with the celestial sphere – approximately the path of the Sun in the sky).

Coordinates: *Celestial longitude* (λ), measured E from the vernal equinox, in degrees; *celestial latitude* (β), measured N (+) or S (−) of the ecliptic, in degrees. See Fig. A.8.

A.2.5 The Galactic Coordinate System

Poles: North, south galactic poles.

Circle: Galactic longitude and latitude circles.

Reference circles: *Galactic longitude circle* through the galactic center (direction of Sagittarius); *galactic equator* (Milky Way).

Coordinates: *Galactic longitude* (ℓ), measured E from galactic center, in degrees; *galactic latitude* (b), measured N (+) or S (−) of the galactic equator, in degrees. See Fig. A.9.

A.3 Transformations Between Coordinate Systems

A.3.1 RA and HA Equatorial Systems

The two equatorial systems are linked by the local sidereal time (LST), which is defined as the hour angle (HA) of the vernal equinox:

$$\text{LST} \pm \text{HA}_\Upsilon = \text{HA*} + \text{RA*} \qquad \text{(A.3)}$$

The asterisk refers to any object. Therefore, the hour angle of any object can be found from the right ascension (RA) and the local sidereal time. Further, the RA of any object can be found from the HA and the LST. Figure A.10 illustrates the relation for an arbitrary local sidereal time.

A.3.2 Equatorial and Altazimuth Systems

The HA depends on the local meridian and so is directly linked to the observer. The *astronomical triangle*, which links the north celestial pole, the zenith, and a celestial object, contains all the quantities needed for the transformations between these two coordinate systems. It is shown on the sphere and enlarged in Figs. A.11 and A.12, respectively.

By application of the law of cosines, we find that:

$$\sin \delta = \sin \phi \cdot \sin h + \cos \phi \cdot \cos h \cdot \cos A \qquad \text{(A.4)}$$

$$\sin H = -\cos h \cdot \sin A / \cos \delta \qquad \text{(A.5)}$$

These equations serve to transform horizon system coordinates into equatorial coordinates H and δ. Equation A.3 can then be used, if necessary, to find the right ascension, α, if the LST is known. (Recall from §A.2.2 and §A.2.1. that $H \pm \text{HA}$, and $\alpha \pm \text{RA}$.) The same astronomical triangle provides the reverse transformation equations:

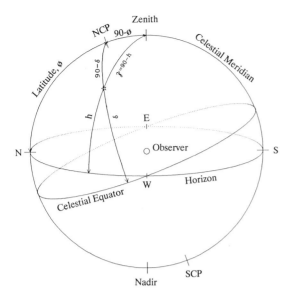

FIGURE A.11. The equatorial and altazimuth systems showing the astronomical triangle.

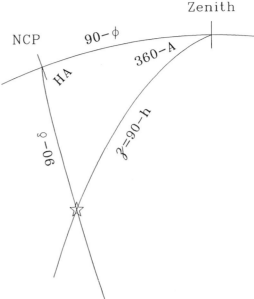

FIGURE A.12. The astronomical triangle.

$$\sin h = \sin \phi \cdot \sin \delta + \cos \phi \cdot \cos \delta \cdot \cos H \tag{A.6}$$

$$\sin A = -\cos \delta \cdot \sin H / \cos h \tag{A.7}$$

Example 1

Find the hour angle and declination of an object at azimuth $A = 120°$ and altitude $h = 60°$, at a site with latitude $\phi = 40°$.

By Eq. A.4,

$$\sin \delta = 0.642788 \cdot 0.866025 + 0.766044 \cdot 0.5 \\ \cdot (-0.5)$$

$$= 0.556670 + (-0.191511) = 0.365159$$

From this,

$$\delta = \arcsin(0.365159) = 21.4174° \quad \text{or} \\ 21°25'02.6''$$

Then, from Eq. A.5,

$$\sin H = 0.5 \cdot 0.866025 / 0.930945 = 0.465132$$

so that

$$H = \arcsin(0.465132) = 27.7188° = 1.8479^h \text{ or} \\ 01^h50^m52.5^s$$

Given the precision of the input data, we would be justified in writing $\delta = 21°$ and $H = 28°$ as the answers in this case. For the purpose of intermediate calculations, however, it is wise not to round off

the data during the steps of the calculation. In the check calculation, we can see the results of retaining "only" six significant figures in the intermediate results. Now the hour angle could have had another solution because the sine and cosine functions are double-valued,[1] and unlike the declination, the hour angle can go from 0° to 360° (i.e., 0^h to 24^h). In this case we do have the correct value for H, but in general, one should also solve for $\cos H$ from Eq. A.6 to confirm the quadrant of H. Solving Eq. A.6 gives

$$\cos H = (\sin h - \sin \phi \cdot \sin \delta) / (\cos \phi \cdot \cos \delta) \tag{A.8}$$

so that

$$\cos H = (0.866025 - 0.642788 \cdot 0.365160)/ \\ (0.766044 \cdot 0.930945)$$

$$= 0.631305 / 0.713145 = 0.885240$$

so that

$$H = +27.7189$$

in agreement in both sign and magnitude with our earlier result to the last decimal point, which has

_____

[1]For any arbitrary angle Θ, $\sin \Theta = \sin(180° - \Theta)$ and $\cos \Theta = \cos(360° - \Theta) = \cos(-\Theta)$.

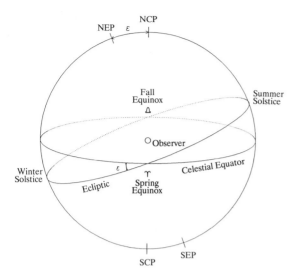

FIGURE A.13. The ecliptic and equatorial coordinate system shown together.

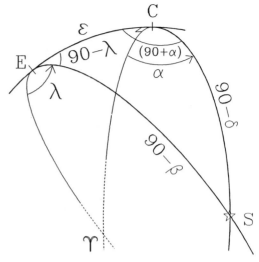

FIGURE A.14. The spherical triangle involving the poles of the ecliptic and equatorial coordinate systems.

been slightly affected by round-off error; the answer is still $H \approx 28°$.

A.3.3 Ecliptic and Equatorial Systems

The two systems (Fig. A.13) are linked by the vernal equinox, which is the zero point for both, and which moves at a rate of $50.2''$/y westward along the ecliptic (precession). Nutation and other periodic effects are not considered in the definition of the ecliptic system. The poles are separated by the *obliquity of the ecliptic* ($\sim 23.5°$), which slowly varies with time. The transformation between the systems can be obtained from the spherical triangle seen in Fig. A.14.

The transformation equations are obtained from applications of Eqs. A.1 and A.2:

$$\sin \beta = \cos \varepsilon \cdot \sin \delta - \sin \varepsilon \cdot \cos \delta \cdot \sin \alpha \tag{A.9}$$

$$\cos \lambda = \cos \alpha \cdot \cos \delta \, / \cos \beta \tag{A.10}$$

which transform the equatorial system to the ecliptic, and

$$\sin \delta = \cos \varepsilon \cdot \sin \beta + \sin \varepsilon \cdot \cos \beta \cdot \sin \lambda \tag{A.11}$$

$$\cos \alpha = \cos \beta \cdot \cos \lambda \, / \cos \delta \tag{A.12}$$

which transform the ecliptic to the equatorial coordinates.

Example 2

Find the ecliptic coordinates for an object at right ascension $22^h02^m43^s = 330.679167°$ and declination $42°16'40'' = 42.277778°$ (at the mean equinox 2000.0).

The obliquity, $\varepsilon = 23.439291$ (see Chapter 14). Therefore, from Eq. A.9,

$$\sin \beta = \cos(23.439291°) \cdot \sin(42.277778°)$$
$$- \sin(23.439291°) \cdot \cos(42.277778°)$$
$$\cdot \sin(330.679167°)$$

or

$$\sin \beta = 0.91748206 \cdot 0.67272560 - 0.39777715$$
$$\cdot 0.73989207 \cdot (-0.48969951)$$

or

$$\sin \beta = 0.61721367 - (-0.14412452)$$
$$= 0.76133819$$

so that

$$\beta = \arcsin (0.76133819) = 49.582312°$$
$$= 49°34'56''.$$

Now from this result and Eq. A.10,

$$\cos \lambda = \frac{\cos(330.679167°) \cdot \cos(42.277778)}{\cos(49.582312°)}$$

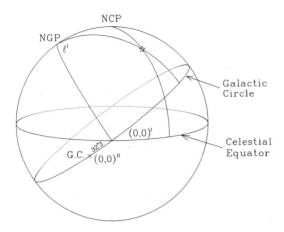

FIGURE A.15. The galactic coordinate systems I and II. System II coordinates are referred to the galactic center and are currently designated (ℓ,b). The position of the galactic center in the (α,δ) system is $\alpha = 17^h42.4^m$, $\delta = -28°55'$ (1950.0). The older coordinates $(\ell^{\mathrm{I}}, b^{\mathrm{I}})$ are referred to the ascending node of the galactic circle on the celestial equator.

or

$$\cos\lambda = \frac{0.87189127 \cdot 0.73989207}{0.64835497}$$

$$= 0.99498803$$

so that

$$\lambda = \arccos(0.99498803)$$

thus

$$\lambda = 5.73883° \quad \text{or} \quad (360° - 5.73883°)$$
$$= 354.26117°$$

Equation A.11 must be used to resolve the ambiguity:

$$\sin\lambda = \frac{\sin\delta - \cos\varepsilon \cdot \sin\beta}{\sin\varepsilon \cdot \cos\beta} \quad (A.13)$$

Therefore,

$$\sin\lambda = \frac{\sin(42.277778) - \cos(23.439291°) \cdot \sin(49.582312°)}{(\sin(23.439291°) \cdot \cos(49.582312°)}$$

or

$$\sin\lambda = \frac{0.67272560 - 0.91748206 \cdot 0.76133819}{0.39777715 \cdot 0.64835497}$$

$$= \frac{-0.02578853}{0.25790079} = -0.09999400$$

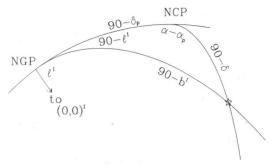

FIGURE A.16. The spherical triangle permitting derivation of the transformation equations between the equatorial and galactic (I) coordinate systems. α_P and δ_P are the coordinates of the North Galactic Pole in the equatorial system, and ℓ^{I} and b^{I} are the coordinates of the longitude and latitude of the object (*) in the galactic (I) system.

The sign of this number shows the celestial longitude cannot be in the first quadrant; indeed,

$$\arcsin(-0.09999400) = -5.73882°$$
$$= 354.26118° = 354°15'40''$$

is the answer.

A.3.4 Galactic and Equatorial Systems

The galactic coordinate system has changed in recent decades as the location of the galactic center has become more certain (thanks to the ability of radio waves to penetrate the interstellar dust that obscures distant regions in the Milky Way). Formerly the zero point of the system was one of the points of intersection between the celestial equator and the galactic equator. Coordinates in this system, which is no longer in use, were designated $(\ell^{\mathrm{I}}, b^{\mathrm{I}})$. In this system, the galactic center has coordinates $\ell^{\mathrm{I}} = 327°41'$ and $b^{\mathrm{I}} = -1°24'$. In the new system, designated for a time $(\ell^{\mathrm{II}}, b^{\mathrm{II}})$, but now written simply as (ℓ,b), the galactic center is the zero point and so has coordinates $(0,0)$. Figure A.15 shows the equatorial and current galactic systems, and Fig. A.16 contains the spherical triangle from which the transformation equations can be obtained.

The procedure chosen here is to compute the galactic coordinates in system $(\ell^{\mathrm{I}}, b^{\mathrm{I}})$ and then convert to (ℓ,b). Let the right ascension and the declination of the North Galactic Pole be α_P and δ_P, respectively. Then, from Fig. A.16 and application of the cosine law,

$$\sin b^{\mathrm{I}} = \sin\delta_P \cdot \sin\delta + \cos\delta_P \cdot \cos\delta \cdot \cos(\alpha - \alpha_P) \quad (A.14)$$

By the law of sines,

$$\cos \ell^{\mathrm{I}} = \frac{\cos \delta \cdot \sin(\alpha - \alpha_P)}{\cos b^{\mathrm{I}}} \qquad (A.15)$$

For the epoch 1950,

$$\alpha_P = 12^{\mathrm{h}}49^{\mathrm{m}} = 12.8167^{\mathrm{h}}, \quad \delta_P = 27°24'$$

From system I, system II coordinates are found (near the galactic plane) by

$$\ell^{\mathrm{II}} = \ell^{\mathrm{I}} + 32°19', \quad b^{\mathrm{II}} = b^{\mathrm{I}} + 1°24' \qquad (A.16)$$

To find the right ascension and declination from galactic coordinates, transformations produce the following equations. First, obtain ℓ^{I}, b^{I} from Eq. A.16. Then

$$\sin \delta = \sin \delta_P \cdot \sin b^{\mathrm{I}} + \cos \delta_P \cdot \cos b \cdot \sin \ell^{\mathrm{I}} \qquad (A.17)$$

and

$$\cos(\alpha - \alpha_P) = \frac{\sin b^{\mathrm{I}} - \sin \delta_P \cdot \sin \delta}{\cos \delta_P \cdot \cos \delta} \qquad (A.18)$$

Example 3

Suppose the 1950 equatorial coordinates of a certain star are $\alpha = 18^{\mathrm{h}}35.2^{\mathrm{m}}$ and $\delta = +38°44'$. Find its galactic coordinates.

We have

$$\alpha_P = 192.2500°, \qquad\qquad \delta_P = 27.4000°$$

$$\alpha = 18.5867^{\mathrm{h}} = 278.8005°, \quad \delta = +38.7333°$$

Thus

$$\alpha - \alpha_P = 86.5505°$$

From Eq. A.14,

$$\sin b^{\mathrm{I}} = \sin(27.4°) \cdot \sin(38.7333°) + \cos(27.4°) \\ \cdot \cos(38.7333°) \cdot \cos(86.5505°)$$

$$= (0.46020) \cdot (0.62570) + (0.88782) \\ \cdot (0.78007) \cdot (0.06017)$$

$$= 0.32962$$

Therefore,

$$b^{\mathrm{I}} = 19.25°$$

From Eq. A.15,

$$\cos \ell^{\mathrm{I}} = \frac{\cos(38.7333°) \cdot \sin(86.5505°)}{\cos(19.25°)}$$

$$= \frac{0.78007 \cdot 0.99819}{0.94409} = 0.82477$$

Therefore,

$$\ell^{\mathrm{I}} = 34.435° \quad \text{or} \quad 325.565°$$

But from Eq. A.17,

$$\sin \ell^{\mathrm{I}} = \frac{\sin \delta - \sin \delta_P \cdot \sin b}{\cos \delta_P \cdot \cos b}$$

we find

$$\sin \ell^{\mathrm{I}} = \frac{\sin(38.7333°) - \sin(27.4°) \cdot \sin(19.25°)}{\cos(27.4°) \cdot \cos(19.25°)}$$

$$= \frac{0.62570 - (0.46020) \cdot (0.32969)}{(0.88782) \cdot (0.94409)}$$

$$= \frac{(0.62570 - 0.15172)}{(0.83818)} = 0.56549$$

Therefore,

$$\ell^{\mathrm{I}} = 34.436°$$

The correct solution for the galactic longitude is the first-quadrant result.

References and Bibliography

Mills, H. R. (1987) *Positional Astronomy and Astronavigation Made Easy: A New Approach Using the Pocket Calculator.* Wiley, New York.

Smart, W. M., and Green, R. M. (1977) *Textbook on Spherical Astronomy*, 6th ed. Cambridge University Press, Cambridge.

Woolard, E. W., and Clemence, G. M. (1966) *Spherical Astronomy.* Academic Press, New York.

Appendix B

A Brief Summary of Error Analysis

Statistics probably engenders more disparate opinions than does any other area of mathematics. The public at large seems to regard statistics as the manipulation of data and to feel that such manipulation can produce any result and bolster any case. This may have been brought about by the widespread practice of presenting incomplete statistical analyses in support of a particular argument, to the exclusion of fuller analyses which may refute it. Partly at least, this attitude is also the cumulative result of the transmission of inaccurate or incomplete statistical results by untrained journalists to an unsuspecting public. Nevertheless, statistics is a highly respected discipline in mathematics and the sciences. The evaluation of the error in an experimental or computational result is an essential requirement for science.

Here we attempt only the briefest summary of the basic notions of error analysis. It is not intended to be a definitive or exhaustive treatment.

B.1 Statistical Quantities and Their Uncertainties

If you carry out a series of measurements of a quantity x under closely comparable circumstances, each measured value, x_i, will almost certainly not be identical. The values will be seen to be distributed about some *average* or *mean* value, $<x>$ (sometimes written $\bar{x}$):

$$<x> = \Sigma\, x_i\, /\, N \qquad (B.1)$$

where N is the total number of measurements and capital sigma, Σ, stands for the summing operation, this time over each value from $i = 1$ to $i = N$.

The N values constitute a *sample population*, a sampling of an infinitely large theoretical parent population. Since you cannot possibly measure an infinite number of values, at best you can get an estimate of the true mean. This is true also of other statistical quantities which are specified, like the mode, the median, and even the uncertainty with which these numbers can be found.

The *mode* is that value which is measured most frequently. The *median* is the midpoint of the distribution: there are as many measured values of x below as above the median. An estimate of the *uncertainty* or *error* is given by a measure of the width of the distribution of the x_i values. Note that these terms have no pejorative meaning in this context. An error in this context still means a departure from the "true" value, but we may not know what, exactly, is the true value that we are trying to measure. Assuming that none of the measurements are biased in some way, and that all are equally reliable, the error or uncertainty, as defined here, gives some idea of how well we are determining the true value. Thus a wider distribution has a larger uncertainty than a narrower distribution, because the data appear to be more scattered. The term to describe the distribution width is called the *standard deviation*[1] (σ) of the distribution, where

$$\sigma = \sqrt{\{\, [\, \Sigma\, (<x> - x_i)^2\,]\, /\, (N - 1)\, \}} \qquad (B.2)$$

where the numerator, $N - 1$, is used instead of N to compensate slightly for the fact that our sample size is limited and that the actual distribution of values is likely to be wider than what is sampled. The standard deviation is a measure of the dispersion of the distribution. The same quantity is used as a measure of the uncertainty in each value of x,

---

[1]In Europe, especially in former years, this quantity is called the *mean standard error*.

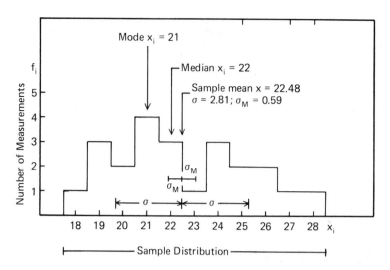

FIGURE B.1. A distribution of measured values of some quantity x. The ordinate, f_i, is the frequency or the number of times that a particular value of x_i was measured. In this case, $<x> = 22.48 \pm 0.59$; $\sigma = 2.81$; the mode = 21 and the median = 22.

so it is sometimes called the standard deviation (or the mean standard error) of a *single* measurement. The uncertainty in the mean itself is called the *standard deviation of the mean* (or the *mean standard error of the mean*):

$$\sigma_m = \sigma / \sqrt{N} \qquad (B.3)$$

When N is very large, $N - 1 \approx N$ and the sample mean approximates the true or expected value of x, while the uncertainties of a single observation and of the mean are essentially those of the expected distribution; here we designate these expected errors as s and s_m, respectively.

Figure B.1 shows how the distribution of measured values can be plotted and illustrates the terms described in this section. The plot is a histogram of the number of times (or *frequency, f_i*) a particular value of a quantity x was measured. The mean, median, and mode are all marked. Equations B.1 to B.3 are basic tools used by every working experimental or observational scientist.

B.2 The Gaussian Distribution

Among the various kinds of distributions of measured data, one of the first to be studied, and one of the most important, is the Gaussian distribution. This is the distribution to be expected from a purely random scattering of error in our measured values. The distribution is independent of the particular cause of the deviation of each individual value from the true value. In other words, it is independent of the type of error affecting each measurement, provided only that this error is random.[2]

The experiment of Galton's pegboard demonstrates dramatically that the total of a large number of such errors (perhaps even undetectably small errors) can approximate a Gaussian distribution. This device, a common sight in university physics demonstration rooms, consists of pegs placed in a series of rows to form the shape of a triangle:

A metal ball bearing is dropped through a slot above the first peg. The position at which it finishes is a result of a number of encounters at each of the rows of pegs. The very small changes in initial conditions that lead to one set of paths instead of another are not readily perceptible, but the probability that a certain path will be followed, in an

---

[2]If all the measurements were biased—because, for example, the *zero point* on the scale of the measuring device was incorrect—a *systematic error* would be seen, and the mean would be shifted. On the other hand, if the *scale* of the measuring device were faulty, so that the error depended on the value measured, the errors would not be random and would show up as a *skewedness* of the distribution of measured values, a departure from the Gaussian distribution.

unbiased set of experiments, is obtainable. We will designate the position just after row i as x_i. Starting at the narrow neck at the top of the apparatus,

$$x_0 = 5$$

then after the first row,

$$x_1 = 4 \text{ or } 6$$

and after the second,

$$x_2 = 3 \text{ or } 5 \quad \text{or} \quad 5 \text{ or } 7$$

and so on. Thus the chances of the ball being in the fifth (or central) position after the second row are twice as great as for either 3 or 7, and are zero for positions other than these three. The probability can be computed in similar fashion for as many rows as one wishes. The results can be displayed in a triangle, called Pascal's triangle.[3]

$$
\begin{array}{c}
1 \\
1 \quad 1 \\
1 \quad 2 \quad 1 \\
1 \quad 3 \quad 3 \quad 1 \\
1 \quad 4 \quad 6 \quad 4 \quad 1 \\
\cdot \\
\cdot \\
\cdot
\end{array}
$$

The results of 256 trials are summarized in Table B.1, where x_8 is the position after the eighth row, f'_i is the frequency or number of ball bearings arriving at position i, and ϕ_i is f'_i / N, where N is the total number of ball bearings used in the experiment. The quantity ϕ_i is a measure of the probability of a ball arriving at the ith position.

Figure B.2 illustrates the distribution of measured end positions tabulated in Table B.1. There are $n = 9$ possible endpoint positions. Despite the step size, $q = 1$, which makes for a rather blocky-looking histogram, the sample distribution closely approximates a Gaussian function. The parameter are as follows: the mean, $<x> = 5 \pm 0.08$, and

TABLE B.1. Results of trials on a Galton's pegboard.

x_8	f'_i	ϕ_i
1	1	0.004
2	8	0.031
3	28	0.109
4	56	0.219
5	70	0.273
6	56	0.219
7	28	0.109
8	8	0.031
9	1	0.004

[3]Each (nth) row of this triangle contains the coefficients of the binomial theorem, i.e., the coefficients of the expansions of the expressions $(1 + x)^n$. Try it and see! Note also that one can obtain the coefficients in the nth or ith row by adding up the number of chances of the $(i - 1)$ row. This triangle is seen in a book from China entitled *Precious Mirror of the Four Elements*, published in A.D. 1303, and the comments accompanying it suggest that it was already well known at that time (J. Needham and C. A. Ronan, *The Shorter Science and Civilisation in China*, Cambridge University Press, Cambridge. 1981, pp. 15 ff especially 54–57).

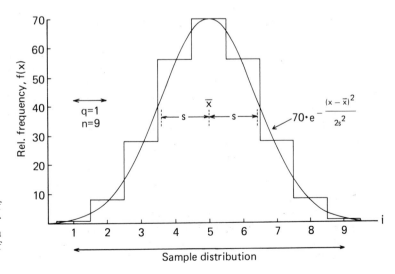

FIGURE B.2. The distribution of final positions of a large number of ball bearings falling through a triangle-shaped configuration of pegs on a Galton pegboard.

the standard deviation, $\sigma \approx s = \sqrt{2}$. The analytic form for the Gaussian function is

$$f(x) = f(<x>) \cdot e^{-[x-<x>]^2/2s^2} \qquad (B.4)$$

where $f(<x>)$ is the frequency at the mean value, here 70. It has a characteristic bell shape and so it is sometimes called the *bell-shaped curve*, or simply the *error curve*. The Gaussian error distribution curve produced by measurements of the solar diameter is discussed in Chapter 27.

B.3 The Poisson Distribution

Next to the Gaussian distribution, probably the Poisson distribution is the most significant for astronomy. Here it is also an example of an asymmetric distribution.

The Poisson distribution describes the frequency of occurrences and so basically deals with dimensionless integers, like the number of raindrops falling on a brick patio during a light drizzle. If the expected average were 2 drops per brick per second, on rare moments 5 or more per second could be expected; however, we could not expect -1 drop per second or fewer, so the distribution is nonsymmetric.

If the number of occurrences is N, recorded in m measurements, the expected number of occurrences per measurement is

$$<x> = \bar{x} = N/m$$

We state without proof that each measured value of $x \geq 0$ occurs with a frequency

$$f(x) = m \cdot (\bar{x}^x / x!) \cdot e^{-<x>} \qquad (B.5)$$

where $x! = 1 \cdot 2 \cdot \ldots \cdot x$, and $0! = 1! = 1$.

The standard deviation of a measurement is given simply by

$$s = \sqrt{<x>} \qquad (B.6)$$

Notice that Eq. B.5 contains only two parameters (N and m), whereas the Gaussian function given by Eq. B.4 requires three [$<x>$, $f(<x>)$, and s]. Moreover, the form of the expected error is quite different. In the case of the Gaussian, $<x>$ and s are independent of each other, but not so for the Poisson case.

Examples of the application of Poisson statistics abound in astronomy. A large number of physical measurements consist of the counting of events; for example, photon counting with a photomultiplier tube. In most cases, the events are Poisson-distributed, and Eqs. B.5 and B.6 apply. Photon statistics essentially involves the Poisson distribution. As a measure of relative precision, the quantity $r = s/<x> = 1/\sqrt{<x>}$ will serve. The inverse of r is the *signal-to-noise ratio*, S/N, so

$$S/N = \sqrt{<x>} \qquad (B.7)$$

If we strive for a precision of 1%, this means that we are looking to have $r = 0.01$ or a signal to noise ratio of 100; to achieve this, we require that $<x> = 10,000$ events or more. The counting requirements for even higher precision, that is, higher S/N and smaller values of r, require much greater numbers of events. For example, achievement of 0.5% precision requires that $S/N = 200$ and $<x> = 40,000$. In other words, a doubling of the relative precision requires a fourfold increase in the numbers counted. If we are observing starlight on a telescope equipped with a photoelectric photometer, the only way to increase the numbers of counts is to observe longer. Therefore, to double the relative precision, the observing time must be increased by a factor of 4. In Figs. B.3 to B.7, the S/N improvement is obvious as the integration time is made increasingly longer.

Figures B.3 to B.7 show the region of the spectrum of the star α Centauri around the Na–D line (580 to 600 nm). The spectra were obtained with a scanning photoelectric spectrophotometer on a telescope at the European Southern Observatory in Chile. The integration times were 0.3, 1.2, 4.8, 19.2, and 76.8 s, respectively. Thus the times were successively increased by a factor of 4, so that the counts for successive panels increase in the ratios 1 : 4 : 16 : 64 : 256. The ordinate scale, which indicates the numbers of counts per scanning step (0.2 nm), reflects this increase. The improvement in S/N is obvious from figure to figure. Notice that in Fig. B.3, the spectral features are indistinguishable from the noise. In Fig. B.4, an absorption feature is suggested, but not convincingly. In Fig. B.5, a spectral line is definitely present, and there is a strong suggestion that a doublet (two close lines) is present. This is confirmed in Fig. B.6. The Na–D doublet is at wavelengths 589.0 and 589.6 nm. In Fig. B.7, much weaker features in the spectrum are suggested, as the precision improves to $\sim 0.4\%$. These figures show the dramatic improvement in the relative error, even though the absolute error ($\approx \sqrt{N}$) increases with observing time. Another application of Poisson statistics can be found in Chapter 29.

Of course, there are practical limits to the increase in observing time, considering the relative

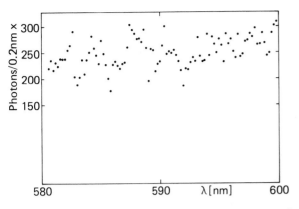

FIGURE B.3. The Na–D line region of the Spectrum of the star α Centauri. Integration time (IT) was 0.3 s.

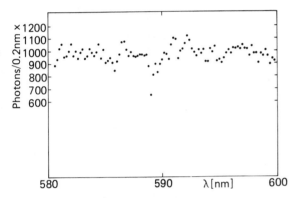

FIGURE B.4. The same as B.3 but with IT = 1.25.

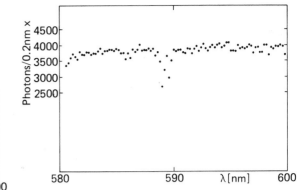

FIGURE B.5. The same as B.3 but with IT = 4.8 s.

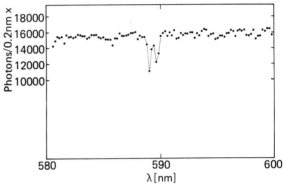

FIGURE B.6. The same as B.3 but with IT = 19.2 s.

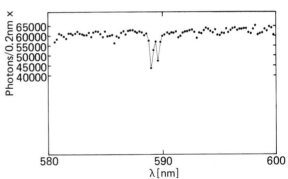

FIGURE B.7. The same as B.3 but with IT = 76.8 s.

improvement to be gained, and sky conditions cannot be expected to remain constant if the observing time is increased too greatly. Variation in the object itself may limit the integration time. There are intrinsic limitations to the improvement of precision, also. Beyond a certain point, systematic error, originating, for example, in the apparatus, may become important. These situations make observing something of an art, but within reasonable limits, the improvement of astronomical knowledge through an increase in the number of photons is the basis for requiring ever larger telescopes and more efficient detectors in exploring the limits of the universe.

Appendix C

Constructing a Cross-Staff

A cross-staff is a device to measure angular separation of objects in the sky. It was widely used by mariners during Europe's age of exploration in the fifteenth century. Although some extant cross-staffs are beautifully elaborated, the basic device is nothing more than a ruled surface held perpendicular to a long shaft. The shaft must be of a particular length, so that the graduations on the ruled surface are easily seen (by flashlight illumination) and can be calibrated. About 55 cm (21.5 in.) is optimum. Figure C.1 illustrates the basic form. The observer places his or her eye at the bottom of the cross, sights along the long shaft, and views the separation of the objects (e.g., a star transiting the celestial meridian and the southern horizon) so that they straddle the center and both read the same value on the graduated scale. The angular separation is then twice the scale reading.

A difficulty with the basic device (even the classical moving-crosspiece described in Chp. 1) is that different points along the crosspiece are at different distances from the observer's eye. This means that the scale is not linear in angular measurement of degrees. The slightly modified design seen in Fig. C.2 remedies this situation by curving the scale along the circular arc centered on the observer's eye. The curve, centered on a point about 2 cm beyond the end of the main shaft, can be marked on a slat of wood (which functions as the scale support) with a piece of chalk tied to a string of appropriate length.

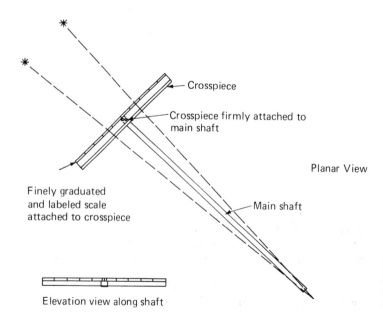

Crosspiece

Crosspiece firmly attached to main shaft

Planar View

Finely graduated and labeled scale attached to crosspiece

Main shaft

Elevation view along shaft

FIGURE C.1. Basic cross-staff design. Note the difference from the movable crosspiece design described in Chp. 1.

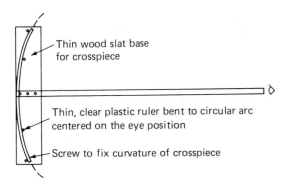

FIGURE C.2. Modified cross-staff design.

Thin wood slat base for crosspiece

Thin, clear plastic ruler bent to circular arc centered on the eye position

Screw to fix curvature of crosspiece

TABLE C.1. Calibration of the cross-staff scale.

$\Theta°$	s / ℓ	y / ℓ	s (cm)
0.5	0.0087	0.0087	0.5
1	0.0175	0.0175	1.0
2	0.0349	0.0349	2.0
3	0.0524	0.0524	3.0
4	0.0698	0.0699	4.0
5	0.0873	0.0875	5.0
6	0.1047	0.1051	6.0
7	0.1222	0.1228	7.0
8	0.1396	0.1405	8.0
9	0.1571	0.1584	9.0
10	0.1745	0.1763	10.0
11	0.1920	0.1944	11.0
12	0.2094	0.2126	12.0
13	0.2269	0.2309	13.0
14	0.2443	0.2493	14.0
15	0.2618	0.2680	15.0
16	0.2793	0.2869	16.0
17	0.2967	0.3057	17.0
18	0.3142	0.3249	18.0
19	0.3316	0.3443	19.0
20	0.3491	0.3640	20.0

Screws can be used to constrain a plastic ruler to fit the curve and hold it next to the slat. The slat is, of course, firmly fixed to the main shaft. The device is used in the same way; the angular separation of any two objects can be read off from any part of the device, but making the objects equidistant from the center position so that they read the same angle has the advantage of providing an automatic check on the reading. To use it in this way, the midway point of the plastic ruler must be exactly on the main shaft axis labeled with a permanent marking pen.

The scale of the traditional square design can be calibrated in the following way. The relationship between the distance from the center along the crosspiece, y, is related to the length of the main shaft, ℓ, and the angle to be measured, Θ, by the relation

$$y = \ell \cdot \tan \Theta \qquad (C.1)$$

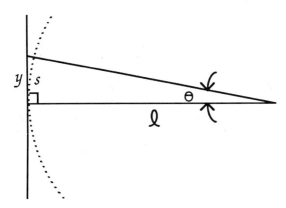

FIGURE C.3. Calibration of cross-staff scales. See Eqs. C.1 and C.2.

With the modified design, the relationship is

$$s = \ell \cdot \Theta \ (\text{rad}) = \ell \cdot \Theta \ (\text{deg}) \ / \ 57.30 \quad (C.2)$$

Figure C.3 illustrates the geometry in the two cases, and Table C.1 lists the angular separations corresponding to divisions on the ruler divided by the distance to the eye along the main shaft (~ 57 cm or 22.5 in.). For a distance $\ell = 57.3$ cm, the arc length value s is also calculated. You can calculate y or s for any value of ℓ in a similar way. Note that s (cm) is identical to Θ (deg) but y is not.

The difference between the tangent function and the angle variable expressed in radians is very small at very small angles and increases as the angle increases. The approximations

$$\Theta \ (\text{rad}) \approx \tan \Theta° \approx \sin \Theta°$$

are used in computing distances to the stars from stellar parallaxes that are all very much less than 1°. The use of a curved scale and of a shaft length to give $\ell = 57.3$ cm has a great advantage when the angles to be measured are more than about 10°, at which point there is about a 1% difference between the two types of cross-staff angle readings. By 20°, the percentage difference is $(20.9 - 20.0) / 20.0 = 4.5\%$.

Good building and good observing!

Appendix D

Julian Day Number Table

The Julian day number, often called the Julian date or Julian day, by astronomers, is one of a continuous count of the number of consecutive days since a particular date in the remote past. The count changes at noon Universal Time (UT). The system of Julian day numbers originated in 1582 (the same year as the Gregorian calendar) and was named for Julius Scaliger by his son, J. J. Scaliger, to whom the system is credited. It has nothing to do with the Julian calendar, which was devised by Sosigenes of Alexandria and introduced by Julius Caesar in 46 B.C. JDN 0 began at noon on the Julian Calendar Date January 1, 4713 B.C.

The table of Julian day numbers is reprinted from the *Astronomical Almanac*. It includes the day numbers over the period 1900 to 2050. To find the JDN corresponding to a particular date, look up the JDN for day zero of the month and year in question, and add the day of the month. As an example, find the JDN corresponding to the afternoon of the Gregorian date June 21, 1975. From the table,

$$JDN \ (June \ 0, \ 1975) = 2,442,564$$

Add 21^d to get

$$JDN \ (June \ 21, \ 1975) = 2,442,585$$

Note that the JDN does not change until the following noon. Decimals of a day are regularly used in astronomy, thus 2,442,585.5001 refers to an instant 8.6 s after midnight on June 22, 1975.

The Julian day numbers and decimals thereof are very useful in the computation of variable star phases as well as lunar phases and planetary positions. For example, the JDN of a new Moon can be found from the following ephemeris: JDN (new Moon) $= 2,441,747 + n \cdot 29.5305882$, where n is an integer. The current JDN can be used to find the current value for n, and the remainder is the phase of the Moon (0.25 for first quarter, 0.5 for full Moon, etc.). The formula with $n + 1$ in place of n then yields the next new Moon JDN, the Gregorian equivalent of which can be found from Table D.1.

The JDN also provides a simple way to find the day of the week. Divide the current JDN beginning at noon UT of the day of interest by the number 7. The remainder (the decimal of the quotient multiplied by 7) reveals the day of the week: 0 for Monday, 1 for Tuesday, . . . , 6 for Sunday. For example, on April 13, 1990 the JDN starting at noon is 2447995. Dividing by 7 gives 349,713.5714. . . . Multiplying 0.5714... by 7 gives 4 as the remainder, which according to our scheme is Friday. This date was in fact Good Friday 1990.

Be careful when using the table to note the correct prefix, which changes from 244 to 245, for example, in the year 1995. This and the other prefix changes are marked by asterisks.

TABLE D.1. Julian Day Number. Days elapsed at Greenwich noon, A.D. 1900–1950.

Year	Jan. 0	Feb. 0	Mar. 0	Apr. 0	May 0	June 0	July 0	Aug. 0	Sept. 0	Oct. 0	Nov. 0	Dec. 0
1900	241 5020	5051	5079	5110	5140	5171	5201	5232	5263	5293	5324	5354
1901	5385	5416	5444	5475	5505	5536	5566	5597	5628	5658	5689	5719
1902	5750	5781	5809	5840	5870	5901	5931	5962	5993	6023	6054	6084
1903	6115	6146	6174	6205	6235	6266	6296	6327	6358	6388	6419	6449
1904	6480	6511	6540	6571	6601	6632	6662	6693	6724	6754	6785	6815
1905	241 6846	6877	6905	6936	6966	6997	7027	7058	7089	7119	7150	7180
1906	7211	7242	7270	7301	7331	7362	7392	7423	7454	7484	7515	7545
1907	7576	7607	7635	7666	7696	7727	7757	7788	7819	7849	7880	7910
1908	7941	7972	8001	8032	8062	8093	8123	8154	8185	8215	8246	8276
1909	8307	8338	8366	8397	8427	8458	8488	8519	8550	8580	8611	8641
1910	241 8672	8703	8731	8762	8792	8823	8853	8884	8915	8945	8976	9006
1911	9037	9068	9096	9127	9157	9188	9218	9249	9280	9310	9341	9371
1912	9402	9433	9462	9493	9523	9554	9584	9615	9646	9676	9707	9737
1913	9768	9799	9827	9858	9888	9919	9949	9980	*0011	*0041	*0072	*0102
1914	242 0133	0164	0192	0223	0253	0284	0314	0345	0376	0406	0437	0467
1915	242 0498	0529	0557	0588	0618	0649	0679	0710	0741	0771	0802	0832
1916	0863	0894	0923	0954	0984	1015	1045	1076	1107	1137	1168	1198
1917	1229	1260	1288	1319	1349	1380	1410	1441	1472	1502	1533	1563
1918	1594	1625	1653	1684	1714	1745	1775	1806	1837	1867	1898	1928
1919	1959	1990	2018	2049	2079	2110	2140	2171	2202	2232	2263	2293
1920	242 2324	2355	2384	2415	2445	2476	2506	2537	2568	2598	2629	2659
1921	2690	2721	2749	2780	2810	2841	2871	2902	2933	2963	2994	3024
1922	3055	3086	3114	3145	3175	3206	3236	3267	3298	3328	3359	3389
1923	3420	3451	3479	3510	3540	3571	3601	3632	3663	3693	3724	3754
1924	3785	3816	3845	3876	3906	3937	3967	3998	4029	4059	4090	4120
1925	242 4151	4182	4210	4241	4271	4302	4332	4363	4394	4424	4455	4485
1926	4516	4547	4575	4606	4636	4667	4697	4728	4759	4789	4820	4850
1927	4881	4912	4940	4971	5001	5032	5062	5093	5124	5154	5185	5215
1928	5246	5277	5306	5337	5367	5398	5428	5459	5490	5520	5551	5581
1929	5612	5643	5671	5702	5732	5763	5793	5824	5855	5885	5916	5946
1930	242 5977	6008	6036	6067	6097	6128	6158	6189	6220	6250	6281	6311
1931	6342	6373	6401	6432	6462	6493	6523	6554	6585	6615	6646	6676
1932	6707	6738	6767	6798	6828	6859	6889	6920	6951	6981	7012	7042
1933	7073	7104	7132	7163	7193	7224	7254	7285	7316	7346	7377	7407
1934	7438	7469	7497	7528	7558	7589	7619	7650	7681	7711	7742	7772
1935	242 7803	7834	7862	7893	7923	7954	7984	8015	8046	8076	8107	8137
1936	8168	8199	8228	8259	8289	8320	8350	8381	8412	8442	8473	8503
1937	8534	8565	8593	8624	8654	8685	8715	8746	8777	8807	8838	8868
1938	8899	8930	8958	8989	9019	9050	9080	9111	9142	9172	9203	9233
1939	9264	9295	9323	9354	9384	9415	9445	9476	9507	9537	9568	9598
1940	242 9629	9660	9689	9720	9750	9781	9811	9842	9873	9903	9934	9964
1941	9995	*0026	*0054	*0085	*0115	*0146	*0176	*0207	*0238	*0268	*0299	*0329
1942	243 0360	0391	0419	0450	0480	0511	0541	0572	0603	0633	0664	0694
1943	0725	0756	0784	0815	0845	0876	0906	0937	0968	0998	1029	1059
1944	1090	1121	1150	1181	1211	1242	1272	1303	1334	1364	1395	1425
1945	243 1456	1487	1515	1546	1576	1607	1637	1668	1699	1729	1760	1790
1946	1821	1852	1880	1911	1941	1972	2002	2033	2064	2094	2125	2155
1947	2186	2217	2245	2276	2306	2337	2367	2398	2429	2459	2490	2520
1948	2551	2582	2611	2642	2672	2703	2733	2764	2795	2825	2856	2886
1949	2917	2948	2976	3007	3037	3068	3098	3129	3160	3190	3221	3251
1950	243 3282	3313	3341	3372	3402	3433	3463	3494	3525	3555	3586	3616

TABLE D.2. Julian Day Number, 1950–2000 of day commencing at Greenwich noon on:

Year	Jan. 0	Feb. 0	Mar. 0	Apr. 0	May 0	June 0	July 0	Aug. 0	Sept. 0	Oct. 0	Nov. 0	Dec. 0
1950	243 3282	3313	3341	3372	3402	3433	3463	3494	3525	3555	3586	3616
1951	3647	3678	3706	3737	3767	3798	3828	3859	3890	3920	3951	3981
1952	4012	4043	4072	4103	4133	4164	4194	4225	4256	4286	4317	4347
1953	4378	4409	4437	4468	4498	4529	4559	4590	4621	4651	4682	4712
1954	4743	4774	4802	4833	4863	4894	4924	4955	4986	5016	5047	5077
1955	243 5108	5139	5167	5198	5228	5259	5289	5320	5351	5381	5412	5442
1956	5473	5504	5533	5564	5594	5625	5655	5686	5717	5747	5778	5808
1957	5839	5870	5898	5929	5959	5990	6020	6051	6082	6112	6143	6173
1958	6204	6235	6263	6294	6324	6355	6385	6416	6447	6477	6508	6538
1959	6569	6600	6628	6659	6689	6720	6750	6781	6812	6842	6873	6903
1960	243 6934	6965	6994	7025	7055	7086	7116	7147	7178	7208	7239	7269
1961	7300	7331	7359	7390	7420	7451	7481	7512	7543	7573	7604	7634
1962	7665	7696	7724	7755	7785	7816	7846	7877	7908	7938	7969	7999
1963	8030	8061	8089	8120	8150	8181	8211	8242	8273	8303	8334	8364
1964	8395	8426	8455	8486	8516	8547	8577	8608	8639	8669	8700	8730
1965	243 8761	8792	8820	8851	8881	8912	8942	8973	9004	9034	9065	9095
1966	9126	9157	9185	9216	9246	9277	9307	9338	9369	9399	9430	9460
1967	9491	9522	9550	9581	9611	9642	9672	9703	9734	9764	9795	9825
1968	9856	9887	9916	9947	9977	*0008	*0038	*0069	*0100	*0130	*0161	*0191
1969	244 0222	0253	0281	0312	0342	0373	0403	0434	0465	0495	0526	0556
1970	244 0587	0618	0646	0677	0707	0738	0768	0799	0830	0860	0891	0921
1971	0952	0983	1011	1042	1072	1103	1133	1164	1195	1225	1256	1286
1972	1317	1348	1377	1408	1438	1469	1499	1530	1561	1591	1622	1652
1973	1683	1714	1742	1773	1803	1834	1864	1895	1926	1956	1987	2017
1974	2048	2079	2107	2138	2168	2199	2229	2260	2291	2321	2352	2382
1975	244 2413	2444	2472	2503	2533	2564	2594	2625	2656	2686	2717	2747
1976	2778	2809	2838	2869	2899	2930	2960	2991	3022	3052	3083	3113
1977	3144	3175	3203	3234	3264	3295	3325	3356	3387	3417	3448	3478
1978	3509	3540	3568	3599	3629	3660	3690	3721	3752	3782	3813	3843
1979	3874	3905	3933	3964	3994	4025	4055	4086	4117	4147	4178	4208
1980	244 4239	4270	4299	4330	4360	4391	4421	4452	4483	4513	4544	4574
1981	4605	4636	4664	4695	4725	4756	4786	4817	4848	4878	4909	4939
1982	4970	5001	5029	5060	5090	5121	5151	5182	5213	5243	5274	5304
1983	5335	5366	5394	5425	5455	5486	5516	5547	5578	5608	5639	5669
1984	5700	5731	5760	5791	5821	5852	5882	5913	5944	5974	6005	6035
1985	244 6066	6097	6125	6156	6186	6217	6247	6278	6309	6339	6370	6400
1986	6431	6462	6490	6521	6551	6582	6612	6643	6674	6704	6735	6765
1987	6796	6827	6855	6886	6916	6947	6977	7008	7039	7069	7100	7130
1988	7161	7192	7221	7252	7282	7313	7343	7374	7405	7435	7466	7496
1989	7527	7558	7586	7617	7647	7678	7708	7739	7770	7800	7831	7861
1990	244 7892	7923	7951	7982	8012	8043	8073	8104	8135	8165	8196	8226
1991	8257	8288	8316	8347	8377	8408	8438	8469	8500	8530	8561	8591
1992	8622	8653	8682	8713	8743	8774	8804	8835	8866	8896	8927	8957
1993	8988	9019	9047	9078	9108	9139	9169	9200	9231	9261	9292	9322
1994	9353	9384	9412	9443	9473	9504	9534	9565	9596	9626	9657	9687
1995	244 9718	9749	9777	9808	9838	9869	9899	9930	9961	9991	*0022	*0052
1996	245 0083	0114	0143	0174	0204	0235	0265	0296	0327	0357	0388	0418
1997	0449	0480	0508	0539	0569	0600	0630	0661	0692	0722	0753	0783
1998	0814	0845	0873	0904	0934	0965	0995	1026	1057	1087	1118	1148
1999	1179	1210	1238	1269	1299	1330	1360	1391	1422	1452	1483	1513
2000	245 1544	1575	1604	1635	1665	1696	1726	1757	1788	1818	1849	1879

TABLE D.3. Julian Day Number, 2000–2050 of day commencing at Greenwich noon on:

Year	Jan. 0	Feb. 0	Mar. 0	Apr. 0	May 0	June 0	July 0	Aug. 0	Sept. 0	Oct. 0	Nov. 0	Dec. 0
2000	245 1544	1575	1604	1635	1665	1696	1726	1757	1788	1818	1849	1879
2001	1910	1941	1969	2000	2030	2061	2091	2122	2153	2183	2214	2244
2002	2275	2306	2334	2365	2395	2426	2456	2487	2518	2548	2579	2609
2003	2640	2671	2699	2730	2760	2791	2821	2852	2883	2913	2944	2974
2004	3005	3036	3065	3096	3126	3157	3187	3218	3249	3279	3310	3340
2005	245 3371	3402	3430	3461	3491	3522	3552	3583	3614	3644	3675	3705
2006	3736	3767	3795	3826	3856	3887	3917	3948	3979	4009	4040	4070
2007	4101	4132	4160	4191	4221	4252	4282	4313	4344	4374	4405	4435
2008	4466	4497	4526	4557	4587	4618	4648	4679	4710	4740	4771	4801
2009	4832	4863	4891	4922	4952	4983	5013	5044	5075	5105	5136	5166
2010	245 5197	5228	5256	5287	5317	5348	5378	5409	5440	5470	5501	5531
2011	5562	5593	5621	5652	5682	5713	5743	5774	5805	5835	5866	5896
2012	5927	5958	5987	6018	6048	6079	6109	6140	6171	6201	6232	6262
2013	6293	6324	6352	6383	6413	6444	6474	6505	6536	6566	6597	6627
2014	6658	6689	6717	6748	6778	6809	6839	6870	6901	6931	6962	6992
2015	245 7023	7054	7082	7113	7143	7174	7204	7235	7266	7296	7327	7357
2016	7388	7419	7448	7479	7509	7540	7570	7601	7632	7662	7693	7723
2017	7754	7785	7813	7844	7874	7905	7935	7966	7997	8027	8058	8088
2018	8119	8150	8178	8209	8239	8270	8300	8331	8362	8392	8423	8453
2019	8484	8515	8543	8574	8604	8635	8665	8696	8727	8757	8788	8818
2020	245 8849	8880	8909	8940	8970	9001	9031	9062	9093	9123	9154	9184
2021	9215	9246	9274	9305	9335	9366	9396	9427	9458	9488	9519	9549
2022	9580	9611	9639	9670	9700	9731	9761	9792	9823	9853	9884	9914
2023	9945	9976	*0004	*0035	*0065	*0096	*0126	*0157	*0188	*0218	*0249	*0279
2024	246 0310	0341	0370	0401	0431	0462	0492	0523	0554	0584	0615	0645
2025	246 0676	0707	0735	0766	0796	0827	0857	0888	0919	0949	0980	1010
2026	1041	1072	1100	1131	1161	1192	1222	1253	1284	1314	1345	1375
2027	1406	1437	1465	1496	1526	1557	1587	1618	1649	1679	1710	1740
2028	1771	1802	1831	1862	1892	1923	1953	1984	2015	2045	2076	2106
2029	2137	2168	2196	2227	2257	2288	2318	2349	2380	2410	2441	2471
2030	246 2502	2533	2561	2592	2622	2653	2683	2714	2745	2775	2806	2836
2031	2867	2898	2926	2957	2987	3018	3048	3079	3110	3140	3171	3201
2032	3232	3263	3292	3323	3353	3384	3414	3445	3476	3506	3537	3567
2033	3598	3629	3657	3688	3718	3749	3779	3810	3841	3871	3902	3932
2034	3963	3994	4022	4053	4083	4114	4144	4175	4206	4236	4267	4297
2035	246 4328	4359	4387	4418	4448	4479	4509	4540	4571	4601	4632	4662
2036	4693	4724	4753	4784	4814	4845	4875	4906	4937	4967	4998	5028
2037	5059	5090	5118	5149	5179	5210	5240	5271	5302	5332	5363	5393
2038	5424	5455	5483	5514	5544	5575	5605	5636	5667	5697	5728	5758
2039	5789	5820	5848	5879	5909	5940	5970	6001	6032	6062	6093	6123
2040	246 6154	6185	6214	6245	6275	6306	6336	6367	6398	6428	6459	6489
2041	6520	6551	6579	6610	6640	6671	6701	6732	6763	6793	6824	6854
2042	6885	6916	6944	6975	7005	7036	7066	7097	7128	7158	7189	7219
2043	7250	7281	7309	7340	7370	7401	7431	7462	7493	7523	7554	7584
2044	7615	7646	7675	7706	7736	7767	7797	7828	7859	7889	7920	7950
2045	246 7981	8012	8040	8071	8101	8132	8162	8193	8224	8254	8285	8315
2046	8346	8377	8405	8436	8466	8497	8527	8558	8589	8619	8650	8680
2047	8711	8742	8770	8801	8831	8862	8892	8923	8954	8984	9015	9045
2048	9076	9107	9136	9167	9197	9228	9258	9289	9320	9350	9381	9411
2049	9442	9473	9501	9532	9562	9593	9623	9654	9685	9715	9746	9776
2050	246 9807	9838	9866	9897	9927	9958	9988	*0019	*0050	*0080	*0111	*0141

Appendix E

Planetary Position Tables

Because of the periodicity of the motions of the planets, the date of a particular perihelion passage, in Julian day numbers (JDN), can be found from an expression of the kind

$$T_0 = a + n \cdot b \qquad (E.1)$$

where n is an integer, a is an epoch (a certain JDN when the planet was at perihelion), and b is the sidereal period in mean solar days. Table E.1 lists the parameters a and b for each of the planets.

In the same way, the time of the planet's location at some other point in the orbit can be computed. For example, at aphelion,

$$T_a = a + (n + 0.5) \cdot b$$

For any other phase, 0.5 can be replaced by the decimal quantity n'. At any instant, T, the phase, n', can be computed from the expression

$$n' = [(T - a) / b] - n \qquad (E.2)$$

Since n is an integer, n' is the decimal part of the quantity in square brackets in Eq. E.2. For circular orbits, the phase angle varies linearly with time. In highly eccentric orbits this is not the case. Since the orbital area must be swept out at a uniform rate (Kepler's second law), the planet must move fastest at perihelion, where the distance to the Sun is smallest, and slowest at aphelion, where the distance is greatest. Accordingly, the positions in the heliocentric equatorial system, X, Y, Z, have been tabulated for each planet as a function of n'. The quantities X, Y, and Z are illustrated in Fig. E.1. The idea is to use Eq. E.2 to calculate n' and the tables to compute X, Y, Z. From X, Y, and Z, the geocentric coordinates are computed:

$$\Delta X = X - X_E, \quad \Delta Y = Y - Y_E, \quad \text{and}$$
$$\Delta Z = Z - Z_E \qquad (E.3)$$

where X_E, Y_E, and Z_E are the terrestrial positions, looked up in the table but using n_E', computed with Earth's elements, not the planet's. The differences, ΔX, ΔY, ΔZ, are referred to as x_g, y_g, and z_g in Chapter 14 (see Eqs. 14.10). From them the equatorial coordinate system coordinates, α and δ, can be found. Defining

$$r = \sqrt{[\Delta X^2 + \Delta Y^2]}$$
$$\delta = \arctan[\Delta Z / r]$$
$$\alpha = A + \arctan[\Delta Y / \Delta X] \qquad (E.4)$$

The quantity A is a correction for quadrant. It has the value 360° if $\Delta X > 0$ and $\Delta Y < 0$, 180° if $\Delta X < 0$, and 0° otherwise. Finally, the geocentric distance, D, can be obtained:

$$D^2 = \sqrt{[\Delta X^2 + \Delta Y^2 + \Delta Z^2]} \qquad (E.5)$$

This is a very simple procedure compared to that described and illustrated in Chapter 14, and the price paid for using it is a slight lack of precision

TABLE E.1. Ephemeris parameters for the major planets.

Planet	a	b
Mercury	244 3936.9	87.970
Venus	244 4098.4	224.71
Earth	244 1320.3	365.26
Mars	244 3951.0	687.00
Jupiter	244 2637.5	4,332.7
Saturn	244 2064.0	10,759.4
Uranus	243 9606.5	30,686
Neptune	240 8719.0	60,192
Pluto	235 7268.3	90,700

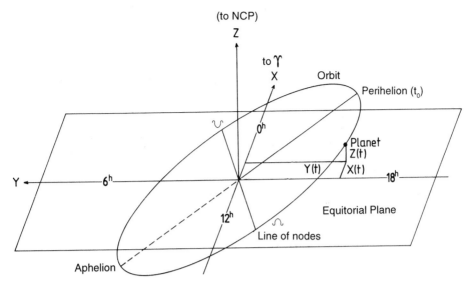

FIGURE E.1. Definitions of the heliocentric equatorial coordinates used in this appendix.

in the positions. As a challenge, compute the position of Mars on Oct. 6, 1988, using the Julian date equivalence (from Appendix D) and Table E.1, to get T, and then Eq. E.2 to get n' (the decimal part of the computed phase). You may check the results against those of §14.6. The results are

$$\alpha = +4.453° = 00^h17.8m, \quad \delta = -2.53°, \quad \text{and}$$
$$D = 0.406 \text{ au}$$

these numbers compare to $00^h17.5^m$, $-2.515°$,

and 0.408 au, computed by the method of Chapter 14. They are precise enough for many purposes.

A bonus to this interesting tabulation is that you can calculate the positions of any planet in the sky of *some other planet*. Just replace X_E, Y_E, and Z_E with the coordinates of the desired planet in Eq. E.3. Mind that the celestial equator with respect to which you will have obtained the positions is not that of your planet. This involves another transformation, and another challenge, if you want to undertake it. Have fun!

TABLE E.2. Rectangular equatorial coordinates of the Sun and planets. Heliocentric coordinates in astronomical units (aus).

		Sun							
n'		X			Y			Z	
		0.00			0.00			0.00	
		Mercury			Venus			Earth	
n'	X	Y	Z	X	Y	Z	X	Y	Z
0.00	0.070	0.267	0.136	−0.473	0.481	0.247	−0.213	0.881	0.382
0.01	0.041	0.271	0.141	−0.506	0.452	0.236	−0.275	0.866	0.376
0.02	0.011	0.272	0.145	−0.538	0.421	0.224	−0.336	0.848	0.368
0.03	−0.019	0.272	0.147	−0.567	0.388	0.211	−0.395	0.826	0.358
0.04	−0.048	0.269	0.149	−0.594	0.354	0.197	−0.453	0.801	0.348
0.05	−0.077	0.265	0.149	−0.618	0.318	0.182	−0.509	0.773	0.335
0.06	−0.106	0.258	0.149	−0.640	0.281	0.167	−0.562	0.741	0.322
0.07	−0.134	0.249	0.147	−0.660	0.243	0.151	−0.614	0.707	0.306
0.08	−0.161	0.239	0.145	−0.676	0.204	0.135	−0.663	0.669	0.290
0.09	−0.187	0.227	0.141	−0.690	0.164	0.117	−0.709	0.629	0.273
0.10	−0.211	0.214	0.136	−0.702	0.123	0.100	−0.752	0.586	0.254
0.11	−0.234	0.199	0.131	−0.710	0.082	0.082	−0.792	0.541	0.234
0.12	−0.256	0.183	0.124	−0.716	0.040	0.064	−0.829	0.493	0.214
0.13	−0.276	0.166	0.117	−0.719	−0.001	0.045	−0.863	0.443	0.192
0.14	−0.295	0.148	0.110	−0.719	−0.043	0.026	−0.893	0.392	0.170
0.15	−0.312	0.129	0.102	−0.716	−0.084	0.007	−0.919	0.339	0.147
0.16	−0.328	0.110	0.093	−0.710	−0.125	−0.012	−0.941	0.285	0.124
0.17	−0.342	0.090	0.083	−0.701	−0.166	−0.030	−0.960	0.229	0.099
0.18	−0.354	0.069	0.074	−0.690	−0.206	−0.049	−0.975	0.173	0.075
0.19	−0.364	0.049	0.064	−0.675	−0.245	−0.068	−0.986	0.116	0.050
0.20	−0.373	0.028	0.053	−0.658	−0.283	−0.086	−0.993	0.058	0.025
0.21	−0.381	0.007	0.043	−0.639	−0.320	−0.104	−0.996	0.000	0.000
0.22	−0.387	−0.015	0.032	−0.617	−0.356	−0.121	−0.995	−0.058	−0.025
0.23	−0.391	−0.036	0.021	−0.592	−0.390	−0.138	−0.990	−0.115	−0.050
0.24	−0.394	−0.057	0.011	−0.566	−0.423	−0.155	−0.981	−0.172	−0.075
0.25	−0.395	−0.077	−0.001	−0.536	−0.454	−0.171	−0.969	−0.229	−0.099
0.26	−0.396	−0.098	−0.011	−0.505	−0.484	−0.186	−0.952	−0.284	−0.123
0.27	−0.394	−0.118	−0.022	−0.472	−0.511	−0.200	−0.932	−0.339	−0.147
0.28	−0.392	−0.138	−0.033	−0.437	−0.537	−0.214	−0.908	−0.392	−0.170
0.29	−0.388	−0.157	−0.044	−0.400	−0.560	−0.227	−0.881	−0.444	−0.192
0.30	−0.383	−0.176	−0.054	−0.362	−0.581	−0.239	−0.850	−0.493	−0.214
0.31	−0.377	−0.194	−0.065	−0.322	−0.600	−0.250	−0.815	−0.541	−0.235
0.32	−0.370	−0.212	−0.075	−0.281	−0.616	−0.260	−0.778	−0.587	−0.255
0.33	−0.361	−0.229	−0.085	−0.239	−0.630	−0.269	−0.738	−0.631	−0.274
0.34	−0.352	−0.246	−0.095	−0.196	−0.642	−0.277	−0.694	−0.672	−0.291
0.35	−0.342	−0.261	−0.104	−0.152	−0.651	−0.284	−0.649	−0.710	−0.308
0.36	−0.331	−0.277	−0.114	−0.108	−0.658	−0.289	−0.600	−0.746	−0.324
0.37	−0.319	−0.291	−0.123	−0.063	−0.662	−0.294	−0.550	−0.779	−0.338
0.38	−0.306	−0.305	−0.131	−0.018	−0.663	−0.297	−0.497	−0.809	−0.351
0.39	−0.292	−0.318	−0.140	0.027	−0.662	−0.300	−0.442	−0.836	−0.363
0.40	−0.278	−0.330	−0.148	0.072	−0.658	−0.301	−0.386	−0.860	−0.373
0.41	−0.263	−0.341	−0.155	0.117	−0.652	−0.301	−0.328	−0.881	−0.382
0.42	−0.248	−0.352	−0.163	0.161	−0.643	−0.300	−0.269	−0.898	−0.389
0.43	−0.231	−0.361	−0.169	0.205	−0.632	−0.297	−0.209	−0.912	−0.395
0.44	−0.215	−0.370	−0.176	0.248	−0.618	−0.294	−0.148	−0.922	−0.400
0.45	−0.198	−0.378	−0.182	0.290	−0.602	−0.289	−0.087	−0.929	−0.403
0.46	−0.180	−0.386	−0.188	0.331	−0.584	−0.284	−0.025	−0.932	−0.404
0.47	−0.162	−0.392	−0.193	0.370	−0.563	−0.277	0.037	−0.932	−0.404
0.48	−0.143	−0.397	−0.198	0.408	−0.540	−0.269	0.099	−0.928	−0.403
0.49	−0.125	−0.402	−0.202	0.445	−0.515	−0.260	0.160	−0.921	−0.399
0.50	−0.106	−0.405	−0.206	0.480	−0.488	−0.250	0.221	−0.911	−0.395
0.51	−0.086	−0.408	−0.209	0.512	−0.459	−0.239	0.280	−0.897	−0.389
0.52	−0.067	−0.410	−0.212	0.543	−0.428	−0.227	0.339	−0.879	−0.381
0.53	−0.047	−0.411	−0.215	0.572	−0.396	−0.215	0.397	−0.859	−0.372
0.54	−0.028	−0.411	−0.217	0.599	−0.362	−0.201	0.453	−0.835	−0.362

TABLE E.2. (*Continued*).

	Sun								
n'	X			Y			Z		
	0.00			0.00			0.00		

	Mercury			Venus			Earth		
n'	X	Y	Z	X	Y	Z	X	Y	Z
0.55	−0.008	−0.410	−0.218	0.623	−0.327	−0.187	0.507	−0.808	−0.350
0.56	0.012	−0.408	−0.219	0.645	−0.290	−0.172	0.560	−0.778	−0.337
0.57	0.032	−0.405	−0.220	0.664	−0.253	−0.156	0.610	−0.745	−0.323
0.58	0.051	−0.401	−0.220	0.681	−0.214	−0.140	0.658	−0.709	−0.307
0.59	0.071	−0.396	−0.219	0.696	−0.175	−0.123	0.704	−0.670	−0.291
0.60	0.090	−0.390	−0.218	0.707	−0.135	−0.105	0.746	−0.629	−0.273
0.61	0.110	−0.384	−0.216	0.716	−0.094	−0.088	0.786	−0.586	−0.254
0.62	0.128	−0.376	−0.214	0.722	−0.053	−0.070	0.824	−0.540	−0.234
0.63	0.147	−0.367	−0.212	0.725	−0.012	−0.051	0.857	−0.493	−0.214
0.64	0.165	−0.358	−0.208	0.725	0.030	−0.033	0.888	−0.443	−0.192
0.65	0.183	−0.348	−0.205	0.723	0.071	−0.014	0.915	−0.392	−0.170
0.66	0.200	−0.336	−0.200	0.717	0.112	0.005	0.939	−0.339	−0.147
0.67	0.217	−0.324	−0.196	0.709	0.152	0.024	0.959	−0.285	−0.124
0.68	0.233	−0.311	−0.190	0.698	0.192	0.042	0.976	−0.230	−0.100
0.69	0.249	−0.297	−0.184	0.685	0.231	0.061	0.988	−0.174	−0.075
0.70	0.264	−0.282	−0.178	0.669	0.269	0.079	0.997	−0.117	−0.051
0.71	0.278	−0.266	−0.171	0.650	0.306	0.097	1.002	−0.060	−0.026
0.72	0.291	−0.250	−0.164	0.628	0.342	0.114	1.003	−0.002	−0.001
0.73	0.303	−0.232	−0.156	0.604	0.377	0.131	1.001	0.055	0.024
0.74	0.314	−0.214	−0.147	0.578	0.410	0.148	0.994	0.112	0.049
0.75	0.324	−0.196	−0.138	0.549	0.441	0.164	0.983	0.169	0.073
0.76	0.334	−0.176	−0.129	0.519	0.471	0.179	0.969	0.225	0.098
0.77	0.341	−0.156	−0.119	0.486	0.499	0.194	0.950	0.280	0.122
0.78	0.348	−0.135	−0.108	0.451	0.525	0.208	0.928	0.334	0.145
0.79	0.353	−0.114	−0.097	0.414	0.549	0.221	0.902	0.387	0.168
0.80	0.357	−0.092	−0.086	0.376	0.570	0.233	0.873	0.438	0.190
0.81	0.359	−0.070	−0.075	0.337	0.590	0.244	0.840	0.488	0.212
0.82	0.360	−0.048	−0.063	0.296	0.607	0.254	0.804	0.535	0.232
0.83	0.359	−0.025	−0.051	0.253	0.621	0.264	0.764	0.581	0.252
0.84	0.356	−0.003	−0.038	0.210	0.634	0.272	0.721	0.624	0.270
0.85	0.352	0.020	−0.026	0.166	0.643	0.279	0.676	0.664	0.288
0.86	0.345	0.043	−0.013	0.122	0.650	0.285	0.628	0.702	0.304
0.87	0.337	0.065	0.000	0.076	0.655	0.290	0.577	0.737	0.320
0.88	0.327	0.087	0.013	0.031	0.657	0.294	0.524	0.769	0.333
0.89	0.315	0.109	0.026	−0.015	0.656	0.296	0.468	0.797	0.346
0.90	0.301	0.130	0.038	−0.060	0.652	0.297	0.411	0.823	0.357
0.91	0.285	0.150	0.051	−0.106	0.646	0.298	0.352	0.845	0.366
0.92	0.268	0.169	0.063	−0.150	0.638	0.297	0.292	0.864	0.375
0.93	0.248	0.187	0.074	−0.195	0.626	0.294	0.230	0.879	0.381
0.94	0.227	0.203	0.085	−0.238	0.613	0.291	0.167	0.890	0.386
0.95	0.204	0.218	0.096	−0.281	0.597	0.286	0.104	0.898	0.389
0.96	0.180	0.232	0.105	−0.322	0.578	0.281	0.040	0.902	0.391
0.97	0.154	0.243	0.114	−0.362	0.557	0.274	−0.024	0.902	0.391
0.98	0.127	0.253	0.122	−0.401	0.534	0.266	−0.087	0.898	0.390
0.99	0.099	0.261	0.129	−0.438	0.509	0.257	−0.151	0.892	0.387

	Mars			Jupiter			Saturn		
n'	X	Y	Z	X	Y	Z	X	Y	Z
0.00	1.258	−0.506	−0.266	4.80	1.13	0.37	−0.39	8.33	3.46
0.01	1.298	−0.417	−0.226	4.71	1.44	0.50	−1.02	8.27	3.46
0.02	1.331	−0.326	−0.185	4.60	1.73	0.63	−1.65	8.18	3.45
0.03	1.357	−0.233	−0.144	4.46	2.02	0.76	−2.27	8.05	3.42
0.04	1.376	−0.139	−0.101	4.30	2.30	0.88	−2.88	7.88	3.38
0.05	1.388	−0.045	−0.058	4.13	2.57	1.00	−3.48	7.67	3.32
0.06	1.392	0.050	−0.015	3.93	2.82	1.12	−4.06	7.43	3.24
0.07	1.390	0.145	0.029	3.72	3.07	1.23	−4.62	7.15	3.15

TABLE E.2. (*Continued*).

	Sun								
n'	X			Y			Z		
	0.00			0.00			0.00		
	Mars			Jupiter			Saturn		
n'	X	Y	Z	X	Y	Z	X	Y	Z
0.08	1.380	0.239	0.072	3.49	3.30	1.33	−5.16	6.84	3.05
0.09	1.363	0.331	0.115	3.15	3.51	1.43	−5.68	6.50	2.93
0.10	1.340	0.422	0.158	2.99	3.71	1.52	−6.17	6.13	2.80
0.11	1.310	0.511	0.199	2.71	3.90	1.61	−6.63	5.73	2.65
0.12	1.273	0.598	0.240	2.43	4.06	1.68	−7.06	5.31	2.50
0.13	1.231	0.681	0.279	2.13	4.21	1.75	−7.46	4.86	2.33
0.14	1.182	0.762	0.318	1.83	4.34	1.82	−7.83	4.39	2.15
0.15	1.128	0.838	0.354	1.52	4.45	1.87	−8.17	3.90	1.97
0.16	1.069	0.911	0.389	1.20	4.54	1.92	−8.46	3.40	1.77
0.17	1.005	0.980	0.423	0.87	4.62	1.96	−8.73	2.88	1.57
0.18	0.936	1.045	0.454	0.55	4.67	1.99	−8.95	2.35	1.36
0.19	0.864	1.105	0.484	0.22	4.70	2.01	−9.14	1.81	1.14
0.20	0.788	1.160	0.511	−0.12	4.72	2.03	−9.29	1.26	0.92
0.21	0.708	1.210	0.536	−0.45	4.71	2.03	−9.40	0.70	0.70
0.22	0.626	1.255	0.559	−0.77	4.69	2.03	−9.47	0.14	0.47
0.23	0.541	1.296	0.580	−1.10	4.64	2.02	−9.50	−0.41	0.24
0.24	0.454	1.331	0.598	−1.42	4.58	2.00	−9.50	−0.97	0.01
0.25	0.365	1.361	0.614	−1.74	4.50	1.97	−9.46	−1.52	−0.22
0.26	0.274	1.385	0.628	−2.05	4.41	1.94	−9.38	−2.07	−0.45
0.27	0.183	1.405	0.640	−2.35	4.29	1.90	−9.26	−2.61	−0.68
0.28	0.091	1.419	0.649	−2.64	4.16	1.85	−9.11	−3.14	−0.90
0.29	−0.002	1.428	0.655	−2.92	4.01	1.79	−8.93	−3.65	−1.12
0.30	−0.094	1.432	0.659	−3.19	3.85	1.73	−8.71	−4.16	−1.34
0.31	−0.186	1.431	0.661	−3.45	3.67	1.66	−8.46	−4.65	−1.55
0.32	−0.277	1.424	0.661	−3.70	3.48	1.58	−8.18	−5.12	−1.76
0.33	−0.368	1.413	0.658	−3.93	3.28	1.50	−7.86	−5.57	−1.96
0.34	−0.457	1.397	0.653	−4.15	3.06	1.41	−7.52	−6.00	−2.15
0.35	−0.545	1.377	0.646	−4.35	2.84	1.32	−7.15	−6.41	−2.34
0.36	−0.631	1.351	0.637	−4.54	2.60	1.23	−6.76	−6.80	−2.52
0.37	−0.714	1.322	0.626	−4.71	2.35	1.12	−6.34	−7.16	−2.68
0.38	−0.796	1.288	0.612	−4.86	2.10	1.02	−5.90	−7.50	−2.84
0.39	−0.875	1.250	0.597	−5.00	1.84	0.91	−5.44	−7.81	−2.99
0.40	−0.951	1.208	0.580	−5.12	1.57	0.80	−4.96	−8.09	−3.13
0.41	−1.024	1.162	0.561	−5.22	1.29	0.68	−4.46	−8.35	−3.26
0.42	−1.094	1.113	0.540	−5.30	1.02	0.56	−3.95	−8.58	−3.37
0.43	−1.161	1.060	0.518	−5.36	0.73	0.45	−3.42	−8.77	−3.48
0.44	−1.224	1.004	0.493	−5.41	0.45	0.32	−2.89	−8.94	−3.57
0.45	−1.283	0.944	0.468	−5.44	0.16	0.20	−2.34	−9.08	−3.65
0.46	−1.338	0.882	0.441	−5.45	−0.12	0.08	−1.78	−9.19	−3.72
0.47	−1.389	0.817	0.413	−5.43	−0.41	−0.04	−1.22	−9.26	−3.77
0.48	−1.436	0.750	0.383	−5.41	−0.70	−0.17	−0.66	−9.31	−3.82
0.49	−1.479	0.681	0.352	−5.36	−0.98	−0.29	−0.09	−9.32	−3.85
0.50	−1.517	0.609	0.320	−5.29	−1.26	−0.41	0.48	−9.30	−3.86
0.51	−1.551	0.535	0.288	−5.21	−1.53	−0.53	1.04	−9.25	−3.87
0.52	−1.580	0.460	0.254	−5.11	−1.80	−0.65	1.61	−9.17	−3.86
0.53	−1.604	0.384	0.219	−4.99	−2.07	−0.76	2.16	−9.06	−3.84
0.54	−1.623	0.306	0.184	−4.85	−2.32	−0.88	2.71	−8.92	−3.80
0.55	−1.638	0.228	0.149	−4.70	−2.57	−0.99	3.26	−8.75	−3.75
0.56	−1.647	0.149	0.113	−4.53	−2.81	−1.09	3.79	−8.55	−3.69
0.57	−1.651	0.069	0.076	−4.34	−3.04	−1.20	4.30	−8.32	−3.62
0.58	−1.651	−0.011	0.039	−4.14	−3.26	−1.30	4.81	−8.06	−3.54
0.59	−1.645	−0.091	0.003	−3.93	−3.47	−1.39	5.29	−7.77	−3.44
0.60	−1.633	−0.177	−0.037	−3.70	−3.66	−1.48	5.76	−7.46	−3.33
0.61	−1.618	−0.250	−0.071	−3.46	−3.85	−1.56	6.21	−7.12	−3.21
0.62	−1.596	−0.328	−0.107	−3.20	−4.01	−1.64	6.63	−6.76	−3.08

TABLE E.2. (*Continued*).

	Sun								
n'	X			Y			Z		
	0.00			0.00			0.00		

	Mars			Jupiter			Saturn		
n'	X	Y	Z	X	Y	Z	X	Y	Z
0.63	−1.570	−0.405	−0.144	−2.94	−4.17	−1.72	7.04	−6.37	−2.93
0.64	−1.538	−0.481	−0.179	−2.66	−4.31	−1.78	7.41	−5.96	−2.78
0.65	−1.502	−0.555	−0.214	−2.38	−4.43	−1.84	7.77	−5.53	−2.62
0.66	−1.460	−0.628	−0.249	−2.08	−4.54	−1.90	8.09	−5.08	−2.45
0.67	−1.414	−0.698	−0.282	−1.78	−4.64	−1.94	8.39	−4.61	−2.27
0.68	−1.363	−0.766	−0.315	−1.47	−4.71	−1.98	8.65	−4.12	−2.08
0.69	−1.307	−0.832	−0.346	−1.16	−4.77	−2.02	8.88	−3.62	−1.88
0.70	−1.246	−0.894	−0.377	−0.84	−4.81	−2.04	9.08	−3.11	−1.68
0.71	−1.181	−0.953	−0.406	−0.52	−4.83	−2.06	9.25	−2.58	−1.47
0.72	−1.112	−1.009	−0.433	−0.20	−4.83	−2.07	9.38	−2.05	−1.25
0.73	−1.039	−1.061	−0.459	0.12	−4.82	−2.07	9.48	−1.50	−1.03
0.74	−0.962	−1.110	−0.483	0.45	−4.79	−2.06	9.54	−0.96	−0.81
0.75	−0.881	−1.154	−0.506	0.77	−4.73	−2.05	9.56	−0.40	−0.58
0.76	−0.797	−1.193	−0.526	1.09	−4.66	−2.03	9.55	0.15	−0.35
0.77	−0.710	−1.228	−0.544	1.40	−4.57	−2.00	9.49	0.70	−0.12
0.78	−0.620	−1.258	−0.561	1.71	−4.47	−1.96	9.40	1.26	0.11
0.79	−0.527	−1.283	−0.574	2.01	−4.34	−1.91	9.27	1.80	0.34
0.80	−0.432	−1.303	−0.586	2.30	−4.20	−1.86	9.11	2.34	0.57
0.81	−0.336	−1.317	−0.595	2.59	−4.04	−1.79	8.90	2.87	0.80
0.82	−0.238	−1.325	−0.602	2.86	−3.86	−1.73	8.66	3.38	1.02
0.83	−0.139	−1.328	−0.606	3.12	−3.67	−1.65	8.38	3.89	1.24
0.84	−0.039	−1.324	−0.607	3.37	−3.46	−1.57	8.07	4.37	1.46
0.85	0.061	−1.316	−0.605	3.60	−3.24	−1.48	7.72	4.84	1.67
0.86	0.160	−1.301	−0.601	3.82	−3.00	−1.38	7.34	5.29	1.87
0.87	0.259	−1.279	−0.594	4.02	−2.75	−1.28	6.93	5.71	2.06
0.88	0.357	−1.252	−0.584	4.21	−2.49	−1.17	6.48	6.11	2.24
0.89	0.452	−1.219	−0.571	4.37	−2.22	−1.06	6.01	6.48	2.42
0.90	0.546	−1.180	−0.556	4.52	−1.93	−0.94	5.51	6.82	2.58
0.91	0.637	−1.135	−0.538	4.64	−1.64	−0.82	4.98	7.13	2.73
0.92	0.725	−1.084	−0.517	4.75	−1.34	−0.69	4.44	7.41	2.87
0.93	0.809	−1.028	−0.493	4.83	−1.04	−0.56	3.87	7.65	2.99
0.94	0.889	−0.966	−0.467	4.90	−0.73	−0.43	3.28	7.86	3.10
0.95	0.964	−0.900	−0.439	4.94	−0.42	−0.30	2.68	8.03	3.20
0.96	1.034	−0.829	−0.408	4.96	−0.10	−0.16	2.07	8.17	3.28
0.97	1.099	−0.753	−0.375	4.95	0.21	−0.03	1.45	8.27	3.35
0.98	1.159	−0.674	−0.340	4.92	0.53	0.11	0.82	8.32	3.40
0.99	1.212	−0.591	−0.304	4.87	0.84	0.24	0.18	8.34	3.44

	Uranus		
n'	X	Y	Z
0.15	−11.74	−13.42	−5.71
0.16	−10.78	−14.13	−6.04
0.17	−9.77	−14.79	−6.34
0.18	−8.72	−15.38	−6.61
0.19	−7.63	−15.90	−6.86
0.20	−6.52	−16.36	−7.08
0.21	−5.37	−16.76	−7.26
0.22	−4.20	−17.08	−7.42
0.23	−3.02	−17.33	−7.55
0.24	−1.83	−17.52	−7.65
0.25	−0.62	−17.64	−7.72
0.26	0.58	−17.68	−7.75
0.27	1.79	−17.66	−7.76
0.28	2.98	−17.57	−7.74
0.29	4.17	−17.41	−7.68

TABLE E.2. (*Continued*).

	Sun		
n'	X	Y	Z
	0.00	0.00	0.00

	Uranus		
n'	X	Y	Z
0.30	5.34	−17.18	−7.60
0.31	6.48	−16.89	−7.49
0.32	7.61	−16.53	−7.35
0.33	8.71	−16.12	−7.18
0.34	9.77	−15.64	−6.99
0.35	10.80	−15.11	−6.77
0.36	11.79	−14.52	−6.53
0.37	12.73	−13.88	−6.26
0.38	13.63	−13.19	−5.97
0.39	14.48	−12.45	−5.66
0.40	15.28	−11.67	−5.33

	Neptune		
n'	X	Y	Z
0.59	−5.38	−27.68	−11.20
0.60	−3.51	−27.91	−11.34
0.61	−1.62	−28.03	−11.44
0.62	0.27	−28.05	−11.49
0.63	2.16	−27.95	−11.50
0.64	4.04	−27.74	−11.46
0.65	5.90	−27.42	−11.37
0.66	7.74	−27.00	−11.25
0.67	9.56	−26.47	−11.07
0.68	11.33	−25.83	−10.86
0.69	13.06	−25.10	−10.60
0.70	14.73	−24.26	−10.30
0.71	16.35	−23.33	−9.96

	Pluto		
n'	X	Y	Z
0.96	−27.08	−12.56	4.24
0.97	−25.67	−15.08	3.02
0.98	−24.03	−17.47	1.78
0.99	−22.17	−19.70	0.52
0.00	−20.10	−21.75	−0.76
0.01	−17.85	−23.60	−2.02
0.02	−15.44	−25.23	−3.27
0.03	−12.89	−26.63	−4.48
0.04	−10.22	−27.80	−5.65

Appendix F

Photographic Targets

The photography of the objects of the heavens is a very satisfying pastime. Although it has largely been replaced as a data-gathering technique—many observatories have converted their darkrooms into other facilities—it is still widely used for instructional purposes (as in this book) and by the public at large, and it is likely to be used for some time to come. Moreover, the use of very fast color film (at this writing 1700 ASA is available) does give a glimpse of the splendors of the sky, although it is not reasonable to expect that it should be able to capture anywhere near the full spectral distribution of the radiation from space. Nevertheless, to study the color of sunsets, lunar eclipses, the green flash, or aurorae, it is a very desirable medium.

Table F.1 lists classes of projects for you to tackle, many of them described in this book, and offers hints for their effective completion. The topics are arranged according to the conditions of illumination: day, twilight, and night. "Technique" refers to the type of device recommended: $C = $ a 35-mm camera with a 50-mm focal lense for direct photography; $t = $ the camera body with a telephoto lens attached; and $T = $ the camera body attached to a telescope. The speed of the film is indicated in ASA numbers which are equivalent, but you may want to experiment with faster films now widely available. The ratio of the focal length of the system, f, to the diameter of the aperture, D, that is, f/D, is called the *focal ratio* or *f/ratio*. Increasing f spreads out the image and gives better magnification; increasing D collects more light. f/D controls your image's brightness: the smaller f/D, the brighter the image. The relation is

$$b \propto 1/(f/D)^2 \qquad (F.1)$$

where b is the image brightness in, say, watts per square meter. The stops on your camera are so arranged that the ratio of apertures from one stop to the next (e.g., from 2.8 to 4) is $\sqrt{2}$; this represents a factor of 2 in area and therefore in light-gathering power. The light gathered is directly proportional to the exposure, too, so you can alter either to achieve the same brightness and to accommodate a different film speed. The sensitivity is directly proportional to the ASA speed, so if you use a film with ASA 400 instead of ASA 200, for example, you can expose for half as long, or increase the f ratio by one stop. The stops are limited, and at the open end (small numbers), the depth of field may be too small to accommodate your subjects at the same focus, so the exposure time is usually easier to adjust.

TABLE F.1. Suggested photographic projects.

Object	Technique	ASA	f/D	t(s)	Objectives
Day Projects					
Sun	t, T + filters, sunscreen projection	25	15	0.01	Limb darkening and reddening; radiation laws; sunspots and active regions; eclipses
Sun	T + Hα filter	400	15	0.03	Prominences
Terrestrial objects	t, T	25	15	0.02	Focal length determination; optics laws
Twilight Projects					
Moon	$(t) + T$	50	15	0.1...0.01	Moonscapes; shadowlengths; angular diameter variations
Lunar parallax.					
Sun	t, T	25	15	Variable	Refraction; atmospheric extinction
Green flash	t, T	50	8	0.2	Differential refraction; atmospheric dispersion anomalies
Earthshine	t, T	50	4	1	Earth's albedo
Satellites (+ chopper)	C	160	2	500	Orbits; velocities with rotating chopper blades in front of camera
Night Projects					
Exposures by moonlight	C	400	2	200	Comparison of moonlight to Sunlight; differences in exposures due to moonlit sky color
Exposures by light of the night sky	C	400	2	10^4	Atmospheric chemistry; differences from Moonlight and sunlight
Night sky	C	400	2	10^3	Atmospheric chemistry
Aurorae	C	400	2	300	Solar terrestrial relations; Earth's magnetic field; plasmas
Star trails extrafocal images	C	160	2	300	The photographic process; photometry; stellar distances, etc.
Star trails on the horizon	C	160	2	10^3	Refraction
Lunar eclipse	t, T[a]	400	15	60	Earth shadow studies; refraction, extinction, and reddening by Earth's atmosphere
Planets	C[a]	160	4	10	Kinematics ; dynamics; orbits; velocities; brightness variation; distances
Planetary surfaces	T[a]	400	150	1	Oblateness of Jupiter, Saturn; red spot and other atmospheric features; satellite phenomena
Comets	C, t, T[a]	400	4...15	5...100	Dust and gas; tail morphology; solar wind
Milky Way	C,[a] t,[a] T[a]	400	2	10^3	Stars, gas and dust; point sources and extended sources; thermal and nonthermal (Hα) radiation

[a]Requires guiding.

Appendix G

Topics Guide

Topic	Chapter														
	1	2	3	4	5	6	7	8	9	10	11	12	13	14	15
Archaeoastronomy	+	+	+	+	+	−	−	−	−	−	−	−	−	−	−
Atoms/molecules	−	−	−	−	−	−	−	−	−	−	−	−	−	−	−
Circular motion	−	−	−	−	−	−	−	−	−	+	+	+	+	−	−
Cosmology	+	−	−	−	−	−	−	−	−	−	−	−	−	−	−
Electronics	−	−	−	−	−	−	−	−	−	−	−	−	+	−	−
Gravitation	−	−	−	−	−	−	−	−	−	−	+	+	−	−	−
Light/color	−	+	−	+	−	+	−	−	−	−	−	−	−	−	+
Observation	+	+	−	+	+	+	+	−	−	−	−	+	+	−	+
Orbital motions	−	−	+	−	+	−	+	+	+	+	+	+	+	+	−
Parallax	−	−	−	−	+	−	+	+	−	+	+	−	−	−	+
Photometry	−	+	−	−	−	−	−	−	−	−	−	−	−	−	+
Radiation	−	−	−	−	−	−	−	−	−	−	−	−	−	−	+
Spectra	−	−	−	−	−	−	−	−	+	−	−	−	−	−	−
Spherical astronomy	+	+	+	+	+	−	+	+	−	−	−	−	−	+	−
Level[a]	1	1	1	1	1	1	1	2	1	2	1	1	2	2/3	2

Topic	Chapter														
	16	17	18	19	20	21	22	23	24	25	26	27	28	29	30
Archaeoastronomy	−	−	−	+	−	−	−	−	−	−	−	−	−	−	−
Atoms/molecules	−	−	+	+	+	−	−	+	−	−	−	−	−	−	−
Circular motion	−	−	−	−	−	−	−	−	−	−	−	+	−	−	−
Cosmology	−	−	−	+	−	−	−	−	−	−	−	−	−	−	+
Electronics	−	−	−	−	−	−	+	−	+	+	+	−	−	−	−
Gravitation	−	−	+	−	−	−	−	−	−	−	+	+	−	−	+
Light/color	+	+	−	+	+	+	+	+	+	−	+	+	+	+	+
Observation	+	+	−	+	+	+	+	+	+	+	+	+	+	−	+
Orbital motions	+	−	−	−	−	−	−	−	+	−	+	−	−	−	−
Parallax	+	−	−	−	−	−	−	−	−	−	+	−	−	+	−
Photometry	−	+	−	+	−	+	+	+	−	+	+	+	+	+	+
Radiation	−	+	+	+	+	+	−	−	+	+	+	+	+	+	+
Spectra	−	−	−	−	+	−	−	−	+	+	+	+	+	+	+
Spherical astronomy	−	−	−	−	−	−	−	−	−	−	−	−	−	−	−
	2	2	2	3	2	2	2/3	2	2	2	2	2	3	1/2	3
Level[a]	1	1	1	1	1	1	1	2	1	2	1	1	2	2/3	2

[a]These are rough guides only. Some "advanced" topics may be easily mastered by an individual who is interested in them. A 1 indicates that some geometry and trigonometry is needed; appropriate for high school students and beginning amateur astronomers. A 2 indicates that some physics and perhaps calculus would be helpful; appropriate for senior high school and beginning university students and more advanced amateur astronomers. A 3 indicates that physics and calculus are necessary; appropriate for upper-level undergraduates and dedicated amateur astronomers.

Index